PURE SCIENCE
AND
THE COSMOS
OF
MORAL REASONING

PURE SCIENCE AND THE COSMOS OF MORAL REASONING

SOME CALCULATIONS OF MEANING

BY JACQUELINE MARY BROWN

Publisher: Jacqueline Mary Brown 2020

ISBN 978-1-9163835-0-0

Table of Contents

DEDICATION

IN DEFENCE

OF

NATURE

THE

"WAR TO END ALL WARS"

WILL BE FOUGHT ON
PHILOSOPHICAL GROUND

WITH

PSYCHOLOGICAL TOOLS

Acknowledgments

The detail in this narrative was accessed and the arrangement as a whole composed by me alone, as I am accustomed to doing when I work as an expert witness. And I take full responsibility for my work as I am required to do when working in that context. The opinion I give is entirely my own. In this case I was not instructed to conduct an enquiry and used my own initiative in doing so. There is no other difference. I use one particular method. But by peer agreed definition this investigation is not the work of an expert witness. I do not claim that this is an enquiry conducted by an expert witness by peer agreement. I have not written this narrative for presentation in a Court as a witness there. But I am by definition an expert in my field. And my field is psychology. I am a forensic psychologist with extensive clinical experience. And I note that determining the terms of reference is a significant factor that determines the direction of an enquiry. This is an independent investigation and as you will see, I define independence in a particular way.

Without the work of others there would be no material on which to take a view. I am grateful to all those whose work I have quoted in this narrative even when I have not agreed with their opinions. All their contributions to the Calculation of Meaning have value. Without Comparative Analysis we cannot progress towards Knowledge.

Some parts of this book will be uncomfortable reading. Perhaps all of it will be for some. The work pushed my own tolerance for discomfort to the limit. But I felt a need to make sense of the doings and happenings I have observed during my lifetime. And it is necessary to look at and deal with things that are difficult. So, I have tried to make this narrative and your journey through it as interesting and comfortable as possible. I will observe the reaction in due course, under the condition that any of you choose to read this book.

PROLOGUE

Meaning travels across time and space. We pick up Meaning from its Representation. Artists have the ability to create Illusions. But they can also portray truths. There are many Ancient Texts from which Meaning is drawn. In the Sixth Century BC a poem, attributed to the poet Theognis of Megara, was written:

> Hope is the only good god remaining among mankind;
> The others have left and gone to Olympus.
> Trust, a mighty god has gone, Restraint has gone from men,
> And the Graces, my friend, have abandoned the earth.
> Men's judicial oaths are no longer to be trusted, nor does anyone
> Revere the immortal gods; the race of pious men has perished and
> Men no longer recognize the rules of conduct or acts of piety.

Poems are open to interpretation and they also deliver particular Meaning. People have always had different Opinions. In this poem hope and trust are among the gods. Only hope was said to be present among mankind at that time. It is said that trust had gone, and coupled with its disappearance was the leaving of restraint and the graces. But gods are immortal and all that is necessary is for men to revere them and recognise the rules of conduct; in this poem acts of piety are associated with those things.

Now, Pandora's Jar and the Trojan Horse are enduring models of things with the False Appearance of Gifts of Great Value that turn out to be a Curse. The consequences of accepting such gifts were portrayed. They brought chaos and wreaked havoc. But as well as having Meaning in common, they are different models with different Meaning. Pandora's Jar is a warning against giving in to Curiosity and the Trojan Horse is a warning to be Curious

Hope was kept confined in Pandora's Jar and was all that was left when the other evils escaped into the world and became Active there. And some poets portrayed hope as the Most Evil of evils. Some said that hope prolongs man's torment. Hope can be raised and dashed. Hope gives the other evils time to do their work. And trust is a companion of hope. Trust is opposed to curiosity. Hope and trust were involved in pulling the Trojan Horse within the walls of the city.

Some poets and painters envisaged an alternative Container holding Gifts of Blessings: - harmony (concordia), peace (pax), security (salus), fairness (aequitas), mercy (clementia), freedom (libertas), happiness (felicitas), worth (virtus) and joy (laetitia). The vessel that contained those qualities was left unopened and was hidden. The Blessings have remained confined across the Centuries. Is it too late to unleash The Blessings now and bring Order to Disorder? Before we can open it and find out, we must find the Container that holds them and has lain hidden across the Centuries.

CURIOSITY IS A WONDERFUL THING

Curiosity is one of The Blessings confined. Curiosity about the Meaning of things is The Most Blessed of The Blessings and can lead to attainment of the others. I invite you to see where Curiosity can lead. Hope is a comfort and sustains us, but hope is clearly not enough to combat the forces that were unleashed and wreaked havoc.

Jacqueline Mary Brown
January 2020

CHAPTER ONE: SCIENCE VIEWED FROM THE ANGLE OF THE COURT

PART ONE of Chapter one

JUSTICE MISCARRIES IN UNKNOWN QUANTITIES

Malfunctioning of the Justice System has been documented by Judges and researchers. It is a Fact that Errors occur in Criminal Trials. That has been proved Beyond Doubt. Garrett (2012), an American Professor of Law, presented a detailed analysis of a selected sample of 250 Single Cases in which retrospective DNA Testing had overturned convictions. The sample consisted of the first exonerations by DNA Evidence in the USA. The serious types of crime for which these individuals had been convicted carried high penalties, including death. The title of his book was "Convicting the Innocent". Convicting the Innocent is one type of Error. A different type of Error occurs when the Guilty Go Free. Both types of Error have Significant Effects and Outcomes. While we cannot distinguish between Guilt and Innocence in Single Cases we cannot know when Justice is served. And Justice is a matter of definition; we need to perceive its Meaning.

It is a Fact that the same Meaning is perceived when a commonality between things is picked up and differences are dropped. That which has the Appearance of being the same in Meaning can have different Meaning. Convictions differ in type. Differences in type are Qualitative Differences in Meaning and Action. The Guilty are not Convicted when the Innocent are Convicted. But Convicting the Innocent creates the False Appearance that the Guilty Do Not Go Free. The difference in the type of Meaning of a Conviction is made visible by the Emergence of True Meaning in Single Cases. The Illusion that the Guilty Are Punished for their crimes is not created when nobody is convicted for a crime.

Meaning is Qualitative. The distinction between Innocent and Guilty is Qualitative. DNA is a Qualitative Test. The type of Test fits the type of distinction. DNA is a product of Pure Science and had provided Verification of Innocence in these Single Cases where previously there was the Appearance that Guilt was Certain. Those who are Innocent cannot be Guilty. That is a difference in type of

Meaning. That which is one type of thing cannot also be a different type of thing. Verification of Innocence emerged through Falsification of Guilt by DNA Evidence. Verification and Falsification are Intrinsically Related and Relative to a particular thing. In a trial the finding is of Not Guilty or Guilty. That is a different Qualitative distinction and reflects the Quality of Uncertainty carried within the Trial Process. A Trial does not search for Innocence. But though Guilt is one type of thing that can be present or absent, its presence and absence are Not Quantifiable. One type of thing has particular Meaning. Meaning cannot be Quantified.

The 250 individuals in Garrett's sample had been found Guilty of crimes on the basis of evidence presented at their trials. The evidence was flawed. I will put that Meaning in a different form. There never was Evidence of Guilt in these cases because Guilt was not there. Garrett describes how DNA Testing evolved following its initial discovery and how the Test became increasingly discriminating as it was developed. Qualitative development enabled more differences to be picked up, not fewer. Advances in Technology are rooted in Knowledge produced by Pure Science. That which becomes possible through the engagement of human agents with Pure Science is not possible prior to that engagement. Garrett states that:

> "DNA exonerations have changed the face of criminal justice in the United States by revealing that wrongful convictions do occur and in the process, altering how judges, lawyers, legislators, the public, and scholar, perceived the system's accuracy."

The word accuracy represents the Power to detect a difference between things. Within the Boundaries of Pure Science, accuracy is defined as the Power to discriminate between particular things using the Qualitative Standard of Truth in relation to their Meaning. Garrett used that Standard for accuracy. Different Standards are applied outside the Process of Pure Science and the word accuracy has a different type of Meaning there. A finding of Not Guilty in relation to a happening is a finding that Guilt was not proved beyond Reasonable Doubt. I do not know how many people who perceive the Structure of the Justice System Believe that the System can find the Truth, and that Trial in Court is part of Action to discover True

Meaning in a Single Case, but some do. And in some Single Cases Truth is found, but we need to be able to tell which Cases those are. Before the advent of DNA, there was widespread Belief that a Method of making a Qualitative distinction between Guilty and Not Guilty was known and in use. And the Belief that the difference between Innocent and Not Guilty is Trivial is Active in the public domain but not in Specialist Contexts. Pure Science can Falsify or Verify a particular Belief, and in doing so transforms Belief into Fiction or Fact. But while the Belief that the Trial distinguished the Guilty from the Innocent has been Falsified in a large number of Single Cases, it was Not Falsified as a general proposition. The theory that evidence in a trial can find Truth in relation to Guilt has withstood many Falsifications in Single Cases. The Fact that making an accurate distinction depends on the type of evidence was not picked up. The difference in type of evidence cannot be accurately picked up while the Process of Pure Science is Unmarked.

In the 250 Cases that Garrett analysed, Uncertainty in relation to Guilt was given the Appearance of Certainty. Once Innocence was Known for Certain in Single Cases, the happenings that had led in those Cases to the Perception of False Meaning and wrongful convictions could be identified. Much can be learned from looking at the details of Errors using Qualitative Analysis. Garrett analysed the trial records and found "that the exonerees' cases were not idiosyncratic". The Errors were Systematic. Given those findings it can be predicted with Certainty that the occurrence of these Errors would extend beyond the 250 cases. The frequency of occurrence of Convicting the Innocent is not known and the number cannot be counted where we cannot tell the difference between Guilt and Innocence in the Single Case. The number of Convictions can be counted and expressed in Statistical terms in relation to frequencies on other variables, but the number of True and False Convictions remains hidden within that number. Nothing can be counted when the type of thing that it is cannot be identified Qualitatively. Qualitative Analysis necessarily takes precedence.

Garrett found that safeguards were lacking at all levels. He comments:

> "One might expect that judges would look for serious mistakes like the ones in these exonerees' cases. However, judges only reluctantly review the evidence after a conviction. After all, most constitutional rights relate to procedure and not accuracy. Most exonerees did not try to challenge flawed trial evidence, and when they did they almost always failed. Judges typically refused to grant a new trial, and some were so sure that these people were guilty that they called the evidence of guilt "overwhelming"."

The Judges saw what the Juries saw. The evidence had been tested by the Adversarial Process and the Illusion of No Reasonable Doubt was created and took hold of the audience. The members of the Judiciary know the Facts of the Law and how it Operates, while members of the Jury do not. That additional information did not change the False Perception of the Meaning of the evidence. It is not designed to do so and is designed to Function in a different way. We assume we know precisely what the purpose of a Trial is. The difference between Not Guilty and Innocence is often not perceived. It is commonly assumed that the purpose of Trials is to distinguish between Guilt and Innocence in each Single Case. But is that the purpose? The evidence was not fit for that purpose in the 250 cases, nor was it fit for the Qualitatively lower Standard of distinguishing between Guilty and Not Guilty. Garrett makes it clear that the Standard he uses for Judgment of the accuracy of a conviction is Truth in Reality, which Focusses Attention on what actually happened. The accuracy of a conviction depends on what actually happened and that is the question asked in each Case. That is not always the question that is answered.

The Premise we all see when viewing the Justice System is that Truth in relation to what happened is the Standard for Judgments of Guilt in that System. Premises have not been Tested for Truth. If they had been then they would be Facts and they are not that. A wrongful conviction is a finding that a particular person is Guilty in relation to a happening when that finding is Not True. In these 250 Cases the Court Process had found something that was not there. It is Reasonable to presume that if we know what we are looking for, then we will recognise it when we find it and know when we have not. So what is looked for by those who administer the Justice System? That question is answered by Judges when we look at the detail of what they say.

Challenges to decisions in Court are made by lawyers on behalf of their clients. Members of the Legal Profession have learned how the System Operates. They know that Truth is not the basis on which a challenge can be mounted. Arguments in Court are accepted as tests of the value of evidence and once evidence has withstood that testing the verdict is assumed to be true and the presumption of Guilt replaces the presumption of Innocence that had been in place. So a different route must be taken to Unsettle the Conviction. Garrett found that on occasions where the evidence in these cases had previously been challenged without DNA Evidence the False conviction had withstood the challenge. It was new Evidence of a different type that changed the verdict.

So, there was the Appearance that the Meaning in the evidence presented at Trial was True when it was False. That type of happening is represented by the word Illusion. An Illusion of Guilt was created by the evidence in these Cases and was accepted by those who saw it. Illusions are Powerful and all human beings are subject to them. Even Judges are Not Immune from entrancement by the Spell of Illusion. The Justice System is designed to produce a binding conclusion and there must be Justification for further testing. The signs indicate the presence of Belief in the accuracy of the Procedure Rules that form the Structure within which evidence is tested and in the accuracy of the Method of testing. In these cases where there was Verification by DNA that the Innocent had been Convicted, the other evidence in the Trial had passed the tests and its Meaning had been perceived by the fact finders. I note that the definition of the word fact covers different types of Meaning. I use the word Fact for Meaning that is True. Truth had not been found in these cases.

It is a Fact that Observation of patterns in happenings is necessary and not sufficient to access Causal Knowledge. Garrett points out that "most constitutional rights relate to procedure and not accuracy". Differences in type of Right are Significant. Since Garrett uses the word accuracy to represent Truth, it follows that most constitutional rights relate to the procedure that is followed and not to the Truth of the Verdict. In other words, there is a right to be tried in a particular way. It could not be otherwise. But surely the Procedure should be related to the Truth of the Verdict. The two Rights should be Unified. A Concept of Justice should be capable of doing that, so it is Reasonable to propose that the Concept is missing, or weak, and that we have in Operation a particular Construct. The differences between Concepts and Constructs will become clear in due course. But I note that I have only added to the range of possible propositions. The proposition that the Concept of Justice is strong still stands and is also a Reasonable proposition. Alternative propositions are set up in the Process of Pure Science in order to guide Testing.

Since Truth in these Cases was Verified by DNA Evidence, only one conclusion can be drawn. There is a Fault somewhere along the route to Conviction. The Fault leading to Error may be in the Justice System or may enter the System from elsewhere. The emphasis on Procedure in the Justice System is Observable. It is Justifiable to have a Structure to which Attention is paid. All parts of the Justice System are tightly Structured; it is a Formal System. There are Procedure Rules for Expert Witnesses and for every other aspect of the proceedings. Rules determine the way things are done and the order in which they are done. On the basis of his analysis, Garrett suggested amendments to the Structure of the Justice System with the aim of changing the Quality of the Judgment. But the Justice System is a Whole Structure with many interrelated parts. One Structure is part of another Structure and the detail of each must be analysed before Attention can turn to discovering the relations between them and the Meaning of the Action as a Whole. Qualitative Analysis is necessary for there to be Causal Knowledge. Garrett sets aside Intention. He found that:

> "The trial records in these cases give us every reason to believe that most of the police, prosecutors, forensic analysts, defense lawyers, judges and jurors acted in good faith. Few likely realised they were targeting the innocent."

I will express that last sentence in a different way in order to draw out a difference. Few likely realised they were targeting a particular person and likely thought they were targeting the question of what actually happened. Given that there is "every reason to believe" the individuals involved across many Cases acted in good faith it is Reasonable to propose that Causation lies elsewhere. With the exception of Jurors, the people in Action in the Cases were all selected from Specialist groups. Garrett uses the term "acted in good faith". The same Meaning can be represented by the term 'put their trust in Beliefs'. That trust was misplaced. Garrett states that:

> "It would be very disturbing if malicious people intended to frame these innocent defendants. But in some ways it would be a relief: those bad actors would also be an isolated problem – a few 'bad apples'."

Well here we see an Assumption that malicious people are rare, or at least have a low frequency of occurrence. Assumptions must be Tested but I will not Test that one in this narrative. That Assumption can stand as a Presumption. The Context for Testing whether or not people are malicious is Single Cases. The word malicious represents a particular Quality of a person in relation to particular Action. Blame is often diverted to single individuals through Judgments of Intention. That does Appear to bring relief, but it does so for those who Blame and not for those who are Blamed. And if Single individuals are Not a Significant Causal Factor then the problem will be repeated. Relief will not last long. If individuals are not responsible for these Errors, then the Source of the problem lies within the Structure that is external to them and it is clear that many different Structures are involved in the Process of the Law. Setting aside notions of Intention and the Blame associated with that does not involve setting aside the Fact that Single Agents Act on the World alone and in groups, and that they do so in particular ways with particular effects and outcomes. Single Agents are parts within Structures and Single Agents are involved in designing Structures to suit particular purposes. It is necessary to gain Knowledge on the topic of Human Action. Knowledge enables understanding of Causation.

When DNA Evidence or any other measure produced by Pure Science is not available, or the conditions under which that type of measure can provide Certainty are not met, then other types of evidence are all there is between conviction and acquittal. Garrett pointed out that:

> "In the vast majority of criminal cases, DNA testing cannot be conducted or cannot answer the question of identity.

It seems to follow that we do not know the accuracy of convictions in the vast majority of criminal cases. That might or might not be the case. It is impossible to count the numbers on either side of a division when both sides look the same. But it is a Fact that Uncertainty can be transformed into Certainty in the absence of Pure Science. That was Verified in 250 Single Cases. It is a Fact that there are different types of Certainty and that the difference in type has been dropped. Different types of Certainty have different Points of Reference. Pure Science leads to the Certainty that comes when Truth is disclosed. DNA is not the only Qualitative Measure produced by Pure Science. Qualitative Measures involve pattern matching, and the match of patterns is Tested for Accuracy by focussing Attention on differences as well as on commonalities. The measures that the Action of Pure Science produces become parts in further Action by human beings. I am not asking the question of how many products of Pure Science are available for use. I am seeking to identify the Action of Pure Science by type. That type of Action leads to the disclosure of Truth in Single Cases and it is a Human Action. And I note that the Quantitative Method picked up the necessity for matching patterns and misused that necessity. The type of thing that is matched makes a Significant Difference to the Quality of the conclusion.

It has been suggested that Science is one type of thing and is Unbiased. Those suggestions have been accepted and are Active in minds. Actions have particular direction. When things move in a particular direction there is a Bias towards reaching a particular destination. If there is No Bias then Chance is in Operation and no particular destination is reached. But Chance is a Hypothetical Construct. Chance Represents Action in the absence of a particular direction as a Whole. Now, Wholes are composed of parts that are Active. And Actions in one direction can cancel the effect of Actions in another direction but that is not the type of Meaning or Action

Represented by the Construct of Chance. I will show Evidence that Pure Science is a Systematic Bias that moves the Perception of Meaning in the direction of Truth. I propose that Pure Science is not based on Chance. And I propose that Chance is a Construct that is not based on Pure Science. I propose that Pure Science is a particular type of Action that leads systematically to good effects. In conjunction with that, I propose that disengaging Pure Science leads to Error and has the Potential to produce bad effects. Action against the discovery of Truth can cancel the effect of Action towards its discovery. That type of Action is not due to Chance and the word random does not cover it. All those propositions are Testable and will be Tested.

Inductive science is based on a model of Chance. Chance is a Construct that does not fit the Action in Social Dynamics. Most evidence is a product of Inductive science. The word Inductive has been dropped and the word science stands alone, thus a difference in type of science is hidden. I propose that there is always a Bias and thus I propose that the Appearance of No Bias in science is False. The Bias in science is hidden and can be Tracked. Whatever the purpose of Inductive science is, it is served by generating Uncertainty. The type of purpose is one Factor among many that determine the Type of Bias in Operation in Human Social Systems. It is generally accepted that if there is a Bias in Operation in the Courts then it is not one of Malicious Intent. It is Reasonable to presume that. If so, then Good Intentions are having bad effects in an Unquantifiable Number of Single Cases. Good Intentions are necessary and not sufficient to produce good effects. We need to discover how Uncertainty can be transformed into Certainty in the absence of Truth. But science is based on a Construct of Chance and Chance leads to Uncertainty, not to Certainty. So, it would Appear that science will be found Not Guilty of leading to Miscarriages of Justice. That Appearance is Deceptive. Appearances must be Tested for Truth.

Garrett described in detail the Errors in the Cases he examined and documented their repeated occurrence across the group of cases. The Errors were collectively referred to as "corrupted evidence, shoddy investigative practices, unsound science, and poor lawyering." He observed that "Judges typically refused to grant a new trial". It is humans who collect and present evidence using Inductive science, humans who conduct investigations, humans who use the products of Pure Science, humans who prosecute and defend, and it is humans who act as Judges and Jurors. The Errors are Human Errors. It is of course also humans who engage in the Process of Pure Science. We cannot gain Causal Knowledge unless we Observe the details of Human Actions. It is humans who are convicted of crimes and find themselves Powerless in the face of Injustice. To be in that situation is Traumatic.

It is humans who commit crimes. To be the victim of a crime is also Traumatic. A pattern of "corrupted evidence, shoddy investigative practices, unsound science, and poor lawyering" also leads to Freedom for the Guilty. Those who Malfunction in Social Dynamics and evade detection can continue to Malfunction. The Guilty are safer after the Innocent are Convicted than they were before that happening, for the official investigation into those crimes ends. We do not look for that which we think we have found. Attention will be focussed in my narratives on telling the difference between "unsound science" and Pure Science for that is Intrinsically Related to the Quality of evidence. I distinguish between Inductive science and Pure Science. I propose that Inductive science is "unsound science". That is not a new distinction nor is it a new proposition. That was the traditional view. Inductive science does not have the Power to detect difference between things. Inductive science can only count differences once they have been distinguished by Qualitative Methods.

Garrett set aside the notion of malign intent and I also set that notion aside. It is possible to Track True Meaning and set aside Intention. But, in setting aside questions of Intent, are we actually setting aside a Causal Variable of Significance? I do not think so, for within the Boundaries of Pure Science things are only set aside until they can be dealt with. Outside those Boundaries, Variables of Significance are set aside and Ignored, thus the potential for accessing Causal Knowledge is removed. Differences in Intention are important. Human Actions are goal directed. But Attributions of Intention are not often Tested for Truth and it is difficult to do that. Whether Attributions are right or wrong, they arouse strong Emotions. Thus, they interfere with Judgments. Attributions of

Intention tied to particular Actions are part of the Error in Social Dynamics. They lead to Blame and Retribution that are misplaced. The fit between Intention and the effects of Action is variable and the fit of Intention to an outcome is even harder to Judge. So until and unless we have Causal Knowledge and can accurately predict where our Actions will lead it is necessary to set aside Intention. Where the True Meaning of particular Actions can be clearly seen in advance, then there is Justification for taking Intention into account. The Ancients knew that. They also knew the importance of Testing for True Meaning.

Before setting the notion aside, I note that Intention is closely related to Motivation. Garrett's analysis bears on the question of the Motivation of people to distinguish between Guilt and Innocence. Before DNA was introduced as Evidence, that which was categorised as evidence in all these cases was not actually Evidence of Guilt. With the developments in DNA Testing, a Method that provided Certainty of Truth in some Single Cases became available. DNA does not always provide Certainty in relation to single happenings, but does so under particular conditions. Garrett demonstrated clearly that there was sometimes reluctance to use DNA Testing to make the distinction between Guilt and Innocence in the types of Case where the Test had the Power to establish the Truth and its potential to do that was known. The conditions that could produce Certainty were met and the financial costs of Testing were covered by the defence, yet DNA Testing was sometimes not permitted by Judges. The Presumption is that these Judges were not malicious and it is a Fact that there is No Evidence that they were. It is also a Fact that there is always a purpose. Human beings are goal directed. From a presumption that these Judges were not malicious, it follows that they had Good Intentions. We recall the possibility that Good Intentions can have bad effects. What could the Good Intentions possibly be?

It Appears to be Reasonable to infer that blocking access to a Test that can Falsify Guilt and thus Verify Innocence is done with the Intention of preventing discovery of the Truth in the Single Case. How could it possibly be otherwise? But now I have asked if there could be a different Intention. It is a Fact that while Attention is focussed on one problem, it is not focussed on another. So, there may not be Intention to Convict the Innocent. That effect might be a by-product of a different Intention. Intentions and Intuitions are related. Intuitions are conclusions with particular Meaning. When we have an Intuition, we clearly see the conclusion as a Whole Meaning and we cannot produce Reasoning leading to that. The lack of fit between the details of the arrangement and the conclusion can be hidden in various ways. We frequently Act on Intuitions. Humans tend to Believe their Intuitions are justified. We have yet to acquire the habit of Testing Intuitions for Justification. We do not know what drives our Actions. In other words, we are not aware of what determines the Responses that we emit.

There is a Fact from Experimental Psychology that I will now draw into the field of view. Fear interferes with the Operation of the Capacity to Reason. You did not actually need professional psychologists to tell you that Fact, for we all have personal experience that leads us to perceive the Truth of that Meaning. The Ancients knew that Fact. But Fear comes in many disguises and it is hard to recognise that type of Emotion even in ourselves. Fear can be suppressed below the level of Awareness. That which is hidden from view can have Powerful effects. And Emotions of all types are transmitted from one person to others. Emotions are sent through signals and signals are received. We respond when we detect Fear in other people and we do not all respond in the same way.

Meaning can be Tracked from Facts and through propositions. There is no value in False Findings of Guilt for those who are punished. Many of those in the audience do not see value in Convicting the Innocent either. So, if there is profit, where could the profit be and who benefits? Where there is profit there is always a beneficiary. Now, set aside the Quantitative Construct of profit and think in Qualitative terms. Just imagine for a moment what the reaction of the audience would be if they were told that in the vast majority of Criminal Cases it is not possible to distinguish accurately between Guilt and Innocence, and that there is rarely the Potential for that happening in any Single Case. Many of those in the audience see that already, but until there is Verification there is always room for Doubt. As the number of Falsifications rises and it becomes clear that there was only an Illusion of Evidence in a multitude of cases, then Belief in the effectiveness of the Justice System as a Whole becomes difficult to maintain.

Now I will recall the strange twist in the Tale that Garret drew to Attention. Permission for obtaining DNA Evidence has on occasions been refused when it is relevant to the Conviction. That type of Evidence is a product of Pure Science. Under certain conditions Pure Science can provide Certainty for a Judgment in relation to Truth. Pure Science can tell the difference between Guilt and Innocence. That has been Verified and is a Fact. Without the Evidence that the Process of Pure Science provides, Guilt cannot be distinguished from Innocence. Guilt remains a possibility when Innocence remains Uncertain. Without the Evidence of Pure Science there could be no successful Appeal in these cases. And in the absence of an Appeal the Conviction stands firm. And Appeals were denied. Also, if Evidence is limited to products of Pure Science and DNA is the only product available then the vast majority of Trials could not take place. But that position is not where Reasoning leads us.

Pure Science leads to technological developments and DNA is not the only form of technology that has been produced, but that is not the Fact I want to demonstrate. A Process is different to a Product. Pure Science produces tools for particular purposes through its Action but that Process is also a tool, and it is a Natural Flexible Tool designed for Processing Sense Data. All types of data can be entered into the Process of Pure Science and Acted upon by the Mechanisms within that. Pure Science is a Process that leads us in the direction of Truth. Truth is the Point of Reference for the Action within that Process. Pure Science enables us to tell the differences in Meaning and Action that distinguish one type of thing from another. Truth is at the beginning and the ending of Pure Science. But the Truth at the beginning is the direction for the Search and the Truth at the end is where the search led. The Point of Reference for the Action in a System determines the type of Functioning of the parts in the System and of the System as a Whole. These are Simple Conclusions. But the details that lead inevitably to particular Conclusions are Not Simple and the Complexities will be clarified in due course.

In order to demonstrate a problem and Track Meaning, let us accept for a moment an extreme proposition that only DNA Evidence is fit for the purpose of finding True Meaning. We accept temporarily the proposition that other Testimony categorised as evidence has a False Appearance of Truth. The Justice System delivers consequences for Actions that are categorised in Law as Offences. Offending is Conduct that is Disorderly. The Perception by an audience that there is no possibility of others gaining control over the Actions of any person including themselves would be a different problem to that of Convicting the Innocent and would be a Significant Problem in its own right. Without consequences being applied for Actions it is not possible to restrain people from doing whatever they wish to do, but that is under the condition there are no Internal Restraints. That is actually a Highly Significant condition. I will in due course provide Evidence that Internal Restraints are a Factor in Human Action that has been neglected. Now, the Appearance of Control over happenings in Social Dynamics disappears if the delivery of consequences is limited to the finding of Truth and if it is impossible to find the Truth of what happens in the Real World and if that impossibility is Visible. Those different Factors working in conjunction would produce particular effects. But Meaning can be hidden. If the impossibility is not seen then there is the Appearance that Truth can be found and there is the Appearance that Consequences are brought to bear in the right place. Those Appearances Act as a Deterrent - or at least that is an underlying theory. The success of that theory depends on the acceptance by an audience that the Meaning Represented in the Appearance is True.

Intention is a Complex Mechanism. The Intention to solve one problem can lead to the Unintentional creation or maintenance of a different problem. I will provide an abundance of Evidence for that statement in due course. It is Reasonable to set aside the Issue of Intention and focus Attention on the type of Action, for Meaning can be Calculated from Actions. It is not necessarily desirable to exchange one problem for a different one. Doing that is Justifiable when the problem substituted is easier to solve and if the solution can then be related to the original problem, thus increasing the possibility of solving that.

Judges are human beings with particular skills and particular experience. They listen to arguments between people who disagree with each other and they are tasked, with or without the presence of a jury, with producing Order where there is Disorder. Here I will give you an example of one possible solution to the problem of the Potential for Disorder if the majority of the population perceive that

True Meaning cannot be found in relation to doings and happenings. This potential solution is Unjustifiable and this example is relevant to the rest of this narrative. If evidence outside the boundaries of Pure Science cannot enable us to tell the difference between Guilt and Innocence, then the conclusion at the end of a Trial may not be the result of a guess and may be the result of a Bias, but the results of Trials cannot be more accurate in the long run than that produced by guessing when a distinction on the Relevant Variable cannot be made. Bias shows as a systematic regularity and the type of Bias determines the type of regularity that is systematic.

If it is True that there is no Method by which Truth can be found then those who commit crimes cannot be dealt with and the number of crimes will escalate. It is a Fact that there is escalation. The frequency of crime is taken as a measure of Disorder in Social Dynamics. Let us presume that the idea that evidence can tell the difference between Guilt and Innocence is an Illusion, the Illusion has been not been Dispelled and the Illusion must be maintained. A Universal Method of finding True Meaning is either not available, not recognised, or not wanted and out of sight. What then? How could the problem of gaining Control and reducing the Potential Increase in Disorder to which the Absence of Control invariably leads be resolved? I will use my imagination.

There is a particular Theory on which much of Social Dynamics has already been constructed. That theory is Probability Theory and it is a theory with which we are all familiar. Probability Theory has increasingly been applied by human beings to problems in all Contexts. Probability Theory is based on Games of Chance. So, let me show you how Probability Theory could apply to the problem of Distributing Consequences in order to gain Control over a population in the absence of finding the difference between Guilt and Innocence. Several different Mechanisms are brought into play in this Apparently Simple Solution that is Without Justification and Totally Unjustifiable.

THE PROBABILISTIC SOLUTION TO THE PROBLEM OF UNCERTAINTYIN RELATION TO GUILT OR INNOCENCE

We cannot tell the difference between Guilt and Innocence in Single Cases, and Pure Science is not there to help us. Crimes are committed and someone is Guilty of each one of them but we cannot work out who did it. Truth is not an Issue for it is impossible to provide Verification. An example must be made of somebody since otherwise everybody will think there are no consequences and crimes will spiral out of control. If that happens, Power will be lost and there will be Chaos until Power is redistributed.

So, let us draw lots from the general population. People will be identified and selected for the draw. Since Selections are Not Justifiable, the criteria for the allocation of tickets will not be revealed, nor will the identity of those who determine the criteria for selection. Those Agents who have gained Power will choose from the group of those who have No Power the ones who will enter the draw. Tickets are free and represent by number those who have attracted Attention. The free tickets cannot be refused, after all a crime has been committed and it will be seen that a criminal must be found.

The numbers will be placed into a container which will be spun. The Action of spinning mixes the numbers up. From the Confusion of Numbers, one will be drawn. The Action of spinning gives everyone with a ticket an equal chance of being drawn. This game is fair because, as you can see, it is ruled by Chance. There is no particular Bias towards any of those entered into the container. The number that is drawn first from the container represents Guilt. The person with that number will receive the sentence that fits the crime. Those in power will agree which punishment fits which crime.

THAT SMALL ENVISIONING IS FICTION.

OR IS IT?

It is Reasonable that punishment is fitted to the crime and is not fitted to the individual, but the individual should fit both those parts and the fit should be aligned to Truth. My short story is a Construct derived from Probability Theory and that Construct does not Appear to have been proposed for Action in the Context of the Justice System. The Absence of Justification is so obvious that it could not be displayed as a possibility. But has it been put into Operation by default? Is the Tale of a Probabilistic Solution to Justice played out in the Real World? It is Reasonable to ask that question. As you will all be aware, I have described a lottery. Lotteries are accepted and welcomed when the prize is a considerable sum of money. They have spread like wildfire for they generate huge amounts of money for the organisers, who have the Power to determine its distribution. Lotteries are a form of gambling. Gambling is addictive. The bookies profit and the gamblers lose. Lotteries are Not Justified and Not Justifiable. The participants have done nothing to merit the conclusion. No skill is involved and the only Action required is that involved in selling or acquiring a ticket.

Just try telling any audience that the prize in a lottery is death or a long term of imprisonment. Just try telling anyone that the possession of particular characteristics independently of Guilt or Innocence will Bias the selection for entry into that type of lottery and that you will not be told what those characteristics actually are because the Bias is Not Justifiable. I think my point is made! Then there would be Outrage. There would also be a high level of Fear. Fear is related to Control. Actions are Relatively Unpredictable once Fear takes hold. That Fact is known. Fear is frequently used as a Mechanism in Social Dynamics and its use is hidden as far as it is possible to do that. But a Balance is hard to Strike and impossible to maintain in Operations based on Fear. Triggering Fear can lead to Disorder and cannot lead to Order. And Disorder serves particular purposes. The Suppression of Fear is merely displacement. That which is displaced moves elsewhere. That which is not properly dealt with remains Unresolved. The Suppression of Fear gives the Appearance that there is Order where there is not.

Outrage would be Justified. It would be Truly Outrageous if that Method of Chances in a Lottery was put into Operation as a Construct of Justice. Of course, my Tale is Not True. Or is it? I have used the Licence of Fiction to express that Meaning. True Meaning can be represented in works of Fiction or not, but we cannot tell the difference for the Quality of Truth is Not Tested in that Context. It is Reasonable to ask if Chance plays a part in the Administration of Justice and whether Truth is an Issue or not. And if Truth is Not an Issue, then we need to ask a question. What actually is the Focus of Attention in the Justice System?

The search for Causal Knowledge can proceed while Intention is set aside. It was accepted only a few Centuries ago that there is No Justification for the sacrifice of one human being by others. Of course that was accepted. That was Universally Accepted. Or was it? Those human beings in Garrett's sample might see that differently. The single word justice covers many different types of Action giving them all the False Appearance of One Common Meaning. That which has the Appearance of Justice is not necessarily what we think Justice is.

The Process of Scapegoating is well known. Scapegoating is an Ancient practice in human societies and is a particular type of Action by some agents upon other agents in the World. A Scapegoat is blamed for doing things they have not actually done and for causing things they have not actually caused. When we convict the Innocent, we engage in the process of Scapegoating. We know it is wrong to do that. We do not grasp that that is actually what is being done. The Process in the Courts has a different Appearance. But my Analysis will disclose that the problem does not lie in all parts of that Process. Finding the solution to the problem of Convicting the Innocent is dependent on understanding the different parts of the Process and how they relate. The solution to the problem of Convicting the Guilty is also dependent on gaining that understanding.

England abolished the death penalty relatively recently, during the 20th Century. America still sometimes applies the death sentence. Seventeen of those in Garrett's sample had been sentenced to death. We cannot count how many Innocent people have been put to death by humans over the Centuries. The number is large and continues to grow. Putting people to death long ago for Witchcraft has the same Meaning as executing people for crimes they did not commit in recent times. There is only the Appearance of a Difference. So, what makes their Meaning the same? Well a common Point of Reference does that. I propose that the Standard for Judgment changed and reliance on Beliefs stayed the same. The Signs of Guilt are different now but that is Trivial if there is the same failure to pay Attention to True Meaning in Reality.

Witchcraft filled a gap in Causal Meaning, and now evidence does that. Curiously the notion of Causation simply disappeared leaving use of Association in its place. The gap in Causal Knowledge seems to be filled but, unless we look closely, we do not clearly see what type of Meaning it is filled with. We do not Believe in Witchcraft and we do Believe in evidence. The type of Belief changed, but the practice of Acting on the basis of Beliefs did not. I repeat that the Common Point of Reference makes their Meaning the same in type. The Mechanism of Association underlies the formation of Beliefs and it underlies the production of evidence by Inductive science. That is a Fact. Beliefs have particular Meaning. Meaning guides Action and Action guides Meaning.

Belief is Not the Point of Reference for Pure Science and Truth is. Beliefs lead to Error in Judgments. Pure Science will not always lead to Verification or Falsification but will not lead to Error. Beliefs cannot be blocked from entering into Judgments in the Courts if nobody knows that Beliefs are there, and once they pass through the gate then they become Active in the Justice System. Garrett analysed Errors in the Justice System and we learn much from observing in detail that which is wrong in a System. We can also learn much from observing in detail that which Functions well in a System. The Process of Pure Science is a particular type of Action in which human beings can engage. Judges are Observers of Human Actions and have a wealth of experience of Social Dynamics. The Judges are in place and in Action as gatekeepers of the Courts. They are watchful. I had particular questions based on experiences in the Forensic Context. I did not know the answers to the questions. I thought I had on occasions glimpsed Pure Science in Action within the Justice System so I decided to look for some of its Meaning there.

Judges form a group of human agents who are peers in a particular Context. Each Judge is different to another Judge and they take different views of the evidence. Judges agree and they disagree and in so doing they leave issues open for further development. Judges document their Observations and the detail that underpins their Judgments. Those details are made public. Both agreement and disagreement are made visible in relation to Single Cases. That has great potential for Tracking Meaning. So, during my search for Pure Science I looked at a sample of published views expressed by Appeal Court Judges in their Judgments of evidence presented in lower Courts. I looked for data within the Common Law System in the United Kingdom and other countries where that is applied. The search of that area was rewarding, as I had predicted it would be. Experts are one Source of Testimony at Trials and it is that Source on which I will focus Attention.

EXPERTS AND THE JUSTICE SYSTEM

There are differences and commonalities in Criminal and Civil Law. The use of experts is a commonality. A New South Wales report (2005) relating to expert witness testimony in civil law proceedings provides some brief historical detail of interest in relation to the use of experts in the Justice System. I have selected the following parts from their narrative:

> "2.1 Since medieval times, the common law has grappled with issues as to when and how to use specialist or expert knowledge to assist it in the resolution of disputes. The law has, at different times and for different reasons, employed three distinct procedures for utilising such knowledge in its work: the expert jury, the assessor and the expert witness."

I note in passing that the word knowledge covers different types of thing as does the word evidence and the word science. In relation to the expert witness it is stated that:

> "2.17 The third way in which English law has taken advantage of specialist knowledge is through the use of "expert witnesses". The earliest records of such a practice date back to the 14th century, and involve cases in which surgeons were summoned to establish such things as whether a wound was fresh. Cases from the 16th and 17th centuries also show that surgeons were summoned to give expert opinion on the cause of death, to advise on whether a child born 41 weeks after a husband's death could

> be legitimate, and even as to whether fits suffered by children could be a result of witchcraft on behalf of a defendant."

Expertise is sought where there is a gap in the Perception of Meaning. That which is thought to be a possible cause varies over time and space. Beliefs are relevant in deciding on which topics and questions experts can give evidence.

> "2.18 .... In Buckley v Rice Thomas [1554], Justice Saunders made what has since become a famous statement: [I]n matters arising in our law which concerned other sciences or faculties, we commonly applied to the aid of that science or faculty which it concerns, which is an honourable and commendable thing in our law. For thereby it appears that we do not despise all other sciences but our own, but we do approve of them and encourage them as things worthy of commendation."

The word science is here applied to the Law and to Contexts beyond that without being defined. Where there is a Premise that we know what science is, a definition is not deemed to be necessary where the word is used. It would be a brave man or woman who admitted they did not know what science is, so the question is not usually asked. But Judges are brave enough to do that. There is no clear answer to that question for science is not defined

The composition of the jury changed over time. Juries became audiences for evidence. At paragraph 2.19, the authors of the report refer to Thayer (1898):

> "[T]he old doctrine of their going on private knowledge began more and more to give way. The jury were told that if any of them knew anything relating to the case, they ought to state it publicly in court. This lay long in the shape of a moral duty of the jurors, not enforceable; but after a time it was enforced and the court assumed that, in general, nothing was known to the jury except what was publicly stated in court – adding to this, under the notion of judicial notice, what they were legally supposed to know and what was known to everybody. This brought matters

> down to the state of things in which we are now living. The jury became merely judges upon evidence."

In relation to the Single Case, the jury became in theory a Blank Slate wiped clean for the hearing of evidence in particular Cases. Then Attention necessarily turned to the question: What constitutes evidence? That question is not an easy one to answer.

> "2.21 One ….. rule appearing in the latter half of the 18th century was what is today known as the opinion rule: "that a witness must have personal knowledge, must state facts, not opinions". The development of this rule was intricately linked to the rise of the rule against hearsay."

The common definition of hearsay is rumour, gossip. That was ruled out. But what witnesses think they know is subject to psychological processes. It is hard to tell the difference between Fact and Opinion in human testimony. Experts differ from those who have personal "knowledge" in that they do not have personal experience in relation to the particular case – they did not see, hear or feel anything around the time of the happening. Then we are told that:

> "2.22 …………… the problem arose as to how to rationalise the use of skilled persons (a practice long established) with the newly developed rule prohibiting opinion evidence. Experts had thus far been sought out specifically to give such opinions. This problem appears to have been settled by Lord Mansfield in the seminal case of Folkes v Chadd … where it was accepted that the evidence of expert witnesses was an exception to the general rule prohibiting mere opinion evidence. The expert witness, as such, became 'a special sort of witness'."

Experts give evidence and their Opinion is defined as evidence. An Opinion from an expert witness is allowed and an Opinion from other witnesses is not. So does that allow hearsay from expert witnesses? Hearsay is not excluded - the development of the rule against taking Opinions as evidence was linked to the rule against that. Hearsay is involved in Opinions. The Ruling against hearsay was Reasonable for hearsay is rumour and rumour is defined as general talk, assertion, and that is of doubtful accuracy. But now we see the word opinion used to describe evidence when it is delivered by experts. Thus general talk and assertion of statements of doubtful accuracy becomes evidence through Association with a skilled person. So, is Opinion actually evidence? The Pocket Oxford dictionary defines opinion as:

opinion: 1. unproven belief. 2. view held as probable. 3. what one thinks about something. 4. a piece of professional advice........

In contrast with that, evidence is defined as Facts and, as I use that word, Facts are True for they are Verified. You can clearly see that Meaning is moving across time and is being actively channelled. Allowing Opinions to be given at all allowed a known problem to continue. But some things cannot be proven and decisions must be made. That was the puzzle for which a solution was sought. All human beings have Opinions and only those from professionals were allowed. Opinions were not distinguished by type and were distinguished on the basis of characteristics of the person giving them. That is a Fault. That is Formal logic, which is Inductive science. There are different types of Opinion. Some Opinions are Justifiable and some are not. The failure to recognise that particular difference in type of Opinion leads to Errors of Judgment. The Judges are aware of the difficulties that arise when Formal logic is in use. The problem that arose from the failure to distinguish one type of Opinion from another in terms of its Qualities was seen and expressed. At paragraph 2.24, the report states that:

> "........ by the mid-19th century, there was evident concern within the judiciary and the general public about the use of partisan expert evidence. There was a perception that "experts could be found who would testify to anything absurd"."

While hearsay is allowed the Problem of Hearsay remains Active. The report quotes as follows from Taylor's Treatise on the Law of Evidence (1885):

> "Perhaps the testimony which least deserves credit with a jury is that of skilled witnesses. These gentlemen are usually required to speak, not of facts, but to opinions: and when this is the case, it is often quite surprising to see with what facility, and to what an extent, their views can be made to correspond with the wishes or the interests of the parties who call them."

Opinion is defined as "unproven belief" and here is said to be usually required from "skilled witnesses" instead of Facts. The word evidence is used for Opinion from professionals. It is likely that the acceptance of Opinion reflects knowing the difficulties of providing proof, to which I will turn in subsequent Chapters. But Facts are Not Variable and Opinions are Variable, thus it is not surprising that views can easily be made to correspond. Now recall the Opinion Rule: "that a witness must have personal knowledge, must state facts, not opinions". Personal knowledge is not necessarily composed of Facts and is not necessarily Knowledge. There is confusion about what Facts and Knowledge are. According to my Model of Meaning (MOM), Facts and Knowledge are composed of parts with Meaning that is aligned to Truth. A Fact is something that is True; it is an item of Verified Information. Knowledge is Facts in their Intrinsic Relations. So, a witness must state Facts and those are statements that are True for they have been Verified, while an expert witness is under no obligation to do that. Or at least it Appears that an expert witness was not placed under any obligation to do so during the Nineteenth Century. That is a Curious state of affairs and if that is the case today then nobody told me. I accepted that Truth should be my guide. But the existence of the different position where Truth is not the guide and the Fact that the difference is not made clear would explain many things I have observed while working as an expert witness. My Observations indicate that some experts picked up Meaning that they were not actually told and was there to be found. But I do not suggest they are aware of the difference either, for they might Believe in evidence.

I am puzzled by the expression of surprise on seeing with what facility, and to what an extent, the views of expert witnesses can be made to correspond with the wishes or the interests of the parties who call them. That Action aligns with the Meaning that experts do not have to state things that are True and do not have to Verify them. They are allowed to give Opinions or Beliefs and name them evidence. That is what the exception to the Rule for other witnesses actually means. If that Meaning is seen then the Action that follows is aligned to that perception. Experts could be confined to telling the Truth and required to Verify their Opinions. The historical details show that the solution has long been seen. However, it is not obvious why the Opinion Rule has not been universally applied to witnesses. Perhaps that which has the Appearance of being Simple is actually difficult to do. Knowledge is Facts in their Intrinsic Relations and we do not often have that. Beyond Facts we have Propositions. It is possible to distinguish between different types of statement. Specialists in philosophy have undermined Verification as a possibility, but that which is propositional is asserted by some experts as truths. The Action of Assertion is supported.

Despite Taylor's observation and description, "skilled witnesses" continued to be in demand in Criminal and Civil proceedings. I have observed in recent times many comments with the same Meaning as that expressed by Taylor. Little seems to have changed over the Centuries. So a question springs to mind. If the testimony of skilled witnesses is opinion and "least deserves credit with a jury", then why did the practice of instructing them to provide Opinion persist? The fact finder, be it Judge or Jury, does give their Opinion the credit which it "least deserves". That is necessarily so if the difference between Opinion and Evidence is dropped.

The report notes another detail that is Significant. The Source of this quote was not given:

> "2.27 For others, however, the situation was attributable not so much to the expert witness as it was to the lawyers: Armed with an hour's reading… the great man [i.e. the lawyer] comes down to court to puzzle, bewilder, and very often to confute men of real ability… A pitiable specimen is that poor man of science, pilloried up in the witness box, and pelted by the flippant ignorance of his examiner! What a contrast between the different caution of the true knowledge, and the bold assurance, the chuckling confidence, the vain-glorious self-satisfaction, and mock triumphant delight of his questioner!"

That is perhaps a satirical view of the Adversarial System, or perhaps it is not an exaggeration and is an accurate description in some cases. There is no doubt that the lawyers do on occasions puzzle and bewilder witnesses and that their reading on the topic is not extensive. Through the questions of his examiner, the man of science can be led into Error. Lawyers are satisfied and delighted when they move closer to winning the case by any means possible for that is their goal. Exaggerated or not, this description of the Adversarial System is actually a description of a particular philosophical strategy. I will name that now and you will recognise the name but you will possibly not recognise the description as being that Method until I deal with that later. The Adversarial Action is based on the Socratic Method. The Socratic Method is practised outside the Boundaries of Pure Science.

I note the term "different caution of the true knowledge". Pure Science is different to science. People who habitually engage in Pure Science cannot be led into Error. Men of Pure Science can tell the difference between Fact and Proposition. Pseudo-science is Ignorance by another name. The 'pseudo' part of the Representation that distinguishes its character was dropped leaving the word science. Another word for pseudo as a description of science is the word inductive. Men of Inductive science do not access True Meaning and thus they do not have Knowledge and sometimes they do not even know Facts. Gaps in Knowledge can easily be revealed. I have not expressed an Untestable and Unproven Belief. I will at length provide Evidence for that statement.

There are two main stages in the evaluation of evidence in the Context of the Court. First, the admissibility of specific testimony is decided by a Judge. Second, once admitted into Court, the weight to be given to testimony is decided by Jurors or Judges. In between its admission through the gate of the Court and the Judgment of its weight, evidence is tested by the Adversarial Process. It is commonly acknowledged that it is difficult to separate the issues of admissibility and weight, but a Boundary is Marked. Boundaries are flexible. In this Chapter Attention is focussed on the approach to admissibility.

THE DEFINITION OF AN EXPERT WITNESS

It is for individual Judges to decide who can Act as an expert witness in a specific case and they do not select the particular individual. The difference between an expert and an agent who is not an expert is Marked at a general level. The selection of a particular expert witness is usually delegated to those Acting for the parties in the case. The Judge tests the fit. Qualifications, standing within a profession and experience are all Factors taken into account in the decision. As described in Science in Court October 2005 Number 248:

> "Courts gauge the witness's expertise by enquiring into relevant matters such as their employment history, membership of professional bodies and experience of similar work."

Those individuals who meet the criteria for admission to Court as experts are allowed to give their opinion under certain conditions. Mr Justice King CJ in the Australian case of Bonython (1984) describes some of the conditions for opinion testimony:

> "The general rule is that a witness may give evidence only as to matters observed by him. His opinions are not admissible. One of the recognized exceptions to this rule is that which relates to the opinions of an expert. This exception is confined to subjects that are not, or are not wholly, within the knowledge and experience of ordinary persons....... On such subjects a witness may be allowed to express opinions if the witness is shown to possess sufficient knowledge or experience in relation to the subject upon which the opinion is sought to render his opinion of assistance to the court. Before allowing a witness to express such opinions, the judge must be satisfied that the witness possesses the necessary qualifications, whether those qualifications be acquired by study or experience or both. But when it is established that the witness is an expert in the relevant field of knowledge, he will be permitted to express his opinion, however unconvincing it might appear to be (Commissioner for Government Transport v. Adamcik 2), subject always, of course, in a criminal trial to the

discretion to exclude evidence whose prejudicial effect is disproportionate to its probative value. The weight to be attached to his opinion is a question for the jury."

Mr Justice King adds that:

"Generally speaking, once the qualifications are established, the methodology will be relevant to the weight of the evidence and not to the competence of the witness to express an opinion. The suitability and adequacy of the methods used may well be themselves a matter of expert opinion."

The distinction between admissibility and weight cannot be sharply drawn. There are some Rules of exclusion on admissibility of method and on the weight to be attached to particular methods, to which I will come in due course. First we will look at the consensus on the general criteria that need to be met in order to present Opinion as an expert. In the United States of America, Rule 702 governs expert testimony and that Rule as stated in the case of Daubert (1993) is:

"If scientific, technical, or other specialized knowledge will assist the trier of fact to understand the evidence or to determine a fact in issue, a witness qualified as an expert by knowledge, skill, experience, training, [509 U.S. 579, 8] or education, may testify thereto in the form of an opinion or otherwise."

In similar vein, The UK Guidance Manual for Expert Witnesses (2010) states:

"36.2 There is no definitive legal definition of an expert. It is a matter for the court to rule upon in each case. However for the purposes of this guidance, an expert is defined as: 'a person whose evidence is intended to be tendered before a court and who has relevant skill or knowledge achieved through research, experience or professional application within a specific field sufficient to entitle them to give evidence of their opinion and upon which the court may require independent impartial assistance'."

The Court, i.e. the Judge or Judges, can exclude an expert proposed in a trial and there are Selections before that, of which professional qualifications and experience are a part. You see in the statement from the Disclosure Manual that an expert witness is to give independent impartial assistance. On the issue of competence to give evidence in Court, that Manual also states: -

i. "36.11 A witness is professionally competent if he/she may lawfully give evidence. Whether a witness is competent to give evidence as an expert is for the judge to determine.

ii. 36.12 A witness is credible if he/she is seen to be trustworthy, reliable and capable of being believed. Whether a witness is credible is for the jury to determine.

iii. 36.13 Whilst, for most purposes, the issue of competence and the issue of credibility of an expert may be kept distinct, there may be issues of credibility that affect the expert's competence, and vice versa. For example, an expert who has been convicted of a dishonesty offence may be regarded as less likely to be wholly objective and balanced in giving expert evidence.

iv. 36.36 An adverse judicial finding is a finding by a civil or criminal court, expressly or by inevitable inference that an expert witness has knowingly, whether on oath or otherwise misled the court.

v. 36.37 With regard to experts this definition may be extended to include unintentional misleading of the court, i.e. incompetence.

vi. 2.1 The overriding objective, which is at the heart of the Criminal Procedure Rules (CrimPR), is that criminal cases be dealt with justly. The presumption of innocence and a robust

> adversarial process are essential features of the legal tradition in England and Wales and of the defendant's right to a fair trial."

You are seen to be incompetent as an expert witness if there is a finding in Court that you have been so in a particular case. If you are not seen to be incompetent you are seen to be competent. The definition of a fair trial depends on the type of Standard in Action. A trial cannot be fair if Truth is Not an Issue. Or can it? Well that depends on what you Mean by fair. The Meaning of the term a "fair trial" is not specified. From the guidelines so far you might draw the conclusion that Truth is the Standard. But is that the right conclusion? Note that the word truth is not in use. And there is a Quantitative definition of fair, as I have demonstrated. You can see the Potential for Confusion.

A presumption is not an assumption or a Premise. The difference in Meaning between those words is important. An assumption is a conclusion. A Premise is an assumption in conjunction with the Assertion of its truth. A presumption is a beginning, it sets a starting point to serve a particular purpose and its truth is not Assumed or Asserted. Assumptions and presumptions both introduce a Bias. An assumption is not set to be tested. A presumption is set in order to be tested. Testing makes a Significant Difference, but the type of difference it makes depends on the type of testing.

In each case in Court, Innocence is the prior setting - a defendant is Innocent until Guilt is proved in Court. The presumption of Innocence introduces a particular Bias in order that the Bias can be tested and testing must be fit for that purpose. While a Bias is tested in Court, it is not Innocence that is tested. The Process of testing for Guilt is presumed to be more demanding when the Bias is towards Innocence than it would be if a presumption of Guilt was set. Where there is Innocence there is absence of Guilt. The absence of Guilt is difficult to prove. In general, it is difficult to prove that something is not there and it is easier to prove what is there. But when Assertion replaces proof an audience may accept that what is not there is there. Assertion only replaces proof if it is allowed to do so and its presence must be identified before it can be prevented. Round and Round we go in Dynamic Systems. Natural Systems are Dynamic Systems that go Round and Round and yet Natural Systems do not stay in one place. Artificial Systems can go Round and Round while remaining fixed in the same position. The Justice System is designed as a Whole by human beings and has some Natural parts.

It is a Fact that differences in starting point have an influence on conclusions. That Fact has been known for Centuries. The positioning of the starting point at Innocent indicates that Innocence must be Falsified. It is not possible to be both Innocent and Guilty in relation to one particular happening. That is a Fact. But the Fact is not as Simple as it Appears to be. In order to understand the underlying Complexity, it is necessary to understand the Structure of Meaning. Curiously, Meaning is rarely mentioned by Specialists in philosophy and Appearances are. A Representation is representational. In the Pocket Oxford Dictionary, the word representational is defined as:

representational: adj. Art depicting a subject as it appears to the eye.

The subject of a Representation is Meaning. The Representation of Meaning in Art is defined as depicting Meaning as it appears to the eye. Appearances are Deceptive more often than not. But Meaning can fit its Representation. And it is a Simple Fact that one single type of Evidence can Verify or Falsify one Whole Meaning as it Appears to be. Pure Science is involved in the Action of Verifying and Falsifying.

So, the Adversarial Process is set to test the offering of proof that Guilt is there presuming that Innocence is there. That setting is Justifiable. According to MOM, there is always a Bias of one type or another in Social Dynamics. The Significant Issue is not whether there is a Bias and is what type of Bias is there. So, in Theory, a Criminal Trial is at the start Biased towards Innocence. However, Garrett showed that a Bias towards Guilt was Active on evidence before the evidence passed through the gate and entered the Court. If the evidence is Biased then that Bias is present during the Trial. In order to Act at all, a Bias must be in a position where it can Act. If Truth is not the Bias in the evidence then the Bias of Truth will not be in that position to Act in Court. And the adversarial process failed to detect Error in these cases.

I will repeat that Meaning in a different way. The presumption of Innocence begins at the gate into Court. If the Bias towards Guilt was Active before the Trial it can be carried into Court in the form of evidence. Once through the gate, that Bias becomes Active in Court. It is a Fact that one type of Bias Acts against another type of Bias. Bias is a weight, and weight counts in the final Judgment. Evidence is necessarily Biased towards either Guilt or Innocence and in Court evidence of Guilt is added to the scales and weighed against a static amount of Innocence. At some point one side of the scale is heavier than the other and Meaning changes. The Defence tries to take off some of the weight on one side.

As evidence related to Guilt is progressively added to the Scales of Justice there comes a point at which there is a Qualitative change in the perception of Meaning by those who are observing the weighing. The fact finders are necessarily Observing as others add the weights to one side of the scale. Their opportunities for questioning are limited. Most of the questioning is done by lawyers who "armed with an hour's reading ...... come down to court to puzzle, bewilder, and very often to confute men" whatever their abilities might be. Fact finders are Not Immune from the effects of that type of Action. Lawyers are Active on both sides of the argument but Attention is mainly focussed on one side of the scales. And it is clear that some men of science stand their ground supported by Impressive Qualifications.

SOME DUTIES AND RESPONSIBILITIES OF EXPERTS IN CIVIL CASES

In the Case of the Ikarian Reefer (1993), Mr Justice Cresswell set out a number of criteria to clarify the role of expert witnesses. For Civil Cases, the duties and responsibilities of experts were summarised as:

1. "Expert evidence presented to the Court should be and should be seen to be the independent product of the expert uninfluenced as to form or content by the exigencies of litigation..."

2. An expert witness should provide independent assistance to the Court by way of objective unbiased opinion in relation to matters within his expertise….An expert witness in the High Court should never assume the role of advocate.

3. An expert witness should state the facts or assumptions on which his opinion is based. He should not omit to consider material facts which detract from his concluded opinion…

4. An expert witness should make it clear when a particular question or issue falls outside his expertise.

5. If an expert's opinion is not properly researched because he considers that insufficient data is available then this must be stated with an indication that the opinion is no more than a provisional one……

6. If after exchange of reports, an expert witness changes his view on a material matter….such change of view should be communicated…to the other side without delay and when appropriate to the Court.

7. Where expert evidence refers to photographs, plans, calculations….survey reports or other similar documents these must be provided to the other party at the same time as the exchange of reports……."

As the criteria are commonly reproduced, the responsibility at part 2 specifies that the role of advocate should never be assumed in the High Court. It is possible given that wording to arrive at a conclusion that the role of advocate can be assumed in other Courts. I have always presumed that these responsibilities apply in all Courts. I have seen it stated in other Judgments that an expert should never assume the role of an advocate. That which was specific became generalised. In this case the generalisation of the Rule is necessary. The Courts of Appeal are higher in the System and test the evidence given in other Courts. Experts have already given the evidence to which the Standards in the Court of Appeal will be applied. If those standards are not applied at the outset, during the preparation and presentation of the expert opinion, then the criteria will not be met wherever that opinion is given. The expert witness will then be subject to criticism. However, it is the few not the many Cases that come under scrutiny of the Judges in the Appeal Court. Those who know the Statistical Equations of Probability could calculate the chances of being among the few who are measured against that High Standard. Gamblers do not know those equations but they do calculate the odds informally.

Lawyers Select experts. "Armed with an hour's reading" they cannot assess the difference between experts in the amount of Knowledge that experts have or the Method they use. At the point of Selection, the lawyers cannot look for the Reasoning that Judges look for in the Single Case, for until the Case has been analysed by the expert the Reasoning is not there. A Qualitative Measure of Reasoning cannot be applied where there is none to be seen. There is no training in Reasoning and no formal qualification in that. Training is provided in Inductive reasoning and that is a different type of thing. Skill in Inductive reasoning leads to Impressive Qualifications. I will disclose the difference between Induction and Reasoning in due course. The lawyer is an advocate. The Selection of experts is determined by the purpose of the lawyer, and that is to win the case for his client.

As Represented in the list of duties, the sixth criterion is not as straightforward as it Appears to be. That criterion suggests experts can change their minds on a material matter and they can, and of course they should do that under certain conditions. But if they do so in the absence of new data then they are admitting to having made an Error. That is frowned upon. Errors are misleading and costly. The effects of Errors are difficult to correct. The original data processing by the expert should have been thorough. Only new evidence should lead to a change of opinion. The Opinion of another expert in their own field is not new evidence for them. And it is against the Rules for parties in the case to ask experts to change their Opinion. Opinion must be Unsettled in order to dislodge it.

I now draw your Attention to a statement I picked up in passing from one of the many "factsheets" at which I have glanced. There was a reference to Mr Justice Garland's Judgment in the Polivitte case. The details of the case are not relevant here and I have not seen the original transcript of that. The following was quoted from his Judgment:

> "I have almost considered the role of an expert to be two-fold: first, to advance the case of the party calling him, so far as it can properly be advanced on the basis of information available to the expert in the professional exercise of his skill and experience; and secondly, to assist the court, which does not possess the relevant skill and experience, in determining where the truth lies."

That Meaning is given in more than one arrangement that I have seen. But that position is not as Simple as it Appears to be. It is said that truth is an Issue in Court and that the expert does not determine where the truth lies. The expert assists the Court to find the truth. And the Court "does not possess the relevant skill and experience" to find the truth without assistance. But it Appears to be the evidence that points the fact finder in the direction of truth. And the evidence is the opinion of the expert so that is Confusing. In order for an expert to assist the Court in determining where truth lies, the purpose of the expert must be the discovery of truth and the evidence of the expert must be aligned to Truth. That is not the Point of Reference of the evidence in an Unquantifiable number of cases. That is not the Point of Reference of Hearsay. The Problem of Hearsay is the absence of Truth. That is Not an Insoluble Problem for Truth can be told and thus Truth can be heard. And I will redefine the Problem of Hearsay as the absence of Certainty of Truth. Hearsay must be Verified or Falsified for the problem to disappear. Pure Science is the only Process that can do that. Pure Science is a Process that can provide Certainty.

Members of the legal profession do not commonly want independence in their witnesses. They are tasked with winning the case. They want advocates or at least support for their side of the argument. The expert is selected to advance the case of the party calling him. If the Bias of the Court is the finding of Truth then there are two Conflicting Biases in Operation in the Expert Witness System. Where there is a Bias there is a particular Point of

Reference. The two different Points of Reference are: To advance the case of the party calling him. And: To assist the Judge in finding the Truth. Different Points of Reference draw different Meaning from arrangements. In theory, the instruction to present the range of possibilities allows the expression of different views from different Points of Reference. But the case of the instructing party is not advanced by showing the view that is aligned to winning their case and then showing the view that comes when Attention is paid to Truth if that Meaning is different. If the expert specifies the difference in Meaning as it aligns to these different Points of Reference then the lawyer's case is moved towards defeat. Also, if the view that advances the case is the one the expert has given as his Opinion and that is not the True Meaning, then the expert would have to argue against himself in Court in order to assist the Judge in the discovery of Truth. That is of course what the expert should do, but the Court is not the place to do that.

Truth cannot be Reconciled with opposing Biases. The expert cannot apply two different Points of Reference and arrive at the same conclusion unless the Points of Reference are the same in type. Things of the same type Align Naturally. An expert cannot change his Opinion in Court. And if he Believes his own Opinion, then he thinks truth is his Point of Reference. Remember, the lawyer selects the expert that will advance his case, so he searches for an expert with Beliefs that fit his argument. The proposition that the criteria for the Selection of witnesses are opposed to the Obligations placed on them by the Court is a Reasonable one. That general proposition is Testable from various angles.

One type of Bias can Act against another type of Bias if it is strong enough. Truth is a Strong Bias if it is present. But the expert has the Certainty of Belief and the fact finder does not have Certainty and needs assistance to find it. The Certainty of Belief would flow from one position to the other. The fact finder would need to distinguish between different types of Certainty in order to prevent that. And if the Point of Reference for the evidence is not Truth then Truth will not be there. If Truth is not sought or has been removed from the Opinion before it reaches the Gate, then evidence with the Quality of Truth cannot be passed through the Gate and cannot become Active to assist the Court. If the Source of the Bias that cannot be reconciled with Truth lies outside Court then it must be dealt with at Source. The lawyer is not the Source of the Beliefs of experts. The lawyer finds what the expert believes and Selects the expert on the basis of the Beliefs he finds there.

AN EXCEPTION ALLOWS OPINION FROM EXPERTS

There are conditions for the exception to the Rule that enables experts to give Opinions in Court. In 1782 in the case of Folkes v. Chadd, Lord Mansfield ruled that an expert's opinion is admissible only if it furnishes the Court with information that is likely to lie outside the common knowledge and experience of the jury. And the Meaning of that statement hinges on the definition of what is common knowledge and experience. Definitions form the foundations for Judgments of Meaning. Normal psychological processes are taken to be within the knowledge and experience of the jury. The opinions of psychiatrists and psychologists are only admissible in respect of some abnormality. That distinction was made clear in R. v. Chard (1972) and R. v. Turner, (1975) and is known as the Turner test. I will return to that test later.

In addition to conditions that relate to professional qualifications and experience, the decision that expert opinion has sufficient basis in its Method to be admitted as evidence is subject to Judicial Discretion. The weight of evidence is a matter for the fact finder once the evidence has been allowed to pass through the gate, and Judges guide the finding of facts. Both qualifications and Method are subject to peer approval within professions. And it is small selected groups of peers within each profession who determine what is approved. Thus, those individuals who govern the professions give their weight to decisions in Court at all points where experts are involved. And decisions in one Court travel through the Justice System. In respect of Method, Mr Justice King C J in Bonython (1984) stated:

> "If the witness has made use of new or unfamiliar techniques or technology, the court may require to be satisfied that such techniques or technology have a sufficient scientific basis to

render results arrived at by that means part of a field of knowledge which is a proper subject of expert evidence."

The ruling of the US Supreme Court in the case of Daubert v Merrell Dow Pharmaceuticals (1993) states some common Rules for expert evidence and here we see stated an underlying assumption:

> "Unlike an ordinary witness, an expert is permitted wide latitude to offer opinions, including those that are not based on first-hand knowledge or observation. See Rules 702 and 703. Presumably, this relaxation of the usual requirement of first-hand knowledge - a rule which represents "a `most pervasive manifestation' of the common law insistence upon `the most reliable sources of information,'" Advisory Committee's Notes on Fed. Rule Evid. 602, 28 U.S.C. App., p. 755 (citation omitted) - is premised on an assumption that the expert's opinion will have a reliable basis in the knowledge and experience of his discipline."

So, the wide latitude in Court to offer Opinions that are not based on first-hand knowledge or observation relies for its standing on Premises. There usually are sets of Premises supporting an argument in Formal logic. Qualifications are a measure of peer approval within a profession for an individual's Actions during training and beyond that. Qualifications are an indicator of knowledge and experience of the methods used in a discipline and of what has been discovered using those methods. So the fit of the expert to the discipline is measured by qualifications. But is the fit of the "knowledge and experience" of the discipline to the task of the Court measured? And can that be done?

If the Court aims to determine where the Truth lies then specialists in the Professions can only assist the Court in that task if the methods in the Profession are governed by a search for Truth and are capable of discovering Truth. We have a question that leads to Testable Propositions. Is Truth the Point of Reference in the Court and in the Professions? That brings us back to the need to distinguish and Mark the Process of Pure Science.

I note the following comment included in Footnote 9 to the Supreme Court Ruling in Daubert:

> “We note that scientists typically distinguish between "validity" (does the principle support what it purports to show?) and "reliability" (does application of the principle produce consistent results?). .... Although "the difference between accuracy, validity, and reliability may be such that each is distinct from the other by no more than a hen's kick ...."

So, here it is noted that a reliable principle is one that produces consistent results. But the type of consistency is not specified. That which is consistent is repeated. A thing that is repeated happens again and again. That which has a high frequency of occurrence can be good or bad, functional or dysfunctional. Application of a particular principle or Method will lead to results that are repeatable but the consistent results could be in Error. It follows using Natural Logic that some things should not be repeated. Repetition is not an indication of the type of Meaning or of its Qualities. Some principles and methods produce consistent Errors. We can rely on them to do that. The Judges will have checked the credentials of the experts in the 250 proven Miscarriages of Justice that Garrett examined. That is necessarily so since Procedure Rules are followed and a failure to do so is grounds for Appeal. In the 250 cases analysed by Garrett, Innocence was reliably transformed into Guilt by Testimony that included evidence prepared and presented by experts who passed with their arrangements through the gate into Court. Their opinions would have been based on the knowledge and experience of their disciplines, or at least will have been perceived as being based on that. So, since nothing is assumed within the Process of Pure Science, we now have some more propositions to Test. One proposition is that the disciplines are not based on Knowledge and are based on something different to that. Another proposition is that Impressive Qualifications do not differentiate between those individuals who have Knowledge in the disciplines and those who do not. The word knowledge covers different types of Meaning and we need to see that difference. Clearly there is a considerable body of experience in a discipline but the type of experience makes a difference to what follows from it. Facts are not discovered using Inductive science. When Facts are discovered through Pure Science and are taught, knowing those does not necessarily lead to Knowledge.

The definition of valid is, as the Judges saw, Intrinsically Related to the definition of reliability as a frequency of occurrence. The difference between validity, reliability and accuracy is no more than a hen's kick in Inductive science and that is where the Judges were looking. Their Observation was True from that point of view. Truth is not the Point of Reference for Inductive science and when the word accuracy is used in that Context its Meaning is not governed by Truth. When accuracy is defined as alignment of Meaning to Truth their Observation was False. When Truth is the Point of Reference for accuracy, the difference between validity and reliability is Trivial while the difference between their Meaning and the Meaning of accuracy is Significant. The Quantitative Construct of reliability stands where a Concept could be. Reliability and accuracy would be related if both words were used in the Context of Pure Science and Truth could be brought to bear on validity. The Significant Difference in Meaning is made by the difference in the Point of Reference. Differences that are Significant in terms of Meaning are rendered Trivial by men of Inductive science. When Judges accept the Meaning given by those men of science, the Authority of the Judges is Associated with that type of Meaning and with those who present it. People rise to positions of Power when Authority is gained through Association. That is an Ancient Fact.

I will Test the common thread of Meaning and Action that lies at the foundations of what is described as "knowledge and experience in disciplines". More than one Tale needs to be told before the Whole Meaning and Action can clearly emerge. In this narrative I begin to provide Evidence that shows the one thing that binds them all in Error. Pure Science has particular Meaning. But the Meaning and Action of Pure Science is different in type to that which bears the name science and it is science that is Active in the social sciences. The social sciences as a Whole are now distinct from each other "by no more than a hen's kick". They have the same Point of Reference and it is Not Truth. I will show you Evidence that differences in type of Meaning and Action have been removed from sight as far as it is possible to do that and as far as it is convenient to do that. The effect of hiding differences in Meaning is to leave Unmarked the

Boundaries where Meaning changes. There is no stopping point. I will show you where the Action of those engaged in the Inductive sciences has led Social Dynamics. I will show you the different type of Action of those who engage in Pure Science. Though they are hidden now, the differences are still there and can be brought into view. It is possible to Mark differences that are of great importance.

PART TWO of Chapter one

PARTS OF THE US APPROACH TO GATEKEEPING

The development of guidelines for admissibility of expert evidence can be Tracked through landmark decisions in Courts. While there are differences in Judicial Opinion in different countries and in different states within them, some US general criteria are often cited and are influential across jurisdictions and in different countries. Meaning is passed on. The detail is interesting and relevant to my search for the Qualities of Pure Science.

The Frye test of "general acceptance" 1923

This test was established on Appeal in a trial for murder. At Trial, Counsel for the defendant offered an expert witness to testify to the result of a deception test applied to a defendant. That test is now a well-known Construct of Deception. That measure has the familiar name of The Lie Detection Test. The test carries Error and that Fact is known. That test remains inadmissible. And the test is used outside the gates. The ruling of the Appeal Court states:

> "The test is described as the systolic blood pressure deception test.........the theory seems to be that truth is spontaneous, and comes without conscious effort, while the utterance of a falsehood requires a conscious effort, which is reflected in the blood pressure. The rise thus produced is easily detected and distinguished from the rise produced by mere fear of the examination itself. In the former instance, the pressure rises higher than in the latter, and is more pronounced as the examination proceeds, while in the latter case, if the subject is telling the truth, the pressure registers highest at the beginning of the examination, and gradually diminishes as the examination proceeds."

The Judges had identified assumptions in the underlying theory. The accuracy of the measure relied on the Truth of the theory and that depended on the Truth of assumptions. But Attention centred on whether the theory was new and generally accepted, rather than on its status as a theory and not Fact. Here, Truth was not the Standard applied in Court to the evidence. A different Standard at the Gate was made clear in delivering judgement to exclude this testimony. So I note that different Standards can lead to the same Judgment. That is a Source of Confusion. The Judge stated:

> "Just when a scientific principle or discovery crosses the line between the experimental and demonstrable stages is difficult to define. Somewhere in this twilight zone the evidential force of the principle must be recognized, and while courts will go a long way in admitting expert testimony deduced from a well-recognized scientific principle or discovery, the thing from which the deduction is made must be sufficiently established to have gained general acceptance in the particular field in which it belongs."

That criterion of "general acceptance in the particular field in which it belongs" places some aspects of admissibility in the hands of groups of active agents in professional bodies outside the Legal Context. Agreement among peers in a professional organisation is said to confer evidential force. It is not safe to take peer agreement as a sign of Justification, for it is not that. Presumably, the Premise is that peers test the principle or discovery and the underlying theory. There are different types of testing. Differences in Type are a Significant Variable and must be taken into account in judging Meaning. A question must be asked of necessity: What is the testing by peers designed to test?

The Judgment on the particular Method in the Frye Case was abstracted as a General Rule. Having thus acquired a measure of vagueness the ruling was variously applied to Single Cases where differences lie. Many different views have been expressed in relation to Frye. I note the following from within the Judgement in the case of Marsh v Valyou (2007):

- "Seventy years after Frye, the United States Supreme Court held that the adoption of the Federal Rules of Evidence superseded the Frye test. See Daubert v. Merrell Dow Pharms., 509 U.S. 579, 587 (1993). Daubert adopted a different test for admissibility under which the Frye test—general acceptance in the scientific community—is simply one factor among several".... "By definition, the Frye standard only applies when an expert attempts to render an opinion that is based upon new or novel scientific techniques." U.S. Sugar Corp."

- "The merits of the Frye test have been much debated, and scholarship on its proper scope and application is legion. .... Petitioners' primary attack, however, is not on the content but on the continuing authority of the rule. They contend that the Frye test was superseded by the adoption of the Federal Rules of Evidence. ..... We agree."

This agreement of the Judges in Marsh v Valyou was in line with the agreement of the Judges in the earlier Daubert Case. Different Judges sitting in Single Cases add notes and riders of their own in the Judgments. Of the additions in Marsh v Valyou (2007) I note the following:

> "While this Court has continued to apply Frye in determining the admissibility of scientific expert opinion testimony after the adoption of the Florida Rules of Evidence, it has done so without confronting the fact that those rules do not mention Frye or the test set out in Frye. Hence, unlike the United States Supreme Court, we have never explained how Frye has survived the adoption of the rules of evidence. Because, like the United States Supreme Court, I find no basis for concluding that Frye has survived Florida's adoption of an evidence code similar to the federal code, I would recede from our cases continuing to apply Frye and hold that the rules of evidence do not include a Frye test for determining the admission of expert testimony. In fact, the adoption of these evidence codes was intended to apply a straightforward relevancy test to expert evidence and, in essence, to establish a rule favoring admissibility once relevancy was established, while leaving it to the fact-finder to determine the credibility and weight of such evidence"

So, the Frye Rule was Unsettled and the Unsettling disturbed but did not disable its use. Sometimes it is accepted that the Frye Rule applies and sometimes not. And the Frye Rule is defined in different ways. There was agreement in Daubert that the Frye Rule was insufficient, but a version of that Rule was included. There was an adaptation. In short, there is agreement and disagreement over the criteria for admissibility and continuing debate. It is noted within Footnote 4 of the Ruling on Daubert that:

> "Indeed, the debates over Frye are such a well-established part of the academic landscape that a distinct term—"Frye-ologist"—has been advanced to describe those who take part. See Behringer, Introduction, Proposals for a Model Rule on the Admissibility of Scientific Evidence, 26 Jurimetrics J. 237, 239 (1986), quoting Lacey, Scientific Evidence, 24 Jurimetrics J. 254, 264 (1984). [N.5]. Like the question of Frye's merit, the dispute over its survival has divided courts and commentators."

Where Meaning is Not Specified, there is Uncertainty and thus room for Opinion and argument. Meaning that is Not Specified is Not Testable. But in essence, Frye passes the decision on admissibility to the professional discipline covering the field and states the general conditions for doing that. Or at least that Appearance is given. Do we know what is actually happening? We need to discover what the terms "general acceptance" and "scientific community" mean in practice. Now we will turn to look at some particular detail of the Daubert case.

The Daubert Judgment

In delivering Judgment in the U.S. Supreme Court (Argued March 30, 1993 Decided June 28, 1993), it was explained that: "In this case, we are called upon to determine the standard for admitting expert scientific testimony in a federal trial." Particular Meaning is necessarily taken from Single Cases. The Significance of the Standard of Frye in matters of litigation becomes clear when the details of the Case from which the Daubert criteria emerged are Observed and Meaning is Tracked from those details. Single Cases are the Context in which the Law is applied and the Law is the Context in which Judgments in Single Cases are made. It is necessary to understand the relation between them. Both Contexts are set out in this Supreme Court Ruling:

> "Petitioners, two minor children and their parents, alleged in their suit against respondent that the children's serious birth defects had been caused by the mothers' prenatal ingestion of Bendectin, a prescription drug marketed by respondent. The District Court granted respondent summary judgment based on a well-credentialed expert's affidavit concluding, upon reviewing the extensive published scientific literature on the subject, that maternal use of Bendectin has not been shown to be a risk factor for human birth defects. Although petitioners had responded with the testimony of eight other well-credentialed experts, who based their conclusion that Bendectin can cause birth defects on animal studies, chemical structure analyses, and the unpublished "reanalysis" of previously published human statistical studies, the court determined that this evidence did not meet the applicable "general acceptance" standard for the admission of expert testimony. The Court of Appeals agreed and affirmed, citing Frye v. United States, 54 App. D.C. 46, 47, 293 F. 1013, 1014, for the rule that expert opinion based on a scientific technique is inadmissible unless the technique is "generally accepted" as reliable in the relevant scientific community."

I am puzzled by the suggestion that animal studies and chemical structure analyses are not "generally accepted" as reliable in the relevant scientific community. The scientific community taken to be relevant is not specified. That detail is missing and it is a Significant Detail. Different communities accept different techniques. A Fact emerges plainly through the description. Given that doubt can be cast on the question of Causation, responsibility for a happening cannot be assigned. In this Single case, the Frye Ruling had enabled Unsettling of a position. It did so in the absence of detail within the Single Case and in the absence of the detail of the Point of Reference for the term general acceptance. The Frye Ruling provides a Structure that Selectively enables the Action of some things and disables others regardless of particular Meaning in Single Cases. Uncertainty in relation to a proposition about Causation had been accepted by the District Court and by the Court of Appeals. A verdict on Bendectin of Not Guilty had been returned at both those levels of the Justice System. A further Appeal against the Verdict to the Supreme Court had been allowed.

We recall that according to Garrett "most constitutional rights relate to procedure and not accuracy". Here we see that the distinction between procedure and accuracy is not as clear as it might appear to be. Rules govern the procedure and the Appeal was allowed on the Issue of the Rule. And the Rule determines the weight given to the type of evidence in this case. The type of evidence admitted into Court and the relative weight given to it are related to the accuracy of a Judgment. And the type of Standard applied is a Significant Factor in the accuracy of a Judgment. The Frye Rule had been used as the Standard for admissibility and for weighing the evidence. Evidence that had been admitted into Court was then restricted in its admissibility within Court. The Justification of the Standard was now the issue to be Judged. The Supreme Court was to decide whether the Standard of Frye was fit for purpose. The Standard of Frye was acceptance in the scientific community. The type of science and the type of community that accepted it were not specified and that is relevant. Different types of communities accept different types of thing. So, did the differences in type become clear?

The Methods to which the Frye test was applied in this Case were shown. I have applied highlights to parts of the next quote from the Supreme Court Judgment. There was one well-credentialed expert instructed for the respondent and he used one Method:

> "After extensive discovery, **respondent** moved for summary judgment, **contending that Bendectin does not cause birth defects in humans and that petitioners would be unable to come forward with any admissible evidence that it does.** In support of its motion, respondent submitted an affidavit of Steven H. Lamm, physician and epidemiologist, who is a well-credentialed expert on the risks from exposure to various [509 U.S. 579, 2] chemical substances. **Doctor Lamm** stated that he **had reviewed all the literature on Bendectin and human birth**

> **defects - more than 30 published studies involving over 130,000 patients. No study had found Bendectin to be a human teratogen (i.e., a substance capable of causing malformations in fetuses).** On the basis of this review, Doctor Lamm concluded that maternal use of Bendectin during the first trimester of pregnancy has not been shown to be a risk factor for human birth defects."

The term "admissible evidence" indicates that there can be evidence that is inadmissible. Admissibility is decided by Judges. The weight to be attached to the admissible evidence is decided by the fact finders in Court. We recall that the petitioner had instructed eight well-credentialed experts whose testimony was composed from mixed methods - animal studies, chemical structure analyses, and the unpublished "reanalysis" of previously published human statistical studies. The eight experts for the prosecution had taken different types of evidence and thus presented different angles of view. The single expert for the defence had taken one type of evidence and presented one angle of view. The Supreme Court noted that the District Court had held that:

> "Given the vast body of epidemiological data concerning Bendectin .... expert opinion which is not based on epidemiological evidence is not admissible to establish causation. .... Thus, the animal cell studies, live animal studies, and chemical structure analyses on which petitioners had relied could not raise, by themselves, a reasonably disputable jury issue regarding causation. .... Petitioners' epidemiological analyses, based as they were on recalculations of data in previously published studies that had found no causal link between the drug and birth defects, were ruled to be inadmissible because they had not been published or subjected to peer review."

So, the evidence that was not epidemiological was allowed through the gate and then its admissibility was restricted in relation to causation. It is said they could not "raise, by themselves, a reasonably disputable jury issue regarding causation" and it was not said that together they could not do that. Or was it? Were these different types of evidence categorised as a single type? That is not made clear. Even together they did not raise a "reasonably disputable jury issue regarding causation" in this case. The District Court had given epidemiological evidence Power over other types of evidence taken alone and together in conjunction. In addition to that, Power had been given to published studies over evidence of the same type that had not been published. In Trials of some other Single Cases the Power to Act against some types of evidence and unpublished analyses was also drawn from the ruling in Frye. It is a Fact that epidemiological data is statistical data. The Frye Rule was used to give weight to Inductive science.

In summary, the Premise set in the lower Courts was that statistical evidence is the only evidence admissible to establish causation. In conjunction with that it was held that the only statistical data deemed acceptable by the Court was data in articles that had met criteria for publication. Publication involves peer review. So that accepts a Premise that peer review is a test of the data. Justification for the Ruling based on Frye depended on the Truth of the Premises and Premises are Assertions that are Not Tested for Truth. The presence of Premises indicates that Formal logic is in Action and their acceptance does too.

Statistics is a Quantitative Method and Publication involves peer review, for which a Quantitative Construct is Active. One Premise is built from others. The Premise that in the absence of epidemiological data there cannot be a reasonably disputable jury issue regarding Causation depends on the Premise that statistical data is relevant to a disputable jury issue regarding Causation. Premises can be turned into Testable Propositions. I will use Single Cases to Test Propositions for Truth and Justification, but I will not take the Cases from the Legal Context. The Law is composed of agreements of peers in particular Contexts. The detail of the Law is beyond my area of expertise. But, Testing the effects of different Methods on the Quality of Judgments is within my area of expertise. A Method is a type of human Action on data.

Statistics is the Action of Juggling with Frequencies through the use of mathematical equations - the Epidemiological Inferences are drawn by Induction. It is peer agreed by Inductive scientists that

Inductive science is not relevant when the issue is regarding Causation. And that is Universally agreed for it is a Fact that is Undeniable. All types of scientific community agree some Facts and that is one of them. That particular agreement was not taken into account in this case. Perhaps the Judges had not seen that one. Or perhaps they knew that and did not bring that into view. Agreements must be Tested for Justification whether or not they suit us in the Sense that they fit our argument as a Whole, so I will do that in due course. But I note that I do not know if drawing that Fact into view would or would not have made a difference in this or any other Single Case in Court. It is for the Judges to decide what is relevant in each Case. Decisions made in Single Cases in Court are Complex Legal Judgments that involve many different Factors. I note that publication involves peer review. The Premises that peer review is relevant to Causation, that peer review is reliable and that peer review is related to accuracy are Testable. Some Judges do not accept these Premises and have drawn their disagreement into view. And in order to understand what is happening it is necessary to know what determines relevance.

The Attention of the Supreme Court was focussed on the use of the Fry Rule as a Standard. Problems with the Frye Rule were picked up and discussed. The Supreme Court reconsidered that Rule and others in relation to the Daubert Case. The Supreme Court favoured a relatively more relaxed approach. Those Judges concluded that:

> "Cross-examination, presentation of contrary evidence, and careful instruction on the burden of proof, rather than wholesale exclusion under an uncompromising "general acceptance" standard, is the appropriate means by which evidence based on valid principles may be challenged. That even limited screening by the trial judge, on occasion, will prevent the jury from hearing of authentic scientific breakthroughs is simply a consequence of the fact that the Rules are not designed to seek cosmic understanding but, rather, to resolve legal disputes. Pp. 15-17. 951 F.2d 1128 (CA9 1991), vacated and remanded. [509 U.S. 579, 3]"

A relatively more relaxed approach is not a different type of approach. I will in due course pick up for analysis the notions of "valid principles", "cosmic understanding" and the use of the word "scientific". For now, let us Observe the Judgment of the Supreme Court on gatekeeping and their ruling on Frye.

The Daubert Criteria

In Daubert the opinion that held was that the Federal Rules of Evidence, not Frye, provide the Standard for admitting expert scientific testimony in a federal trial. The ruling states:

> "Frye's "general acceptance" test was superseded by the Rules' adoption. The Rules occupy the field, United States v. Abel, 469 U.S. 45, 49, and, although the common law of evidence may serve as an aid to their application, id., at 51-52, respondent's assertion that they somehow assimilated Frye is unconvincing. Nothing in the Rules as a [509 U.S. 579, 2] whole or in the text and drafting history of Rule 702, which specifically governs expert testimony, gives any indication that "general acceptance" is a necessary precondition to the admissibility of scientific evidence. Moreover, such a rigid standard would be at odds with the Rules' liberal thrust and their general approach of relaxing the traditional barriers to "opinion" testimony."

The Rules' liberal thrust relaxes the traditional barriers to Hearsay. That allows more Hearsay. Briefly, four criteria were set out in Daubert:

a) Whether the theory or technique can be (and has been) tested

b) Whether the theory or technique has been subjected to peer review and publication. It is stated in the Daubert ruling that "Publication (which is but one element of peer review) is not a sine qua non of admissibility; it does not necessarily correlate with reliability. But submission to the scrutiny of the scientific community is a component of "good science," in part because it increases the likelihood that substantive flaws in methodology will be detected."

c) The known or potential rate of error.

d) General acceptance - It is stated in the Daubert ruling that '"general acceptance" can yet have a bearing on the inquiry. A "**reliability assessment does not require, although [509 U.S. 579, 14] it does permit, explicit identification of a relevant scientific community and an express determination of a particular degree of acceptance within that community**." United States v. Downing, 753 F.2d, at 1238. See also 3 Weinstein Berger 70203., pp. 702-41 to 702-42. Widespread acceptance can be an important factor in ruling particular evidence admissible, and "a known technique which has been able to attract only minimal support within the community," Downing, 753 F.2d, at 1238, may properly be viewed with skepticism.'

By now you will immediately grasp the Significance of the part I highlighted within paragraph d. Those criteria set the scene for that which is to come in my narrative. I started my search for Pure Science while I was full of questions about particular happenings in Social Dynamics. I needed answers and now I saw clearly where they could be found and thus where I should look for Meaning. I saw a need to Test that which is accepted by the Recognized Authorities in order to discover its True Meaning, and to do that I needed to engage the Process of Pure Science. Psychology is an Inductive science. Pure Science is not a generally accepted Method in the scientific community of which I am part nor is it in general a recognised Method. It is not possible to accept something when we do not know what it is. I have to Mark that type of Method and prove that it should be accepted. A Method should be accepted if it produces Facts and Causal Knowledge. A known technique which has been able to attract only minimal support within the scientific community is by definition viewed with scepticism. But the amount of support attracted is not an indicator of the Qualities of a technique. And Pure Science is not actually a known technique. That Fact should arouse Curiosity. The assumption that submission to the scrutiny of the scientific community increases the likelihood that substantive flaws in methodology will be detected can be Tested for Truth and Justification. And the type of community that is defined as scientific can be identified - that is a Significant Variable.

The influence of Specialist philosophers on Legal Judgments is made clear in the Daubert Ruling:

> "Ordinarily, a key question to be answered in determining whether a theory or technique is scientific knowledge that will assist the trier of fact will be whether it can be (and has been) tested. "Scientific methodology today is based on generating hypotheses and testing them to see if they can be falsified; indeed, this methodology is what distinguishes science from other fields of human inquiry." Green, 645. See also C. Hempel, Philosophy of Natural Science 49 (1966) ("[T]he statements constituting a scientific explanation must be capable of empirical test"); K. Popper, Conjectures and Refutations: The Growth of Scientific Knowledge 37 (5th ed. [509 U.S. 579, 13] 1989) ("[T]he criterion of the scientific status of a theory is its falsifiability, or refutability, or testability")."

Within the Boundaries of Pure Science, the trier of Fact is a trier of Truth in relation to a particular thing or happening. There is the Appearance of Truth in the statement that science is distinguished from other fields of human inquiry by generating hypotheses and testing them to see if they can be falsified. It is a Fact that Inductive science has been given the Appearance of Pure Science through the common use of words and does not have its particular Meaning. That which has the Appearance of Science is not Pure Science, and the Judges know the difficulty of making a distinction between Meaning as it Appears to be and True Meaning. We all face that difficulty. That is a Universal Problem in Social Dynamics and has the Appearance of being an Insoluble Problem, but it is not that. The problem has a Solution. It is made clear in the Daubert Appeal that the Judges take gifts of Meaning from Specialists in philosophy. They are not alone in doing that. The Action of Taking Meaning that is Gifted is different in type to that of Testing Meaning for Truth.

Gifts are attractive on the surface. As far as I recall, in the Ancient Tale of the Trojan Horse once the gift had been accepted and taken into the city, the human contents emerged from hiding. The city had been invaded effortlessly and was taken over. I will turn my Attention to Specialist philosophy during the course of this narrative. But in this Chapter I seek to learn from the experience of Judges in the Justice System in order to formulate particular questions and propositions that will take my searches for True Meaning in the right direction.

The 1999 Kumho Tire Judgement extended the potential applicability of the Daubert criteria to all expert testimony.

Kumho Tire Co. v. Carmichael, 526 U.S. 137 (1999)

> "The district court had to determine whether the tire expert's methods could reliably determine what had caused the tire on the Carmichaels' van to explode. The expert's experience as a tire engineer wasn't the problem—the expert had worked for ten years at Michelin. The fact that visual inspection of tires was generally a reliable method wasn't an issue either, because the issue before the court was specific to the tire on the Carmichaels' van. But **the expert said that his inspection of the tire led to the conclusion that a defect caused the tire to explode because he did not see evidence of other causes. "Nothing in either Daubert or the Federal Rules of Evidence requires a district court to admit opinion evidence that is connected to existing data only by the ipse dixit of the expert [i.e., only by the statement of the expert himself]."** The district court acted within its discretion to exclude the evidence proffered by the tire expert in light of these concerns. Accordingly, the Supreme Court reversed the Eleventh Circuit's decision to overrule the district court."

In other words: - ruling one thing out does not rule another thing in – showing that some things were not the cause does not inform us as to the cause. Ruling things out narrows the search field. Or at least that is the Meaning I see here. What is stated is that the connection of the opinion to the data must consist of more than the statement of the expert. I would interpret that as recognition that Opinion in isolation is not evidence. The term 'opinion evidence' suggests that it is. Thus the term 'opinion evidence' is a Source of Confusion. There are different types of Opinion. Different types of thing relate to other things in different ways. Some Opinions are Justifiable and some are not. In order for an audience to discover whether an Opinion is Justifiable, the Justification must be shown and Independently Tested.

In the case of Kumho Tire, it was held that: "The Daubert factors may apply to the testimony of engineers and other experts who are not scientists":

a. "The Daubert "gatekeeping" obligation applies not only to "scientific" testimony, but to all expert testimony. ……….Daubert referred only to "scientific" knowledge because that was the nature of the expertise there at issue. …….Finally, it would prove difficult, if not impossible, for judges to administer evidentiary rules under which a "gatekeeping" obligation depended upon a distinction between "scientific" knowledge and "technical" or "other specialized" knowledge, since there is no clear line dividing the one from the others and no convincing need to make such distinctions."

b. "………The Daubert factors do not constitute a definitive checklist or test, id., at 593, and the gatekeeping inquiry must be tied to the particular facts……. Those factors may or may not be pertinent in assessing reliability, depending on the nature of the issue, the expert's particular expertise, and the subject of his testimony."

Now, the statement that "there is no clear line dividing the one from the others" in relation to "scientific" knowledge and "technical" or "other specialized" knowledge, is a statement of Fact. There is no clear line drawn between them and one cannot be drawn. But the distinctions that are Significant are those between Pure Science and Inductive science and between Common-Sense and the lack of it in relation to Meaning and Operational matters. I will take two statements as propositions for Testing. Pure Science and Common-

Sense are Intrinsically Related. Inductive science is Dissociated from Common-Sense.

When the threads of Meaning bind two things closely together in Intrinsic Relations a failure to tell the difference between them is not a problem. Meaning and Action are of a single type when they have the same Point of Reference. There is "no convincing need to make such distinctions" when things have the same type of Meaning. The development of Technology is part of the Process of Pure Science. While Technology and Pure Science are governed by the same Point of Reference then there is no need to make a distinction for the difference is Trivial then. But, once it has been developed through engagement with the Process of Pure Science, Technology is taken outside the Boundaries of Pure Science and then different Points of Reference govern its use. I will argue that is a Fault leading to problems in Social Dynamics. It is a Fact that once Technology has been developed and Pure Science has been disengaged from its Operation, there is a change in the type of Meaning of Technology. I propose that there is a convincing need to identify the particular Context in which the Technology is being used. That is a different distinction. In other words, there is a convincing need to distinguish between different types of Meaning. Meaning is clarified by Context and the Context is determined by the Point of Reference. Technology can be used well and can be misused. Ideas can be taken out of the Context of Pure Science and Operationalised – then there can be misuse. Ideas Operationalised in the Context of Pure Science serve a particular purpose that is distinctive – in other words, there are different types of Operation. When ideas are Operationalised in the Context of Pure Science, they are used well. The Point of Reference determines the type of Meaning of all things to which it is applied. Common-Sense is aligned to Truth and has the same type of Meaning as Pure Science.

There is a problem with general statements of the type that lack particular Meaning. That was explicitly stated in the Daubert Judgment:

> "Chief Justice Rehnquist, with whom Justice Stevens joins, concurring in part and dissenting in part", comments ""General

> observations" by this Court customarily carry great weight with lower federal courts, but the ones offered here suffer from the flaw common to most such observations - they are not applied to deciding whether particular testimony was or was not admissible, and therefore they tend to be not only general, but vague and abstract."

That part of the Judgment clearly shows the importance of Single Cases to Judgments of Meaning. Meaning is moved backwards and forward from the particular detail in Single Cases to general abstractions and from general abstractions to the particular detail in Single Cases. Vagueness increases the amount of movement that is possible. All types of general Meaning are abstract but some are not vague. Different types of Abstraction have different Qualities and Function in different ways. That difference has been dropped.

PART THREE of Chapter one

THE TURNER RULE: COMMON KNOWLEDGE AND EXPERIENCE

The Turner Rule is based on both Facts and Premises. In R.v. Turner (1975) it was noted that the Fact that an expert witness has impressive qualifications does not by that Fact alone make his opinion on matters of human nature better than the opinion of jurors on matters of human nature, but there is a danger that they may think it does. The general Rule was derived from a Single Case. In brief, the Trial was for an offence of murder and the defence aimed to submit the testimony of a psychiatrist to support a defence of provocation to murder. The testimony was reviewed and rejected by the Judge, and the Judgment that all psychological processes except those involving some form of mental abnormality are part and parcel of the common knowledge and experience of a jury became a general rule. Psychiatric and psychological Opinion is excluded from Court by the Turner Rule unless the Opinion is related to some form of mental abnormality. The Turner Rule is a Significant Rule that Functions in a particular way.

The Turner Rule is designed for the purpose of safeguarding, in the Context of the Court, a particular topic from Specialists. The topic being safeguarded is the Meaning of Human Action. The Courts Universally deal with Human Actions and the Courts are Operated by human beings. Rumour and gossip can enter Court from scientific communities since hearsay is allowed in evidence from them. If hearsay is allowed on the topic of human Action there is no restriction on hearsay in Court. Experts are allowed to give Opinions on subjects that are not, or are not wholly, within the knowledge and experience of ordinary persons. Thus the Turner Rule is supported by the Premise that the topic of Human Action is within the knowledge and experience of ordinary persons. It cannot be disputed that ordinary persons have experience of human Actions and that the experience is extensive. But do ordinary persons have knowledge on that subject? The answer to that question depends in part on the definition of knowledge. If knowledge is defined as facts then ordinary persons have knowledge that varies Quantitatively;

some persons know more facts than others but that gap can be filled in relation to the Single Case. Facts must be related in order to arrive at a conclusion. If knowledge is defined as relating facts and reaching conclusions based on facts then ordinary persons can reach conclusions when they know facts. Thus far the Turner Rule has the Appearance of Justification. But Appearances can be Deceptive. Facts are given Meaning through relating them. There are different types of relation and I note that the word fact covers different types of Meaning. Differences in type are Significant. According to MOM, Knowledge is defined as Facts, which are particular Truths, in their Intrinsic Relations. And Intrinsic Relations are Relations that are True.

Within the Boundaries of Pure Science, Facts and Knowledge are Meaning that is aligned to Truth. It is a Fact that ordinary persons reach conclusions that are Not True and are Not Justifiable. It is a Fact that skilled persons do the same. If a conclusion is Not True and is Not Justified it really does not matter who reaches it for that does not change the type of conclusion that it is. There is No Evidence that skilled persons are more likely than ordinary persons to reach conclusions that are True and Justified and there is abundant Evidence that experts do not aim to do that. If "the testimony which least deserves credit with a jury is that of skilled witnesses" the Turner Rule is safe and cannot be Unsettled. So, it is a Fact that human beings make Errors of Judgment about other human beings and it is a Fact that those Errors are not restricted to Specialists. Those Facts have been known as far back in time as we can see.

Now, general Rules are necessarily vague. Therefore, it is not difficult to find a problem in interpreting the Ruling on common knowledge and experience. The argument on interpretation takes place at the general level and also in the Context of Single Cases. At a general level I see a problem in relation to the Turner Rule, but this problem could not Unsettle the Rule and is not aimed at doing so. The Turner Rule does not exclude testimony from psychiatrists and psychologists on some form of mental abnormality and that leaves a weakness in the defences. I will phrase the problem as Questions. On what basis is it decided that there is mental abnormality? And who makes that Judgment? Psychologists and psychiatrists make the Judgments and thus they decide the basis for the distinction between what Normal is and what Abnormal is. Or at least they Appear to do so and Appearances can be Deceptive. The Justification for Judgments on Diagnosis depends on the Method used. Inductive science has No Justification. The Quantitative Method is extensively used in the diagnosis of mental abnormality. There is no particular point at which a Normal State of Mind changes to an Abnormal State of Mind. That will never be Quantifiable. But such a change can occur and does occur in Single Cases – that is a Fact for which I will provide Evidence in due course. With Causal Knowledge it is possible to predict that happening. We do not currently have that Knowledge. The Factors that determine that happening are Not Verified, but they can be.

Judges are trained to deal with arguments and they gain considerable experience in doing so. The Court is the last resort when an argument cannot be resolved. Most Diagnoses are not tested in Court and when they are it is a Battle between experts in psychiatry and/or psychology who have been Selected by lawyers. The lawyers are armed with a small amount of information and aim to secure a particular Bias in the Opinion and have it accepted by the audience. The experts are usually "men" of science. We will observe in due course a Single Case of an argument by men using science where we can observe the type of philosophy that guides the Action and its effect. If the expert happens to be a "man" inclined to use Pure Science who knows the type of Action that Pure Science involves, then the argument is of a different type and takes a different course.

Peer agreement is threaded through all the Daubert Criteria, and the suggestion is otherwise. Peer approval in the scientific community is allowed as an influence and the type of community from which approval is drawn is not specified. I am puzzled there is no requirement to do that. That lack of specification is a Fault. The Point of Reference that gives particular Meaning to the Action is hidden. The weight given to the Quantitative Method is visible in the details of the case of Daubert and is not visible in the general criteria. It Appears that any type of Method can be agreed by peers and the terms for agreement are not visible. The words tested, correlate, reliability, scientific, and the term peer review are in the Daubert criteria. The word justification is not there. The criteria set a Standard that is not governed by a requirement for Justification and

Truth. Curiouser and Curiouser said I. Judges use the word science for their own Actions and they search for Justification in their arguments - or at least some of them do and the presumption can stand at a general level until it is Tested in Single Cases. But all Judges understand the Significance of Justification. And perhaps there is a presumption that science is Justified. Only Testing can disclose the presence or absence of Justification for agreements. Epidemiological evidence is statistical and that type of evidence has been given Power over other types of evidence in all Contexts. Statistics is Inductive logic and it is a Fact that Induction has No Justification. Using Inductive logic, we can quickly and automatically arrive at a conclusion that is False. That Fact is known.

Qualifications are a criterion for the Selection of expert witnesses. Qualifications do impress audiences. It is a Fact that specialist qualifications do not make an Opinion on matters of human nature better than the opinion of Judges and Jurors on human nature. When Jurors are fact finders, they are likely to think experts know facts and that they can find the facts in that place. After all, the word 'expert' is attached to their evidence and that signifies the witness has special knowledge of, or skill in, the area on which they testify. The members of the Jury do not know the Facts that Judges know about expert testimony and they do know the common Meaning of the word expert. Of course, Judges can explain Facts to Juries and that makes a difference, but we do not know how much difference that actually makes. Illusions are difficult to Dispel. Some human agents do engage the Process of Pure Science in forming Judgments. If that is True then it follows that some individuals actually do know more about Normal Psychological Processes than others. It is simply not possible to engage in the Process of Pure Science and fail to gain Knowledge. The problem is that we do not Mark who those individuals are or where they are within the Professions or outside them. We cannot identify people who have Knowledge while we do not know what the Process of Pure Science is. We also do not know whether the Facts gained through the Process of Pure Science are well used or misused. Facts and Knowledge are bought and sold. And Knowledge is not a personal characteristic, thus people should not be used as a Mark of Knowledge.

The word specialist is applied to people who have studied a particular topic in depth in a particular way and met particular criteria. Impressive Qualifications are Marks that distinguish those who have reached a particular level in their Formal studies. Experts come from groups of specialists and they are ordinary persons with particular skills. The Turner Rule states that impressive qualifications alone do not make a person's Opinion better on matters of human nature, but it does not state, and does not even propose, what it is that does make such Opinion better. That is something left in need of specification. It is a Fact that Hearsay does not make Opinion better. Sometimes Hearsay makes Opinion worse.

TESTING

The Daubert ruling states:

> "Scientific methodology today is based on generating hypotheses and testing them to see if they can be falsified; indeed, this methodology is what distinguishes science from other fields of human inquiry."

Falsification was set as a criterion of science by Popper. The Judges mentioned him. Popper was a Specialist philosopher. Specialist philosophers channelled Attention towards Quantitative Methods. It is claimed that the Quantitative Methodology is scientific methodology and as we have seen in the Daubert Case that claim was not only accepted by Judges, but it also displaced the claim that different types of Method are science. That claim is Testable. When Pure Science is engaged, Attention is paid to the disclosure of Meaning through Qualitative Analysis. The Concepts and Constructs of Pure Science are different in type to those of Inductive science. I propose that Inductive science is pseudo-science. Pure Science does involve generating hypotheses for Testing. Using the word science as a common name transposes the Meaning of one type of thing to another type of thing. The same words are used for different types of Action within different Methods. Where there is pseudo-science there is pseudo-testing of pseudo-hypotheses. Testing must be fit for purpose and pseudo-testing is fit for the purpose of pseudo-science. The Inductive Method leads to Uncertainty.

The Action of Testing within the Process of Pure Science is part of a search for True Meaning. Testing within the Boundaries of Pure Science leads to Justified Conclusions. Justification is a particular Quality of Meaning and is produced by a particular type of Action. Specialists in philosophy have issued a licence to deliver Meaning and Action without Justification. The Appearance that finding True Meaning and Justification is an Insoluble Problem has been constructed by Specialist philosophers and Statisticians. The word justification is not used in a different way by Specialists in philosophy. That word is not used at all as a noun. They do not produce Justification and they do not claim that they do so. But they do use that word as an adjective to describe some of the things they do. The Suggestion that their Actions are Justifiable is put there for the taking. Resemblances are characteristic of a Masquerade. And if you draw the conclusion that Justification is there when it is not, then that Error is your Fault for they have avoided lying. But the False Meaning is put there for the taking, and Specialists in philosophy are at least responsible for doing that.

The word science is used for Inductive science and for Pure Science but those are different types of Action by human beings. Inductive science is only peer agreed methodology in the scientific community that uses Induction. I can show you a clear statement of where Inductive science leads. In a paper titled "Expert Witnesses: Recent Developments in NSW" Justice Peter Biscoe (2006) illustrated the fact that "conflicts of expert opinion" can be "particularly difficult to resolve or intractable" by quoting Justice Sperling, "now retired from the Supreme Court of NSW, who was involved with the Law Reform Commission report". I note at this point that Hearsay can be used well and that the condition for good use is that Hearsay is Tested for Justification. Justice Biscoe quotes Justice Sperling as saying:

> "As a judge I heard a case in which the critical issue was whether a surgeon had left a radioactive substance in the lungs of a patient. If he had, the plaintiff won, if he had not, the plaintiff lost. Two experts gave evidence. Their evidence was based on

> the same x-ray of the patient's lungs. One said it was obvious that the substance was in the lungs. It clearly appeared from the x-ray. The other said that the x-ray showed only common deviations within the norm. Now what is a judge to do with that?"

Well he could use scientific methodology. That is Statistical Methodology and involves Simple Operations. Having specified which Opinion is Represented by each side of a coin, he could toss a coin and see where it lands but he would have to do that many times and count the frequencies of landing on either side in order to meet the peer agreed scientific standard of probability. Or he could do a survey of many experts, which would be a costly way of doing the same thing. That also involves counting the frequencies of landing on either side in order to meet the peer agreed scientific standard of probability. In the case of a survey it is also Opinion that lands on either side. Attaching Weight by Numbers is a Universal Principle governing Action under Rule by the Quantitative Method. If the Judge used that Method then the Judgment might have the Appearance of Justification, under the condition that he hid the Action of Tossing the Coin. A mathematical formula could be part of that doing. But I will take his point seriously as I am sure he intended it should be taken and I will apply it Universally. The term "common deviations within the norm" indicates that the Quantitative Method is in use. The Quantitative Method is based on Probability Theory and that is based on tossing a coin. In one way or another, Statisticians are tossers. One thing is clear. If you are searching for True Meaning then you cannot find that by Tossing a Coin. Pure Science is not engaged when Chance is in Operation.

Frankly what are any of us to do with single Opinions, contradictory Opinions or peer agreed Opinions when there is no Method to Independently Test them for True Meaning? Under that condition there is nothing we can do. The Judge's Observation identifies both a problem and a gap in Method that leaves Meaning Uncertain. Gaps in Meaning tend to be filled. Given that experts disagree and the Judges have no Method to access Truth on the Issue before them, the Court has to find a different way of reaching a decision. This detail shows us again the nature of the problem referred to in Daubert as "cosmic understanding". When the Truth cannot be found, then Attention turns to resolving an Issue in the absence of Truth. We are constantly presented with different Opinions in all Contexts in Social Dynamics. All Opinions are necessarily held to be equal when we cannot tell the difference between what is True and what is Not True. It is a Fact that Opinions are Not Equal when the Standard of Truth is applied. We would need a Concept of Truth to enable the emergence of a Justified Standard. We do not have such a Concept. There is no agreement that there should be one. We have a Construct of Truth for which the word truth is misused. That Construct is Truth by Agreement among Peers.

Justice Biscoe also quotes a reaction to the question posed by Justice Sperling. Justice Downes, the President of the Australian Administrative Appeals Tribunal, responded in a paper delivered to the NSW Bar Association Administrative Law section on 22 March 2006 entitled "Expert witnesses in proceedings in the Administrative Appeal Tribunal". The reaction was that what a Judge should not do with that is to ask a single expert to decide. Justice Peter Biscoe continues:

> "A staunch critic of the idea of the single expert, Justice Downes argues that the most satisfactory way to resolve the difference, for a judge, part of whose expertise should lie in being able to detect where the truth lies, is to resolve the dispute by reference to its context and the criteria identified by the experts. The problem with one expert in a situation such as that which Justice Sperling described, is that the expert might be either of the experts who actually gave evidence. That person may honestly strive to identify the competing experts' views but will undoubtedly settle on the expert's own opinion. The result is that the case will be determined by the identity of the expert selected. There is no adequate way of testing whether the single expert's opinion is correct. Also, it is said to be fallacious to assume that in fields of expert knowledge there is only one answer. By way of comparison, look at the level of disagreement between appellate judges."

From that Opinion, I will pick up two parts of Meaning that are relevant and related. The parts are: "there is no adequate way of testing whether the single expert's opinion is correct." And: "in fields of expert knowledge there is more than one answer." Both those statements are Premises supporting the Formal philosophical position from which they came. While the Judge picked up the Formal philosophical Premise that it is fallacious to assume that in fields of expert knowledge there is only one answer, he did not state that he agreed with that Premise. But he does add a familiar Context for comparison. He picks up the Fact that Judges disagree. It is Reasonable to pick up that commonality, but it is differences in Meaning that are Significant and those are difficult to pick up. In the absence of picking up a difference in Meaning there is the Appearance that the Action has the same Meaning. That Appearance may be True or it may be False. The Meaning is different when Points of Reference are different. And we have yet to discover what those are.

It is True that experts disagree. But the Fact that experts disagree is Evidence that there is more than one answer that can be given, and it is not Evidence that there is more than one answer that is True. Facts are a Single answer to a question. Truth is Fact. When we do not have Truth, we do not have a Fact. Then we have more than one possible answer. We only access Facts by Testing for True Meaning. The Court is not the place where the Premises that lie at the foundations of specialisms can be Tested. But all Premises can be Independently Tested and once the results are available, they can be used in any Context.

There is no adequate way of Testing whether the single expert's opinion is correct if the Premise is accepted that in fields of expert knowledge there is more than one answer and that different answers are equal measures of the Meaning of the data. Then you have Uncertainty. But Meaning can be Calculated and Truth can be our guide. The need for an adequate way of testing the relative weight of expert opinion led to a proposed solution in the Legal Context that has gained in popularity. The Action is described by Justice Biscoe as "the taking of concurrent expert evidence, sometimes irreverently called 'hot-tubbing' ". He continues: "The concurrent evidence procedure in the Land and Environment Court of NSW is not fixed in stone but typically is as follows":

> "Before giving evidence, experts of the same discipline confer and produce a joint report which sets out the matters on which they agree, the matters on which they disagree and their reasons for disagreement. This enables the Court to identify the differences which remain between them and which require resolution through their oral evidence. At trial, the experts are sworn in and give evidence at the same time. It is often useful to have a written agenda of matters to be dealt with in oral evidence. The experts have an opportunity to explain their position on an issue and to question the other witness or witnesses about their position. Questions are also asked by counsel for the parties and the judge. In effect, the evidence is given through discussion in which the experts, the advocates and the judge participate. Questions and discussion on a particular issue by all experts can be completed before moving on to the next issue."

Will that type of Action access True Meaning and lead to Justified conclusions? Or will the argument be settled in the absence of the Calculation of True Meaning? Now, recall the description of the lawyer that I quoted earlier in conjunction with a Fact. It is a Fact that there are people in all fields who with "bold assurance, chuckling confidence, vain-glorious self-satisfaction, coupled with ignorance" will dispatch the caution of true knowledge if they possibly can. That Method of dealing with the opposition is not confined to lawyers. So hot-tubbing sounds like fun for some though not for others. But all participants are paid for their arguments so they can focus on that Representation while under fire and in hot water, as Specialists in philosophy have advised them to do. Or as a colleague acting in a supervisory capacity once said to me, focus on getting out of the situation unscathed and on what you will have for tea when you get home. That is not advice I would give. That is not advice I would accept. Attention is not focussed on the discovery of Truth when it is focussed elsewhere. Recall Taylor's observation that "these gentlemen are usually required to speak, not of facts, but to opinions and when this is the case, it is often quite surprising to see with what facility, and to what an extent, their views can be made to correspond with the wishes or the interests of the parties who call them." The Court calls them into the hot tub and the interests of the Court are served by the experts reaching agreement while they are there. Agreement will be reached. But unless there is a change in the requirement to speak not of Facts but to Opinions it is difficult to see how True Meaning will emerge.

The emphasis on Testing Reasoning is sound. But before you can Test it you have to find the Reasoning. That is impossible to do when there is none to be found. The Appearance of Reasoning is covered by the word reasoning and that Appearance can be strong enough for the Inductive Action that the word reasoning covers to be mistaken for the Real thing. The importance of observing the Reasoning that underpins an opinion has been emphasised time and time again by Judges in Single Case Judgments. For example:

> In Risk v Northern Territory of Australia [2006] FCA 404 per Mansfield J at [469]: "The important thing in any expert's report, in my view, is that the intellectual processes of the expert can be readily exposed. That involves identifying in a transparent way what are the primary facts assumed or understood. It also involves making the process of reasoning transparent, and where there are premises upon which the reasoning depends, identifying them."

Here there is a Premise that the word fact can be applied to something that is assumed or understood. That is in line with the common definition of the word fact which is influenced by Formal logic. The word fact is misused in Formal logics for statements without the Quality of Truth. Within the Boundaries of Pure Science, Facts are True and anything that is Not True is Not Fact. Inductive logics are incapable of producing Truths. It is not sufficient to identify Premises. It is necessary to Test Premises for Truth and Justification once they are found. The Judges know that.

PEER REVIEW AND PUBLICATION

I note the following comment from a Parliamentary Office of Science and Technology Postnote October 2005:

> "Peer review, as a method of assessing fitness for registration, is open to criticism. Experts who wish to see their field grow in size may be unduly lenient; conversely, experts may wish to exclude rivals within their own specialism. Peer review is regarded as inherently conservative and might deprive the courts of novel expert evidence which could be sufficient to cause reasonable doubt. However, few alternatives to peer review exist."

Points of Reference are commonly hidden or misrepresented. That which is conservative is restrictive of some things and the type of thing that is restricted should be specified when that word is used. One thing is conserved while another thing is not. When one thing is restricted its Power to Act is disabled and the Power of different types of thing to Act on the World is enabled. The Construct of Peer Review is not a Justifiable option. Peer review as it is currently constructed selectively gives Power to Opinions in the professions in the absence of paying Attention to their Truth and Justification. I propose that a powerful alternative to Peer Review does exist and has always existed. The Fact that the alternative is not seen has long puzzled me. Puzzles are only solved by Testing for True Meaning in relation to Reality. Much that is hidden can emerge into view once Pure Science is engaged.

Judges are individual human beings with different Points of Reference; they take different views. The perils of relying on peer agreements have been noted by Judges in Court. I quote the following comments from the successful appeal (2004) against the conviction of Angela Cannings:

i. The Judge notes that: "The word "dogma" was used in his evidence by Dr Rushton, a consultant paediatric and perinatal pathologist, to record his unhappiness, and that of some of his colleagues at "a set of rules being laid down", and, because they were felt to be over-restrictive, followed with reluctance.

In evidence, Dr Rushton repeated part of the contents of his report. It included this passage: "However, in a family with a history of this type, current dogma is that an unnatural cause has been established unless it is possible to demonstrate an alternative natural explanation for these events." The judge states: "That dogma encapsulates what in a criminal case we have described as the first, and we would emphasise, the erroneous approach."

ii. The judgment continues: "By way of linking the word "dogma" with SIDS, Dr Rushton went on: "The definition of SIDS, for instance, ... says usually the definitions include babies up to one year of age; it used to be two years of age but it has been decided by experts in the field meeting that one year is the limitation. I would suggest that biology does not behave like this and [in] any event, sudden unexplained deaths occur throughout life – they also occur in adults. So there is not a magical cut-off point at one year of age."

iii. "Professor Golding, professor of paediatric and perinatal epidemiology, made the same point in a different way when she said in evidence: "I think, to put it in context, there is a fashion nowadays that if you have more than one sudden infant death the next one must have been killed deliberately, and that is something that people within the paediatric profession have taken on board without sufficient evidence. Certainly, obviously, there are cases where it happens, but (in the vast majority) there is no evidence of that."

The judge says firmly:

iv. "If that is the fashion, it must now cease." And later in the Ruling adds: "We cannot avoid the thought that some of the honest views expressed with reasonable confidence in the present case (on both sides of the argument) will have to be revised in years to come, when the fruits of continuing medical research, both here and internationally, become available. What may be unexplained today may be perfectly

> well understood tomorrow. Until then, any tendency to dogmatise should be met with an answering challenge."

It is difficult to challenge peers who have Authority in a profession, or indeed those who have Authority in any Context. When a set of rules has been laid down, pressure to follow them is applied. Rules are designed for particular purposes and Compliance in terms of Actions can be enforced. It is not as easy to enforce Cognitive Compliance but Meaning follows Action and thus the Ideology of individuals will move in the direction of the group rules more often than not. There are penalties for breaking rules. Authority should not be allowed to override Reasoning, but it would take a Higher Authority to prevent that and there is none in Action. Once Power is assigned, there is virtually No Control over individuals in positions of Authority. The Structures in Social Dynamics enable Strong Will to take precedence over Reasoning. But pressure is sometimes resisted and some agents within professions do use Reasoning in searches for True Meaning as was shown in the Appeal just cited. Knowledge would not be produced without those who use Reasoning, and Knowledge is produced in the professions. And Structures can be changed.

Observation is necessary and not sufficient to find Causal Knowledge. Theory goes beyond Observation. It is a Fact that within the Boundaries of Pure Science, Meaning is Tested through asking questions and formulating propositions. That Fact is known. Theories are not the ultimate product of Pure Science. Knowledge is the Ultimate Goal and the Ultimate Product. When we find True Meaning, then we have Knowledge. Facts are True Meaning in relation to something particular and I would define them as the Units of Knowledge as a Whole. A Whole is different to the parts of which it is composed. Thus Knowledge is different to Facts. Within the Boundaries of Pure Science, theoretical development is part of a Search for True Meaning. Theories are Tested when the Process of Pure Science is engaged. Testing takes us to different Theories and beyond them lays Knowledge. Uncertainty can be transformed into Certainty through Testing Meaning that is propositional within the Process of Pure Science. Outside the Process of Pure Science, theories are Not Tested and serve a different purpose. Theories become the fashion and that which is fashionable is dogma. That which is Asserted as being necessary and true is dogmatic. Premises are a Faulty Mechanism leading to Errors in Judgment.

PURE SCIENCE AND THE COSMOS

I will pick up the word cosmos from the Daubert Judgment. Pure Science does seek “cosmic understanding”. We have it on good Authority that the Justice System does not seek that. It is not only the Observations of the Judges that I refer to as the Authority for the Truth of that statement. I refer to Pure Science as the Authority. DNA Testing has Verified that the Justice System operates by reaching agreements in the absence of Truth in an Unquantifiable number of Single Cases. The Process of Pure Science enabled Truth to be found in some Cases. The finding of True Meaning does require understanding of the Cosmos. So what Meaning does that word represent? According to Collins dictionary, cosmos is defined in two ways as:

cosmos[1]: n [Gr. kosmos, universe, harmony] 1. the universe considered as a harmonious and orderly system. 2. any complete and orderly system.

cosmos[2]: n [Mod L < Gr. kosmos, ornament] a tropical American plant of the composite family with white, pink, or purple flower heads.

In passing please note that the colour of the flower heads is Trivial in relation to the type of thing that it is. That Meaning should be Abstracted for use. There is a relation between the two definitions. Flowers are a particular type of Whole thing and the flower named cosmos is a harmonious and orderly system that is different in Appearance to other flowers. A flower is a small universe; it is a world of its own. One world depends on other worlds. Having stated the relation, it is clear that we are not usually referring to a flower when we use the word cosmos. It is the first definition that concerns us. A cosmos is simply something as a Whole. To avoid

Confusion, we should specify the particular Whole to which we are referring when we use that word.

The word universe is part of the definition of cosmos and the word cosmos is part of the definition of universe. From Collins dictionary:

universe: [L. universum < unus, one + pp. of vertare, to turn] 1. the totality of all the things that exist; the cosmos. 2. the world.

We can use the words universe, cosmos or whole to represent a system that is complete, harmonious and orderly. A Whole is a totality. A totality is a Single Entity. One totality relates to another totality. Single words are just building blocks for Meaning. Words are single parts that are small Structures used to build larger Structures. Each Structure is a Whole and every Whole is part of a larger universe with many other Wholes within it. All Wholes within a universe are bound together by invisible forces that are the threads governing movements. Relations are composed of invisible forces. As we build the larger Structures with words in relations, particular Meaning takes shape. Words are my building blocks for Meaning. All agents use those building blocks for Meaning. Natural Language is a highly developed Structure through which Meaning is represented. We can acquire the skill of building Meaning precisely with Natural Language. Specialists in philosophy deny that Fact. That Denial Interferes with development of the skill. Attention has been turned to the development of Formal Representations that do not fit the Structure of Meaning accurately. The Quantitative Method of Statistics is the fashion in research and that fashion spread across the professions. "What may be unexplained today may be perfectly well understood tomorrow. Until then, any tendency to dogmatise should be met with an answering challenge." I can do that.

There are different types of System. Some Systems are complete and orderly but could not be described as harmonious. I see Pure Science as an active, complete, harmonious and orderly System of a particular type. Pure Science is a Methodological Cosmos composed of many parts of different types, all of which are co-ordinated in their Action by the search for True Meaning in Reality. In order to put Pure Science into Action, it is necessary to have the right parts in the right places and in full working order. If we are ever to understand the way any Dynamic System works as a Whole, we must understand which parts are necessary, how the parts fit together, and how they Act together once assembled. One way to find out how a System works is to disassemble that system and put it back together. Deconstruction and reconstruction can take place in the Real World and in the World of Imagination. Operations on Meaning in the imagination of human agents can be given material form as models. The Reasoning presented in the Chapters that follow provides Evidence that Pure Science is the only type of Action that could reveal more in relation to psychological processes than is commonly known through experience. In order to distinguish the Whole that Pure Science is, it is necessary to observe in detail that which it is not. This particular narrative is a small cosmos that relates to other narratives.

Now a thought has just sprung to mind. A cosmos is a harmonious and orderly system. Natural Systems are harmonious and orderly. Human beings are Natural Systems. But are they Functioning as such? How many human beings can be described as a cosmos and how can you tell which ones they are? Where there is not harmony and order then there is disorder. Disorder is Not a Natural State in the long-term. The Formal philosophical position is: Order tends to Disorder. That gives Disorder the Appearance of Justification. But I have a different proposition. I propose: Order becomes Disorder then necessarily becomes a different Order. But I will not argue about where to draw lines in Formal Terms for those are Artificial and cannot be meaningfully drawn. Nor will I argue about whether Disorder came first for there is no Order of Precedence in a Dynamic System and that is the type of System in existence. Order and Disorder are particular States of Being that are distinct and Intrinsically Related.

The Turner Rule does not refer to Natural Psychological Processes. Nor does it refer to Artificial Processes. Thus I am Not Restricted by that Rule. I will show you there are Facts about different types of Process and where there are Facts there is no need for Opinion until there is a need to go beyond them. When we do go beyond Facts there are Natural Rules we should follow in order to preserve the alignment of Meaning to Truth.

THE GIFT OF A PARTICULAR PHILOSOPHICAL BIAS WAS ACCEPTED

The Frye Test is a Rule that does not just take peer agreement into account in Judgments of admissibility and weight. The Frye Rule gives Power to the Opinion of small groups of people with Authority in Professional Organisations. The Daubert Ruling did not change that but it does make the role of peer agreement less visible and gives Judges in Single Cases more freedom to Act. The Law is open to interpretation. In a Quantitative System, Opinions are held to be equal and counting determines which Opinion prevails. The Constructs of Selection and Agreement by Numbers are threaded as Mechanisms through all Social Structures. The type of part Selected by Numbers determines the type of Bias that becomes Operational. The People who have been Selected by Numbers then Select other things and they use numbers to reach agreements when agreement is necessary. I will ask a question: Is Selection by Numbers a Reasonable Mechanism to use in Social Systems? And I will define that which is Reasonable as that which is Justifiable. The definition determines the answer to my question. The Quantitative Method is the Action of Artificial Induction and Induction is Not Justifiable. Selection by Numbers involves Judgments based on Frequencies of Occurrence. That is Unjustifiable.

Specialists in philosophy do not make a Personal Appearance in Court, but the Influence of Formal philosophy is clearly Observable in Action in the Courts. It is Reasonable to presume that Judges look for Justification. Where they look determines the Potential for its finding. In Marsh v Valyou (2007) an attempt was made to interpret the Frye Rule in philosophical terms:

> "Testimony is "pure opinion" only when it is based solely on experience and training, and does not rely on a novel scientific principle, test, or methodology: "Pure opinion" refers to expert opinion developed from inductive reasoning based on the experts' own experience, observation, or research, whereas the Frye test applies when an expert witness reaches a conclusion by deduction, from applying new and novel scientific principle, formula, or procedure developed by others."

I propose that we are caught by the Illusion that Artificial philosophy can guide us to Knowledge and the Certainty that comes with that. I Test that proposition for Truth in this narrative. I did not begin with that proposition or with questions directed at Specialist philosophy. In searching for the Mark of Pure Science I found that all paths led in that direction. I saw that the Judges looked at Specialist philosophy and took Meaning from that. Formal philosophy is Art and it is Not Pure Art. All forms of Art are Projective in Action and it is Meaning they project. Different people see different Meaning in arrangements of Art. One thing is Certain. Justification for statements of any type will not be found in a work of Art. Justification does not lay in the Specialist philosophical literature. If Justification is to be found the search must be directed elsewhere. But the Appearance that Formal philosophy is the right place to look for Meaning and Justification is strong and pulls all eyes towards it. True Meaning and Justification cannot be drawn from a Source that has none - that is a Fact.

In the coming Chapters of this particular narrative and in other narratives to come, I will show you in detail the Fault that threads through all Contexts of Human Interaction and I will Track its origin. I will Test the nature of the gifts that have been accepted and now dominate the Action in Social Dynamics. The Evidence I present is the Truth as I found it to be from Testing. I found the Truth for me. But when I Act as an expert witness, I present Evidence before Judges in order to assist them in discovering Truth. I do not find the Truth for them. I tell them what I found and show them the Evidence. They find Truth for themselves. That is a Natural Philosophical position. That is my position in writing this narrative. This book is not written for presentation in Court. But I have applied the same Standard. You will be the judges of this work.

There is only one Point of Reference for Pure Science. Pure Science is a Natural type of Action that is independent of Context and can be engaged Universally in searching for True Meaning. Taking the Formal philosophical view blocks the Perception of the type of thing that Pure Science actually is. The Evidence I present Verifies that Inductive science is pseudo-science. Inductive scientists claim that Verification is not possible and I claim that it is. Competing claims can be Independently Tested. Another term that I have seen used for pseudo-science is junk science. Junk called science presented by a well-credentialed agent who has been categorised as an "expert" does not look like junk. Junk science is a designer product. I discovered there are many parts to the Whole Illusion that has been designed.

CHAPTER TWO: A SELECTION OF OPINION WITH A PHILOSOPHICAL EDGE

PART ONE of Chapter two

CALCULATIONS BEGIN SOMEWHERE

Single Cases are the Wholes from which Meaning is drawn. That Meaning is related to Meaning drawn from other Single Cases. A Cosmos is composed of parts that are united through relations. Every harmonious and orderly narrative is a Cosmos with particular Meaning. There are many fitting ways in which I could start this Chapter. I find myself spoilt for choice in the Sources from which I can draw particular Meaning into Action. All that I choose I will use in my own way. Some narratives are harmonious and orderly systems with precious gifts of Meaning attractively represented. Once their Meaning has been Independently Tested for Truth, I can accept the gift of Representation and use it well. I have selected a small, easily recognisable part from Tolkien's narrative of the Lord of the Rings. That narrative is categorised as Fiction.

One Ring was Forged to draw them in.
One Ring to Bid them all and
In Darkness Bind them.

My Representation may not be precisely the same but that does not matter for Meaning is the same. I will provide Evidence that the Meaning in that small arrangement is True. I will display the thing that is symbolised as the One Ring. I take the One Ring as an Analogy for Inductive science. Tolkien's story of The Lord of the Rings was Fiction as a Whole and in all parts. But while the Representation was in terms of Fiction, the Meaning he represented is True in part and perhaps could be True as a Whole.

Fiction is Art. Art is heavily influenced by the World of Imagination. Those who write Fiction do not present Evidence for the Meaning they represent. The Standard of Truth is not applied to Judge Meaning in Fiction. But that does not mean that Truth is Not Represented in Fiction. That just means that the difference between True and False Meaning is Not Marked. There is No Claim to be representing Truth in Fiction. Or if there is such a claim it is thought to be part of the Fiction, which it is. When an arrangement is categorised as Fact there is a claim to be representing Truth. That claim is only Justified if the Standard of Truth was applied to Judge Meaning and the Standard was met. The presence or absence of Testing makes the difference to the Quality of the Meaning that emerges. True Meaning is found through Experiments that are in search of that type of Meaning. Testing within the Process of Pure Science increases the Potential for the Emergence of True Meaning and reduces the Potential for the Emergence of False Meaning. Realisation of the Potentials depends on how accurate the Test is in discovering Truth.

Meaning and Action are Intrinsically Related. Action leads to things coming into existence and things that exist are in Action. Representations are the products of Action and are Active for they show Meaning. Meaning and Action are Intrinsically Related and the difference between them is no more than a hen's kick. There is no clear line dividing the one from the other and no convincing need to make such a distinction. But, there is a convincing need to make distinctions based on differences in type of Meaning and on the fit between Meaning and its Representation. One type of Meaning can be represented as a different type of Meaning. Some types of Meaning should be prevented from becoming Active. But should they and can they be prevented from coming into existence? In order to answer that question, it is necessary to ask many different questions and find answers to those.

Different types of thing have different Meaning. In order to tell if there is a difference in type of Meaning it is necessary to Test Reactions. Meaning is disclosed by Reactions. A Reaction as a Whole is composed of many parts. Reactions occur when things relate to each other and while there is at a general level the Potential for things to relate in different ways, at the particular level they relate in one way at a time. A Reaction is an Action in response to another Action. A Reaction shows the Relation between two things. A Reaction can show that Meaning is the same or that Meaning is different. Meaning that is the same is of one type. Things that have the same type of Meaning React in the same way to other types of thing and things that have a different type of Meaning react in different ways. It is easier to perceive True Meaning through viewing different Representations in conjunction, but taking different angles of view of a Single Representation can work as well. Something has to change in order for there to be a Reaction and the change can be small.

Representations take Meaning from somewhere. I do not know where Tolkien took Meaning from, but he did not conjure up ideas out of a Blank Space. He took threads of Meaning from the World that surrounded him and wove them into stories within his imagination. Tolkien captured particular Meaning and gave it particular form. That same Meaning has been expressed by others in many different forms. I take Meaning from the Real World. Meaning picked up in the Real World can be represented in terms of Fiction without Evidence. I Act on Meaning in a different way. I represent Meaning picked up in the Real World in terms of questions and propositions while gathering Evidence from Testing their Meaning. There is no point in representing Meaning if it is not noticed. So, on occasions I use the tools of Pure Art to draw Attention to the Process of Pure Science and to the Meaning that emerges from that Process. Pure Art can serve Pure Science well. Pure Science can help to produce Pure Art and can employ it, but Pure Science always governs Art of any type when it is in Action. Art cannot govern Pure Science; it does not have the strength to do that. Unless Pure Science is Master of all it surveys, then it is not there.

Formal logics are Inductive. The threads with which I weave the material in this Chapter of my narrative have been drawn from the work of Specialist philosophers and others who have been influenced by Formal logics. Many people in all Contexts in Social Dynamics have succumbed to their Persuasions. Philosophers work on any Meaning that comes their way. We all do. It is part of a Natural philosophical position that Pure Art can be useful within the Process of Pure Science for the Representation of Meaning. A little dramatic effect draws the Attention of an audience. Aristophanes was a great Poet who lived between 445 and 385 BC. He wrote many plays and some survive. I use the Ancient technique of threading a Chorus through the material of this narrative, though not systematically or invariably for this is an Informal narrative. I will use different arrangements as a Chorus, tailored to fit different parts. It is time to introduce one of several Choruses, for which I owe thanks to Aristophanes. I have selected a small part from his play The Frogs (405 BC). You are invited to join in the Choruses as they occur. With the help of Aristophanes, I can quickly and easily show you the Meaning and Action of those who use Artificial philosophy for their work on Meaning. Philosophers can deal with Meaning in whatever form it comes and can deal it out in the same form or in different form. They use a range of strategies. Sometimes Meaning comes to them clad in heavy armour:

Chorus: "Huge are the words that he hurls, great compounds with rivets and bolts in, And epithets hewn out of stone. Now 'tis the challenger's turn to reply to this verbal bombardment: Neatly each phrase he dissects, with intelligence crafty and keen; Harmless around him the adjectives fall as he ducks into cover, And squeaks, 'it depends what you mean!"

Even with strong defences, Meaning cannot usually stand in the way of Specialists in Disarmament however weak they may Appear to be. Now that gift from Aristophanes is not as straightforward as it might Appear to be. Different people use gifts in different ways and this one is a double-edged sword. Either part of that quotation can apply to either side in a battle. In this case, that which I use against others could be used against me. It really does depend what you mean. Meaning depends on which Point of Reference is applied and it is that which is variable. I know what Meaning I take from that quotation and I also know that different Meaning can be taken. So, for now just appreciate the humour Aristophanes gifted. I want you to relax as you join me in discovering Meaning. You will by the end of the journey understand clearly how different interpretations Function. So, smile and watch as a war of words commences.

A few famous names from the recent past will briefly make an Appearance in this Chapter and the next. No search for Pure Science would be complete without some reference to Einstein. He is often quoted. I will pick up some of his words here. I take one small sentence from an article by Einstein in 1936 as quoted by Howard (2005, 2017):

> "At a time like the present, when experience forces us to seek a newer and more solid foundation, the physicist cannot simply surrender to the philosopher the critical contemplation of the theoretical foundations; for, he himself knows best, and feels more surely where the shoe pinches."

I will transfer Meaning from that to the Context of Social Dynamics and to time now. In order to do so I need to modify the data pattern. In other words, I have seen particular Meaning in Einstein's words and wish to capture and use the same Meaning in a different Context. So, I will borrow from his arrangement and change the parts thus:

> At a time like the present, when experience forces us to seek a newer and more solid foundation for Social Dynamics, the Pure Scientist cannot simply surrender to the philosopher, or to any other agents, the critical contemplation of the theoretical foundations; for she (or he) can feel more surely where the shoe pinches and can find a way of easing the strain.

I would add:

> The only way to ease any strain is to discover the Cause of that strain and deal with it.

I will also add that Physics as a Special–ism is an Inductive science. Einstein was a Physicist but he also engaged in Pure Science. Pure Science is not the deductive part of Induction.

Each Single thing large and small is a Cosmos. Different Worlds are Intrinsically Related. Where there is an Intrinsic Relation there is mutual dependency. Human beings affect all things within reach and strive to affect the totality of all the things that exist, and they do so through Social Structures. The Meaning that humans perceive drives their Actions. Meaning is abstracted by human beings from happenings. Philosophers gift Meaning they abstract and distribute their gifts widely. Formal logics were brought into existence through the Actions of particular agents long ago and are products of the

Action of their Imagination on Observations during their time. Formal logics grew and became Invasive. Academics specialising in Formal logics are called philosophers. The Action of Artificial philosophy is not restricted to Academics now specialising in philosophy. Specialists are given Power to assert Authority in their areas of interest without Justification. Those who specialise in philosophy take an interest in every area of enquiry and specialise in none of them. The results of Testing will show that philosophy is not improved by the use of Formal logic. Those who specialise in every area take an interest in philosophy and ideas are exchanged. I found arguments without endings in the philosophical literature and little Real Meaning, so my Curiosity was aroused. I observed that Meaning disappears into a black hole never to reappear. What happens to it? Where does it go? Those questions can be answered. The difference in type between Artificial logic and Natural Logic can be Marked. We do not know where things go with their Energy when they enter Black Holes in Space, but we can be Certain that Meaning absorbed into the Mind gushes out in different form into the Dimension from whence it came.

Abstract thinking is a Universal human activity. We are all philosophers. Meaning in abstract notions meets Meaning in particular things and both can change following their meeting. Abstract thinking about the World is different to experiencing the World and the two different things are related. Thinking is Intrinsically Related to experiencing. The obvious intimacy between abstract thoughts and experiencing is sometimes denied. That Denial is Puzzling. Many Puzzles have been set across the Centuries and the challenge is to solve them. There is now an urgent need to do that.

Throughout history, pioneers in all areas of enquiry naturally moved across different terrains in their search for Knowledge. Until recently, there were no barriers to cut off their searches. There are now. Curiously those who seek to remove Boundaries between Countries have erected barriers to prevent crossings between Contexts by those who search for Knowledge through Reasoning and those crossings are necessary if Knowledge is to be gained. I aim to take those barriers down and replace them with Boundaries across which explorers can move back and forth, while retaining the differences on either side. The pioneers in past ages explored Meaning wherever its threads and their Curiosity took them. I looked back at their ideas as I searched for the Mark of that which is Pure Science and that which is not. In seeking to represent some of their ideas in this part of my narrative I have used as a broad framework the generally agreed categories into which views have been placed. But as you will observe, Meaning does not sit easily within them. The categories are poorly defined. You will see that commonalities as well as differences have been dropped in order to force an Appearance of sharp divisions. That Appearance is Artificial. You will see that Meaning flows Naturally across those divides.

If it is to arrive at a particular destination the flow of Meaning must be channelled in particular directions, not freed from all constraints. Where there is sharp division there is conflict. Where there is no division there is conflict. We do not currently have the Knowledge that would enable us to prevent conflict. Or do we? I propose that many of the parts we need in order to do so are there, and that we just need to find their Intrinsic Relations for the True Meaning of the Human Cosmos to emerge. Once True Meaning does emerge then the Action that follows will be different, unless True Meaning is ignored.

POSITIVISM, LOGICAL POSITIVISM AND AN APPEARANCE OF OPPOSITION

The words reason and logic are interchangeable and both cover different types of Action. The word logic is used for Formal reasoning, which is Artificial, and also covers a different type of Action in Natural Reasoning. Descriptions are necessary to tell us what happens and adjectives are words that describe the Meaning of the words to which they are attached. Dropping the description drops particular Meaning. When a word that covers mixed types of Meaning is used as an adjective it fails to draw forth particular Meaning from the word to which it is attached. Vague descriptions are unhelpful and misleading. It is hard to tell the difference between Positivism and Logical Positivism. Positivism is a Specialist philosophical approach most clearly associated with a group of philosophers who regularly met in Vienna for discussions during

1924 to 1936. They took their name from the place where they met. This group is known as the Vienna Circle and its influence is still felt. However, it was not one approach and was a constantly changing arrangement of various Beliefs and Intuitions. Thornton (2011) states:

> "Two facts must be clearly recognized if a proper evaluation of the Vienna Circle is to be attempted. The first is that, despite its relatively short existence, even some of the most central theses of the Vienna Circle underwent radical changes. The second is that its members were by no means of one mind in all important matters; occasionally they espoused perspectives so radically at variance with each other that even their ostensive agreements cannot remain wholly unquestioned. Behind the rather thin public front, then, quite different philosophical projects were being pursued by the leading participants with, moreover, changing alliances. One way of taking account of this is by speaking (as above) explicitly of the philosophies (in the plural) of the Vienna Circle (and to avoid the singular definite description) while using the expression "Vienna Circle philosophy" (without an article) in a neutral generic sense."

If the last sentence there is aimed at simplifying Meaning it fails to do that and is particularly confusing, but that does not matter. Meaning can be simplified. The members of the Circle were Single Individuals, each with different Beliefs and Opinions and their gathering together was determined by a common purpose. While the purpose united them in Meaning and Action of one type and held the group together, that purpose did not reconcile their differences. Positivism is one type of approach that took many different forms. Positivism in all its forms mounted opposition to the type of Ideology to which the word Metaphysical is applied. Therefore they all attracted opposition from Metaphysicians. They also attracted opposition from a person who had much in common with their views. The philosopher Karl Popper (1902 to 1994) was not part of the Vienna Circle and his attempts to join the Circle had been rejected by some members of that group. He became a satellite circling around the mass that was the planet on which he could not land.

Popper attacked Metaphysics and Pure Science. Popper succeeded in Unsettling part of a Principle that is actually of Fundamental Significance to Pure Science. That is a Curious Fact among many Facts that are puzzling. Popper became accepted as an Authority on science. He was that, but the Meaning of science is not perceived. Popper dropped a difference in type of Action and his Authority was generalised without Justification to all the types of thing that the word science was used to cover. Popper was an Authority on Inductive sciences and those are one type of thing in various forms. The Inductive sciences are pseudo-science. I regard that as a Fact and will retake that as a proposition for Testing in order to show you the Tests. I did not regard that as a Fact prior to Testing. Popper was skilled in marketing that Artificial Product. The term "logic of science" does Not Mean what it Appears to Mean.

So, what were the qualifications that served as a basis for his recognition as an expert on science? Brief details of the Vienna Circle and some personal Context for Popper are given by Thornton (2011), who informs us that:

> Popper "obtained a primary school teaching diploma in 1925 took a Ph.D. in philosophy in 1928 under the supervision of Karl Bühler and qualified to teach mathematics and physics in secondary school in 1929. The dominant philosophical group in Vienna at the time was the Wiener Kreis, the circle of 'scientifically-minded' intellectuals focused around Moritz Schlick, who had been appointed Professor of the Philosophy of the Inductive sciences at Vienna University in 1922. This included Rudolf Carnap, Otto Neurath, Viktor Kraft, Hans Hahn and Herbert Feigl. The principal objective of the members of the Circle was to unify the sciences, which carried with it, in their view, the need to eliminate metaphysics once and for all by showing that metaphysical propositions are meaningless—a project which Schlick in particular saw as deriving from the account of the proposition given in Ludwig Wittgenstein's Tractatus."

Mathematics within the Process of Pure Science has different Meaning to mathematics within the Inductive sciences but the difference has been dropped and is not seen. Physics is considered by many to be one of the Natural Sciences and is overwhelmed by Inductive science. The topic on which Physicists focus Attention is the Natural World and the dominant Method in the discipline is Not Natural. Physics has a mathematical Bias. Formal philosophy is aligned to mathematics and physics. Popper did not specialise in physics. Popper specialised in the Inductive science of philosophy.

Things that are aligned in Meaning are also different in Meaning; it is the common Meaning that draws them together. But differences can force things apart. Emotion is not the Focus of Attention in this book, but the Capacity for Emotion that active agents have is important and is Intrinsically Related to the Capacity for Reason. Emotion is one form of Response when Meaning is perceived. Emotion is a variable that intervenes between the perception of Meaning and the Action that relates to it. Operations on Emotion are a constant thread through Social Dynamics. Thornton allows a glimpse of the effect of the Factor of Human Emotion during the Inductive "unification" of science:

> "Although he was friendly with some of the Circle's members and shared their esteem for science, the acerbity of Popper's manner and his hostility towards Wittgenstein alienated Schlick, and he was never invited to become a member of the group. For his part, Popper became increasingly critical of the main tenets of logical positivism, especially of what he considered to be its misplaced focus on the theory of meaning in philosophy and upon verification in scientific methodology, and revelled in the title 'the official opposition' which was bestowed upon him by Neurath. He articulated his own view of science, and his criticisms of the positivists, in his first work, published under the title Logik der Forschung in 1934. The book—which he was later to claim rang the death knell for positivism—attracted significant levels of attention."

Popper was excluded by not being included. He took exception to that and Reacted in a particular way. Exclusion worked well for Popper. He worked on the same material and played to a larger audience. There was much in common between Popper and the members of the Vienna Circle. Meaning was a common enemy. They aligned in attacking Metaphysics. Bear in mind that we do not yet know what Metaphysics is or what the differences are between that approach and science. The word Metaphysician can be used for those who engage in Metaphysics, and taken separately the Meaning of physician is Associated with the Action of healing. But a physician intervenes and not all interventions heal, so do not associate the quality of healing with the word Metaphysics for that would lead you into Error. The Vienna Circle and Popper used Induction and called it science. The problem for the Positivists, Popper and all who follow their lead is that Verification is an Insoluble Problem for the Method of Induction. Thus, they had to Act against Metaphysics without being able to Verify the Qualities of their own position.

Promotion of the Inductive sciences was the Focus of Attention of the group and that is True of all Specialist philosophers. As Hume observed, the Universal and Insoluble Problem of Induction is the lack of Justification for conclusions. There can be No Verification in the absence of finding True Meaning, and Justification is aligned to Truth. Truth is a Problem for Induction in more ways than one. Verification is the showing of Truth. It is a Fact that the movement, emergence and development of Meaning are not Justifiable when Induction dominates those Operations. It is a Fact that Verification is an Insoluble Problem for Induction. Whether or not the Problem of Verification is Insoluble for Metaphysics depends on how Metaphysics is defined. If the definition of Metaphysics is restricted to the Action of Induction, then Justification is an Insoluble Problem for that too.

The Vienna Circle was a group of people who were concerned with the philosophy of the Inductive sciences and were not concerned with Pure Science. They merely desired to acquire the Appearance of Pure Science. I propose that Pure Science is in essence Natural Qualitative Logic. Inductive logics are Not Qualitative or Natural. Uebel (2012) states that:

> ".... in the 1920s, Schlick (1926) convicted metaphysics for falsely trying to express as logically structured cognition what is but the inexpressible qualitative content of experience……. Carnap (1928....) edged towards a formal criterion by requiring

> empirically significant statements to be such that experiential support for them or for their negation is at least conceivable."

It is a Fact that the Qualitative content of experience can be expressed and I would be surprised if that was not considered to be a Fact by the members of the Vienna Circle. Experience is composed of Sensations and is Qualitative. While its Representation may be True or False in relation to particular things, its expression is a constant Action of human beings. So presumably it was the difficulty in Representing the Meaning of experience using Formal logic that occupied the Attention of the group. Meaning is the Qualitative Content of Experience. Their final solution to their difficulty was to Ignore Meaning. Schlick and the others in the group saw that Justification is a problem for Induction and did not accept that the problem could not be solved. Nobody Appears to have seen that the problem for Social Dynamics is reliance on the Method of Induction for Judgments, or rather let us presume they did not see that. Popper suggested his own approach was not Inductive and was deductive. He also suggested he knew better than the members of the Vienna Circle what science is and what it is not. Or perhaps the word asserted fits better there. According to Thornton (2011):

> "Popper is unusual amongst contemporary philosophers in that he accepts the validity of the Humean critique of induction, and indeed, goes beyond it in arguing that induction is never actually used by the scientist. However, he does not concede that this entails the scepticism which is associated with Hume, and argues that the Baconian/Newtonian insistence on the primacy of 'pure' observation, as the initial step in the formation of theories, is completely misguided: all observation is selective and theory-laden—there are no pure or theory-free observations. In this way he destabilises the traditional view that science can be distinguished from non-science on the basis of its inductive methodology; in contradistinction to this, Popper holds that there is no unique methodology specific to science. Science, like virtually every other human, and indeed organic, activity, Popper believes, consists largely of problem-solving."

So, Popper did accept that the Problem of Justification for Induction was Insoluble. But skill in Induction was his only claim to expertise. He picked up Meaning from a trick used by Plato and Aristotle, and ran further with that. Popper used the Mechanism of Denial to solve the Insoluble Problem of Induction. He did not deny that there was a problem. He simply denied using Induction. The traditional view is that Science is Not Inductive. That view is accurate. Pure Science is Not Inductive. The Inductive "sciences" are not Science and the difference in type was traditionally seen. Schlick wanted to change the traditional view and so did Popper. But they did not want to change to Science and did not want to be ruled by the Process of Pure Science. They were not equipped by training with skills of the type necessary to engage effectively in that particular Process. They had skill in Artificial Inductive logic. Inductive logic has the False Appearance of Logic. Justification and Pure Science are Intrinsically Related for they have the same Point of Reference - Truth is the Ruler in that Cosmos. The Process of Pure Science has one Ruler only and Natural Deduction serves that Ruler well. Those who want to Rule and do not want to Serve use Artificial Induction.

Members of the Vienna Circle and Popper were united by a common Goal; they just could not tolerate being together in the same room. They wanted audiences to perceive that the Method of Acting on Meaning that they had in common is Justifiable for that perception leads to acceptance of Actions related to that. Dropping differences in Meaning and presenting a Single Whole Appearance of science would enable that perception. They fixed the visible differences in Meaning through redefining the word science. Definitions are an Operationalization of Meaning. The Concept of Science had not been well developed, so that was easily made to disappear. The Inductive Construct of Science took its place. The Representation stayed the same and its Meaning was changed.

I would be puzzled if any philosopher failed to accept that there can never be Justification for the products of Induction. That Fact is easily Tested in any particular instance of Induction. And it is not necessary to Test all particular instances to prove a Universal proposition, but that is said to be necessary by those who use Artificial logics. That suggestion creates Uncertainty. If you accept that suggestion, you can never be sure that Induction is Not Justifiable. Specialists in philosophy have just accepted the teachings of their forebears. The Premise that no matter how many Single Cases are True or False there is always the possibility that there remains one that has not been tested and will be different in Quality was enshrined in an Ancient puzzle. Puzzles defy solution and serve particular purposes. The particular effect of this particular puzzle is the creation of the Illusion of Justification for Uncertainty. Through that Illusion, Meaning from one Single Case is selectively blocked from having a Significant Effect on Meaning in another Single Case simply because all Single Cases are unique.

The competing Method of reaching conclusions is Natural Deduction and that takes Meaning from Single Cases. By suggesting that no conclusion is Certain until and unless every particular instance has been taken into account, clever fellows in the Ancient World Acted to remove Certainty and Justification from Judgments of Meaning. Specialists in philosophy have been taught their craft, so we cannot blame them. But the way their Actions Function is clear. There is only one possible conclusion. Some of our Ancestors made it impossible for others to do that which they would not do. They made room for Doubt and passed it on. That is a type of Equality and that type of Equality is Not Justifiable and never can be. In directing Attention to doing that which it is impossible to do, Attention was turned away from doing that which it is possible to do.

Popper is not unusual among Specialist philosophers in his Operations on Meaning. He does Not Justify his opinions and he uses Vagueness as a cover. To say that science consists largely of problem solving does not get us far in distinguishing the type of thing that Pure Science is. I will add some Meaning to that. Pure Science consists of searches for the correct solution to a problem, using the word correct in the sense of being accurate. An accurate solution solves the problem and does not create other problems that take its place. We have to use the type of Method by which the correct solution can be found in order have the Potential to accurately solve problems.

THE UNOFFICIAL OPPOSITION IS STRONG

Let me show you some Real Opposition over the course of this narrative. Meaning can be Justified. True Meaning can be found. Certainty can be reached. I do not argue that Pure Science is a Single particular Method, for I see it as different to that. Pure Science is an Active Process and all parts of that Process have particular Meaning that is distinctive in its type of Action. I take the traditional view - Pure Science is Not Inductive. Natural Deduction is different in type to deduction within Formal logic. The traditional view that Observation within the Boundaries of Pure Science is the initial step in the formation of theories is not "completely misguided" but the nature and role of Observation is perhaps misunderstood by some and not fully expressed by others. Observation is a necessary part of Testing. Observation is a Complex Active Mechanism. Knowledge can only be gained by Testing Meaning in Observations, and Abstracting Meaning using Natural Deduction is part of doing that. The distinctive character of the Process of Pure Science is determined by the constraints of Truth which exert an influence over all its parts.

Popper used the Fact that Observations are Variable and are Selected to introduce a Negative Bias in Meaning to the Action of Observing. It will become clear during the course of my narratives how useful that is for agents to whom Truth is an inconvenience. Popper was Active against the Action of Observation and against the Action of Pure Science. The False Premise that Popper is an Authority on Science has been accepted. If we are influenced by his Opinions without Testing them, then we are misguided. I propose that if we Test them for True Meaning we will not be guided by them for we will see them as the Misjudgements they actually are.

Selection is a Natural tool. Selection is a Universal Natural Mechanism and part of the Process of Pure Science. But within the Process of Pure Science the Meaning of the word selection is different to its Meaning outside the Boundaries of that Process. The common definition of selection refers to choices without specifying the difference in type of principle on which choices are based or the difference that the difference in principle makes to the outcome. Principles are Active Points of Reference and Points of Reference Govern Meaning and Action throughout arrangements to which they are applied. I would not use the term-theory laden in relation to Observations and I would use the word Biased. I agree that Observations are Biased. I propose that Bias is Natural. Where there is Selection there is necessarily Bias. But I do not agree that Bias is a Universal Problem. Only certain types of Bias are a Problem. The reduction of many parts to fewer parts is the Function and Purpose of the Natural Mechanism of Selection. The parts that are Selected then multiply and differentiate. Natural Systems go Round and Round.

So, according to MOM, Bias of one type or another is Universal in Social Dynamics. It is the differences in type of Bias that are Significant to Meaning and Action. I presume that Popper did not understand the nature of Bias and it is a Fact that he was Biased. Popper Unsettled the traditional view that pseudo-science can be distinguished from Natural Science by its Inductive Method. He used several different Factors in conjunction in order to Unsettle the traditional position. Artificial Induction is distinctive of pseudo-science. The prefix pseudo indicates the taking and presentation of a False Identity.

Within the Process of Pure Science, theories are part of the route to finding Facts through Testing. Now, Popper did pick up that Meaning but could not use it well. Inductive theories are Not Tested and whether or not they are Testable remains to be seen. The deductive part of Induction gives the False Appearance of Testing to the Method of Induction. Within the Process of Pure Science, Testing is geared to the finding of True Meaning. Specialists in philosophy know that Fact. Popper indicated that theories must be Testable or at least must have the Appearance of being Testable at some point. It must Appear to the audience that True Meaning is sought. The Emergence of True Meaning can be delayed. There are Opportunities for profitable Action in the intervening period of time and that effect of the criterion of testability within Induction will be shown later through a particular example. Another effect is to turn Attention away from Concepts and towards Constructs. That is a complicated manoeuvre and is part of a Whole Persuasion that has taken humans away from Natural Systems of Judgment and fostered dependency on Artificial Systems.

Selections that are guided by Theories are not guided by Facts. Popper suggested that there is no "unique methodology" specific to science. That may be True of Inductive science but it is Not True of Pure Science. Perhaps that Error in Judgment results from a failure to understand the Structure of Meaning. Intention is set aside in this narrative, and I will presume that Specialists in philosophy misunderstand the type of thing that Meaning and Pure Science are. They pass their Ignorance to others. There are many types of Mechanism within the Process of Pure Science. While the Action in each part is different to that of the other parts, the Action is the same as a Whole in terms of its Meaning. Pure Science is an alignment of many parts of different types by a single purpose. The purpose of the Action of Pure Science is to discover True Meaning.

So, Pure Science is a Process with particular Meaning. There are different routes through that Process. Popper may have seen the flexibility and failed to understand the necessity of that and what it achieves. Specialists in philosophy may or may not have perceived flexibility as Vagueness. But it is a Fact that they substituted a Construct of Vagueness for a Concept of Flexibility.

Observations by particular individuals take place in particular Contexts. Different Biases are Active in different Contexts. A Bias is a particular Alignment of Meaning and vice versa. The Meaning drawn from all parts during Observation in a particular Context is governed by a Point of Reference. Different Meaning aligns when different Points of Reference are applied. I will show you that Pure Science is a particular and distinctive type of Bias in the Action of human agents upon the World. The Attention of agents engaged in that Process is focussed on Truth in relation to what happens in the World. Agents who Observe Action for the purpose of finding True Meaning are engaging in part of the Process of Pure Science. In order to find True Meaning, Points of Reference must be Tracked and specified. A term with the same Meaning as Point of Reference is Focus of Attention. An Alignment of Meaning happens when there is a Point of Reference on which Attention is Focussed. True Meaning is hidden when the Focus of Attention is not visible. Meaning can be Calculated from Observations under certain conditions and is difficult to Calculate under different conditions. Attention is a Complex Mechanism and an Operational Variable. Facts about Attention are known and misused in Social Dynamics. Facts can be used well. Causal Knowledge can only be discovered when True Meaning is discovered. But the Process of Pure Science does not end with the discovery of True Meaning in Single Cases. True Meaning in Single Cases is the core on which Causal Knowledge is built.

ASPECTS OF THE DEVELOPMENT OF POSITIVISM

Meaning is different in different Contexts. The Specialism of philosophy is a particular Context. We need to find and specify the type of thing in which agents within that Context are specialising. The question for which we need an answer is: Where is the Attention of Specialists in philosophy focussed? In other words, what is their particular Bias? Bias is a Complex Phenomenon. Meaning can be found though Actions in Single Cases. Then that Meaning can be followed along the channels through which it flows. Members of the Vienna Circle picked up Meaning from Ernst Mach (1838 – 1916). Pojman (2011) tells us that:

> "In philosophy, [Mach] is best known for his influence upon the Vienna Circle (a predecessor of which was named the Ernst Mach Verein), for his famous anti-metaphysical attitude (which developed into the verifiability theory of meaning), for his anti-realist stance in opposition to atomism, and in general for his positivist-empiricist approach to epistemology"

I will use as a Chorus part of a Representation of Meaning you have seen before:

Chorus: Huge are the words that he hurls, great compounds with rivets and bolts in, And epithets hewn out of stone.

But let us not squeak weakly in response. Let us boldly Track Meaning that is not given freely. Mach's views were many, varied and sometimes proved to be wrong. His thinking was fruitful nonetheless. All types of Meaning are useful. There must be Meaning to Test before there can be any Discoveries. Mach was acknowledged by Einstein as being the philosophical forerunner of relativity theory. Mach fore-ran much else as well. Here I am concerned with distinguishing those parts of Meaning that are True from those parts of Meaning that are False, not in relation to Mach's Opinions but in relation to the development of the Methodology of Positivism. Where there is an Ideology, there is a Methodology for its implementation. An Ideology is a particular Structure of ideas governed by a Point of Reference and a Methodology is the Action that is Intrinsically Related to that Structure. Meaning and Action are governed by a common Point of Reference. In order to Test Meaning and show its Qualities, we must have particular Meaning in a form that can be Tested. Popper knew that Fact. Apart from being opposed to Metaphysics, which is a word with vague Meaning, there is no specification of the type of thing that Positivism and Logical Positivism are. According to Pojman:

> "Mach is generally considered a 'positivist'. ....Outside of the social sciences (where it is alive and well), the word 'positivism' seems to be used mostly by post-modernists, deconstructionists, and the Frankfurt school as a word to represent that branch of the enlightenment tradition that failed; it has become the 'other' which we all know is wrong without having to know what it is."

Here we have a description of the Operation of Induction. We know things simply because we know them. We have no idea how or why we know what we know, all we actually have is a conclusion and a feeling of Certainty that the conclusion is true. Induction leads to Beliefs, but only if it is allowed to do so. Please note two Facts at this point. Positivism failed and Positivism is alive, well and Active in the social sciences. Those are statements of Fact. Curiouser and Curiouser said I. Positivism is easily seen in the social sciences. But it is not possible to prove that it should not be in Action if we do not know what that is. It follows using Natural Logic that if it is Active then Positivism Certainly did not fail. Positivism was rejected as agents tried to enter with it into some Contexts and, in Reaction to that its Appearance was changed to enable Positivism to take hold. People learn from experience and Rejection is an experience that many humans learn how to avoid. Avoidance is a Response and a Response is a Reaction.

I will use some of the same terms that Pojman used while expressing different particular Meaning using them. And I will show you a small detail that Pojman did not include. There is a word that represents "that branch of the enlightenment tradition" that succeeded. That word is science. It is science that has become the 'other' that we all know is right without having to know what it is. The State of Ignorance in relation to its Meaning protects science from the perception that it is actually the 'other' which is wrong. The audience has no idea of the type of Meaning they have accepted. All clues lead nowhere in particular when they are followed. The audience is kept in a State of Belief that science leads to factual knowledge. They are actually worshipping a False Idol. That is a statement of a conclusion for which you have yet to see Evidence. Ideologies are theories with the Appearance of Facts and they characterise thinking. Cognitions must have Structure if they are to have Meaning. The type of Structure makes a difference to Functioning. I propose that Inductive science is an Ideology and that proposition is Testable through Observations of that which bears the name science in Action. Ideologies Function in particular ways. Meaning can be Calculated from Functioning. And Functioning can be Calculated from Observations of Reactions to other things.

Pure Science is Not an Ideology. Pure Science is the search for True Meaning. When True Meaning is found we have Knowledge. As part of their manoeuvres, the Vienna Circle dropped the adjective Inductive that had showed the particular Meaning of their type of "science". The term "enlightenment tradition" was applied to Inductive science and we know that to be enlightened is to find True Meaning, so Inductive science automatically picks up that particular Meaning by Association. Where it took hold, Inductive science displaced Pure Science. There was no Unification. The word inductive was dropped and the difference in type disappeared with that Simple Action. The Mechanisms of Association and Dissociation are more Powerful Acting in Conjunction than either Mechanism in Action alone. When True Meaning is hidden the audience is subject to Deception. That is not an Unjustifiable Opinion on the Action of these clever fellows. That is a Fact. I will add a Chorus.

Chorus: They Know Not What They Do.

While I am well aware of the Ancient Source of this Chorus, I do not take that Source as an Authority for I have no Religious Beliefs. I do not take Aristophanes as an Authority either. The Choruses in my narratives represent Meaning that can be Independently Justified in the Context of use. I cannot restrict my use of Representations simply because others used them in a different Context. To do that would disable me. Meaning is transferrable and some transfers from one Context to a different Context are Justifiable. The Association of the words science and induction is Not Justifiable. But there is no requirement in Social Dynamics to Justify Meaning and its transfer. I found that puzzling. There is a requirement to present an Appearance of Justification. It follows that the Significance of Justification is known.

According to Uebel (2012):

> "The Vienna Circle's theories were constantly changing. In spite (or perhaps because) of this, they helped to provide the blueprint for analytical philosophy of science as meta-theory—a "second-order" reflection of "first-order" sciences. While the Vienna Circle's early form of logical empiricism (or logical positivism or neopositivism: these labels will be used interchangeably here) no longer represents an active research program, recent history of philosophy of science has unearthed much previously neglected variety and depth in the doctrines of the Circle's protagonists, some of whose positions retain relevance for contemporary analytical philosophy."

All reflections are Illusory. Positivism, empiricism and Formal logic in general are a vague reflection of Science and are not a mirror-image of Science, for mirror-images copy particular detail. Some Formal logics are closer in their Appearance than others to mirror-images of Pure Science. Vague images and mirror-images can Confuse Actions. The direction of movement in a reflection is reversed. It is a Fact that even accurate reflections are not the Real thing, and science is not even an accurate reflection of the Meaning and Action of Pure Science. Empiricism is merely a vague image of a

Method that does exist. Specialist philosophers seem to be aware of that for here it is acknowledged that there is a difference between Inductive science and the traditional Sciences. But beyond being told that Inductive science is a reflection, we are not told what the difference is. Differences can be told.

The Traditional Sciences relate to the Real World and humans are part of the Real World. Pure Science can be applied to the study of that part. Specialist philosophers are human beings who do not analyse the Real World and they do not Appear to pay Attention to Meaning within the World. But Appearances are Deceptive. If you have ever walked into a room of distorting mirrors in a fairground, you will have had an experience that is the same as looking at the literature of Formal philosophy. The experience is not similar to that; it has the same Meaning. The "second-order" reflection of "first-order" sciences is distorted.

I note that the terms logical empiricism and logical positivism are used interchangeably, and where that is done Meaning is the same. The Ideology named science that developed from the early form of logical empiricism is not a reflection of the Real World and is a poor reflection of the Actions of agents who engage in Pure Science. Although Modern Empiricism looks different to Positivism, the foundations of Empiricism are Positivism and that is the branch of the "enlightenment tradition" that "failed". The philosophical foundations of Modern Empiricism are unsound. So, Empiricism succeeded while its foundations failed. The Natural conclusion is that there is more than Positivism holding the Structure up. I propose that all of the foundations of Inductive science are Unstable but some are yet to give way. Any Structure built on Unstable Foundations has the Quality of Instability. If the proposition that the theoretical Structure of Modern Empiricism is inherently Unstable is True, then the Structure will collapse when the conditions for Destabilisation are met. I ask the question: What precisely has Empiricism succeeded in doing? I will show that Empiricism has succeeded in leading us into Uncertainty with the Appearance of Certainty. Empiricism is the philosophy of Inductive science in Action on the World. Empiricism allows Selected Beliefs to triumph over different Beliefs. Another question I ask is: How does that System work? That question is a Simple one of Functioning and the answer is also Simple but finding and providing Verification is not Simple. Many Calculations are necessary to find an accurate answer on Causation and a Potential Solution to the problems we face.

Having understood that Justification is an Insoluble Problem for Induction, Popper Unsettled the Verifiability Theory of Meaning in Inductive science. The Inductive sciences could never possess the Qualities of Pure Science so there was no point in trying to acquire them. In order for the takeover to succeed, the value of the Qualities that Induction actually has, must be accepted instead. So, Popper argued that there is no such thing as Proof. Thus Meaning is always Uncertain. It was Asserted that there was only one type of science and that Induction was it. That was backed up by a different Assertion. Popper Asserted that he used deduction not Induction and in doing so the Illusion that deduction is Uncertain was strengthened. He changed the name of Induction leaving its Meaning intact – that is an Action of Definition. These were not novel Actions and had been modelled Centuries before his time. Popper had a PhD in philosophy. He had studied the teachings of a particular group of Ancients and their followers.

The failure to understand the Structure of Meaning enables continued misuse of Natural Language. Small details are significant and changing them can lead to large changes in Meaning. It is Reasonable to presume for now that there has been no Malign Intent to frame Natural Language as being Guilty of misleading the Perception of Meaning. But there has been Intention. Intentions are variable and the type of Intention must be Verified by Testing after Misconduct has been Verified by Testing. Good Intentions can lead to bad effects. But the Fact is that a Conviction followed the Actions of those involved in the gathering of evidence against Natural Language and the Trial. Natural Language is on Death Row. And books that have not yet been consigned to the flames are not read; the effect is the same. The Conviction is a Miscarriage of Justice that has never been overturned. It will take the Action of Pure Science to Verify Innocence in this Case. Continued misuse of Natural Language is used as evidence of its Dysfunction. That is misuse of the word evidence. Round and Round we go in circles when it is not known what actually determines happenings. Clever fellows across the Centuries have Acted against the Discovery of Causal

Knowledge. It is not wise to do that. Or is it? The Judgment hinges on definitions and the words clever and wise have been undermined. So, I will propose that it is not Intelligent to do that. And I will define Intelligence within MOM as skill in Reasoning and define Reasoning as Natural Deduction aimed at the discovery of True Meaning in relation to Reality and define Reality as what actually happens in the Natural World. Natural Deduction is Intrinsically Related to Common-Sense. But it is not a formalisation of Common-Sense for both Actions are Natural in type and are therefore informal and flexible.

The Process of Pure Science is a Process that is geared towards the finding of True Meaning in Reality. The Action in Pure Science leads to Verification thus providing Certainty, or at least it leads in that direction without deviating from the path. Meaning that is Verified is True. When Meaning is False that can be Verified and the word Falsification is used for that happening. The Meaning of Verification and that of Falsification are Intrinsically Related. If you accept that there can be Falsification then you necessarily accept that there can be Verification. In the absence of Verification there is Uncertainty. When Uncertainty has the Appearance of Certainty there is Belief. Both the words Premise and Belief represent Meaning that is Not True and is not Certain but has the False Appearance of being so.

There is a Natural form of Induction but that is a different type of thing to Artificial Induction and does not concern us here. It is Artificial Induction that is not part of Pure Science. Artificial Induction merely bears some resemblance to the Natural type. Formal logic is a form of Induction that is Not Natural. That is the type of Induction that concerns us here. Mathematics is a Formal logic. The Vienna Circle and Popper were Biased towards the use of mathematics. They had developed that skill. These philosophers were interested in drawing particular conclusions using the Code of Numbers. A False Premise was set. It was Asserted that if Formal logic in general and the Quantitative Method in particular cannot produce Verification then Verification can never be produced. The Vienna Circle and Popper set about Asserting their Authority over the Manipulation of Meaning by Acting against any expression of Meaning that they classified as Metaphysical. Uebel (2012) informs us that the Anti-Metaphysical position of the Vienna Circle was supported by:

> "..... an empiricist criterion of meaning and a broadly logistic conception of mathematics. ....To begin with, its members styled themselves as conceptual revolutionaries who cleared the stables of academic philosophy by showing metaphysics not simply to be false, but to be cognitively empty and meaningless."

Academic philosophy was not the only target of these Specialists in philosophy. They aimed to Unsettle every theory and Belief that did not accord with their own ideas. Philosophers Act on cognitions - the Action is psychological. If the Turner Rule is applied, they know no more and no less about normal psychological processes than any other human being. But the Premise that they are experts on that topic is Active. The Opinions of Specialists in philosophy on normal psychological processes are Active in Court and have been accepted as being authoritative by people in every Context. Thus it is normal in Quantitative terms to accept that these philosophers are experts in the psychology of normal human behaviour, but I challenge that view. A Quantitative Construct of the word normal stands firmly where a Concept of Normal should be. Quantitative Constructs are Active as definitions. I propose that the Opinions of Specialists in Artificial philosophy are abnormal by a Natural Qualitative Standard. Specialists in philosophy are not present as human beings in Court. The Source of the psychological Opinion that is in Action there is hidden from view. But employing experts in psychology who rely on Induction would not make a difference to the Quality of Judgments. Artificial Induction leads to Judgments without the Quality of Truth and Justification. The field of Specialist psychology accepted the Bias of Inductive science. Curiouser and Curiouser said I.

The discussions and ideas of the Vienna Circle were centred on the Meaning of things and on how Formal logic in general and mathematics in particular could work with Meaning of all types. Revolutions change the Balance of Power. The overthrow of Metaphysics changed the Balance of Power to those who have skill in mathematics. An "empiricist criterion of meaning" is a Quantitative Construct of Meaning. Constructs are an Operationalization of

Meaning. Constructs are Representations. A Concept of Meaning was not developed and it is a Fact that the group did not set out to understand the type of thing that Meaning actually is. They simply set out to bring the Representation of Meaning under the control of those Academic philosophers who use the Quantitative Method and under the control of mathematicians in every Context. That is a Fault. Faults lead to Errors. Having gained Power, the clever fellows who use Formal logics feel secure for the Code of Numbers is designed to be hard to break. It is impossible to Unsettle Meaning when Meaning is not there. When Meaning is not in one place, then it is somewhere else. I invite you to join in a modified Chorus:

Chorus: Huge are the equations they hurl, great compounds with rivets and bolts in, And numbers hewn out of stone.

The audience has long squeaked weakly: We do not understand what you mean! Of course we have not understood the messages that they pass between them in code. That is how codes Function. So far, nobody has succeeded in breaking open the code of numbers to see what is there. Meekly we accept any Meaning they give us from their formulae for we can do nothing else. Meekly we accept that the equations are the Source of the changes in Meaning that these Specialists deliver. But it is a Fact that human beings are the Source of the Meaning delivered through mathematical equations. The mathematical equations are a front for Action that takes place elsewhere. The Quantitative Method displaced the Qualitative Method. There is No Evidence that Specialists in philosophy are malicious. But there is Evidence of the effects of their Actions and the outcome from those is fast approaching. And as the outcome gets closer its type becomes clear. Surely it is time for a Revolution in favour of a different type of Action.

At an early stage, the idea took hold that support can replace the Proof of Truth that they could not realise using Inductive science. Uebel states:

> "Note that even though this early Wittgensteinian version of the meaning criterion required conclusive verifiability (which Carnap's of 1928 did not), it also allowed for verifiability in

> principle only (and did not demand actual verifiability). Like Carnap's notion of experiential support, this criterion worked with the mere conceivability of verifiability. (The demand for conclusive verifiability was discussed in the meetings with Wittgenstein.) Around 1931, however, it had become clear to some that this would not do. What Carnap later called the "liberalization of empiricism" was underway and different camps became discernible within the Circle."

The term "verifiability in principle only" bears at least a strong resemblance to the word testability as it is used by Popper. The Circle and the Satellite moved in the same direction as a Whole. The "liberalization of empiricism" can be viewed as a euphemism for lowering the Standard of proof to a level that was possible for Inductive science to meet. Mach had said that "where neither confirmation nor refutation is possible, science is not concerned". Perhaps that has a double Meaning and perhaps Mach did not see that. Where neither confirmation nor refutation is possible, Pure Science is Disabled and Truth is not concerned. When Pure Science is Disabled, Uncertainty can take hold. The Vienna Circle used the words verification and falsification. There is Meaning in common. By definition, to confirm or verify is to establish the truth, correctness, or validity of a proposition by examination. That definition represents different types of Meaning by single words. Specialists in philosophy cannot establish the truth or correctness of a proposition. Observation of any sample of their work will disclose the Truth of that statement. Verification is not possible for Inductive science. The relevant group of peers have agreed that statement is Fact. Thus, the criterion of Verification was abandoned. They can establish validity by agreeing that a proposition is true or correct when they cannot confirm the Truth. The word validity was already inserted in the definition of verification and thus the link to truth and correctness was ready to became Active in our minds. The difference between proof and support was dropped and those words were gifted the same Meaning by the Premise that where there was support there was proof. I do not know at what point in time these definitions were set. The movement away from Pure Science had begun long before the Nineteenth and Twentieth Centuries. Meaning moves continually and there will have been no particular point at which different types of Meaning were mixed together - there has been a gradual progression.

In short, it was recognised that proof was impossible, without recognising the Cause of that, and at some point it was held that the Appearance of travelling some distance towards proof could be given without ever getting there. Popper contributed to the change in the criteria for Science. Having agreed to drop the difference that traditionally distinguished Pure Science from pseudo-science, these clever fellows then agreed that the distance they travelled towards proof could be measured Quantitatively. They also agreed that Metaphysics should be abolished. That is a lot of agreement and there was more to come.

Popper and members of The Vienna Circle, in common with Specialists in philosophy before and since their time, built Meaning from a particular theoretical base, while Asserting their Belief that the base is sound without Testing its Qualities. They criticised Metaphysics for doing precisely that. They ignored the Fact that their Action was of the same type. Having criticised Metaphysics for its lack of Evidence of Truth and found they could not correct the Fault in their own theories they simply decided to agree that the absence of Truth was not a problem for them. They also agreed that the absence of Truth was a problem for Metaphysis. I am surprised they got away with doing that but it is clear that they did. I will in due course show you where they picked up the Psychological Strategies that enable them to hide True Meaning and escape Detection. The inconsistency was not noticed and the Premise that the absence of Truth is only a problem for Metaphysics remained in place thereafter. If there is no Verification then there can be no Facts. It is fallacious to take your own failure to produce Facts as evidence that no Facts can be produced. Many Facts have been produced. Facts are not sufficient for the emergence of Knowledge but they are necessary for that happening. And I note that the word metaphysics covers different types of Meaning and Action.

There was a counterattack and also criticism from within the ranks. Uebel describes the Defences constructed by these clever fellows in the Circle, who were presumably full of good intentions but had a Bias against Truth and Justification:

> "Notwithstanding the metaphysicians" thunder, however, the most telling criticisms of the criterion came from within the Circle or broadly sympathetic philosophers. When it was protested that failure to meet an empiricist criterion of significance did not make philosophical statements meaningless, members of the Circle simply asked for an account of what this non-empirical and presumably non-emotive meaning consisted in and typically received no convincing answer. The weakness of their position was rather that their own criterion of empirical significance resisted an acceptable formal characterization."

Later you will see this same tactic repeated by Popper, and you may already be familiar with that Ancient strategy. The defence of the members of the Vienna Circle was to attack. Popper attacked as a Defence and also attacked when he did not get what he wanted. That is characteristic of the Formal philosophical position and that type of Action is Aggressive. The Formal philosophical position is an Ancient position with Forceful Strategies that have been in use for Centuries. I propose that the idea that Aggression is Universally Necessary is an Illusion. Aggression is neither necessary nor sufficient for Defence. I allow that opponents must be taken out of the equation but there are ways of doing that without resorting to Aggression. I also allow that it is hard to tell the difference between Aggression and Defence that is Non-Aggressive. Defence must be strong enough to defeat those who attack and the weapons must be fit for that purpose. It is the purpose that makes the difference in Meaning between Defence and Attack. There are different types of Force. A Construct of Defence as Aggression stands in place of a Concept of Defence as a Force. The Construct is Quantitative and that type of Construct extends to infinity. There is no Natural stopping point to Aggression once you take steps along that path. There is a Chain Reaction and an Escalation. I propose that Aggression is an Indicator of the Absence of Justification. And that an Absence of Justification leads to Aggression. There is an Intrinsic Relation between them.

Questions turn Attention in particular directions. The Vienna Circle and the Official Opposition used the same strategies in dealing with those who opposed them. Asking for an account of "this non-empirical and presumably non-emotive meaning" gifts the problem of Justification to the opponent. It also ties Empiricism to a lack of influence by Emotion and ties Metaphysics to Emotion. Empiricism has been defined as objective and objective has been defined in part as Not Biased by Emotion and as a Whole as being Unbiased. Dissociation from Emotion is part of science. I will provide Evidence that Dissociation from Emotion is part of a Fault and that the Fault is Selective Dissociation. Dissociation is a Mechanism that is operated in conjunction with the Mechanism of Association. These are Complex Mechanisms. That which is Dissociated becomes Associated with something else. Emotion involves feelings of different types and differences in type are Universally Significant in terms of Meaning and Action. Feeling Emotion is Natural, but in relation to particular things the Emotions of other people are not always of the type we assume them to be. Differences outweigh commonalities and Emotion is no exception to that. There are Universal Natural Principles of Emotion, which are yet to be identified. But many Facts are known. During the course of my narrative it will emerge from the Evidence that particular types of Emotion are Intrinsically Related to Pure Science.

Before I searched for the Meaning of Pure Science, I was forced to accept that there was one type of science and had to silence the questions that welled within me. Nobody had the answers to relevant questions. The questions were dismissed when I asked them. I had accepted the Premise that science was one thing that can be described as Objective without grasping that the Representation was part of the Assertions of those who have agreed to the Formal philosophical position. Agents in positions as designers and educators have forced pseudo-science upon us all. It is a Principle of my Model of Meaning (MOM) that Premises must be Tested. I will provide Evidence that a lack of Emotion is one indicator that Pure Science is Not in Operation. A lack of Control over Emotion is another indicator but that is not distinctive. Emotion Functions in a particular way within the Process of Pure Science. Engagement with that Process requires agents to use the Capacity for Emotion in conjunction with the Capacity for Reason, and the Action of one Capacity constrains the Action of the other. I will of course show you the route I took in arriving at that conclusion, but it is an extensive route so you will have to bear with me on a long haul through this narrative and possibly through several others.

The word science is positioned in the field of view for audiences to see while the Meaning assigned to that word is actually False in all its aspects. False Meaning Functions in particular ways and serves particular purposes. The lack of popularity of the "conceptual revolution" proposed by the Vienna Circle was predictable. Unpopularity is not a problem for Specialists in Formal philosophy and those aligned to that position. They Ignore Indicators that they and their views are Not Popular and carry on regardless for as long as the Rules, which were designed by Specialists in philosophy, allow. Nobody seems to notice that Action of that type is inconsistent with a Concept of Democracy. A Quantitative Construct of Democracy was designed long ago and stands where a Concept of Democracy should be. The Art of Persuasion involves the swaying of Emotion, but those who use Formal logic are Selectively Dissociated from Emotion. Agents who use Formal logic seek to persuade others and will not be persuaded by others. That position is only possible under the condition that you know Facts about the tools of the Art of Persuasion and how they can be related to achieve particular effects. Resistance is possible once True Meaning is perceived.

Uebel (2012) tells us:

> "Not surprisingly, it was the Circle's rejection of metaphysics by means of their seemingly devastating criterion of cognitive significance that attracted immediate opposition. (That they did not deny all meaning to statements thus ruled out of court was freely admitted from early on, but this "expressive" surplus was considered secondary to so-called "cognitive" meaning and discountable in science"

It is clear that they were dealing with Meaning and knew that Fact. Mathematics does not have the flexibility of Natural Language that renders it expressive and so the utility of the property was dismissed. It is True that the expressive property of Natural Language in the Representation of Meaning is misused. The tool was blamed for the misdeeds of its users. Indeed the tool was misused in order that blame could be attached to that. Misattribution is both constructive and destructive. The term "cognitive significance" is clearly a valuation of Meaning. How was the value of Meaning to be judged? Uebel informs us that:

> "The synthetic statements of the empirical sciences meanwhile were held to be cognitively meaningful if and only if they were empirically testable in some sense. They derived their justification as knowledge claims from successful tests. Here the Circle appealed to a meaning criterion the correct formulation of which was problematical and much debated .... **Roughly, if synthetic statements failed testability in principle they were considered to be cognitively meaningless and to give rise only to pseudo-problems.** No third category of significance besides that of a priori analytical and a posteriori synthetic statements was admitted: in particular, Kant's synthetic a priori was banned as having been refuted by the progress of science itself. (The theory of relativity showed what had been held to be an example of the synthetic a priori, namely Euclidean geometry, to be false as the geometry of physical space.) Thus the Circle rejected the knowledge claims of metaphysics as being neither analytic and a priori nor empirical and synthetic. (On related but different grounds, they also rejected the knowledge claims of normative ethics: whereas conditional norms could be grounded in means-ends relations, unconditional norms remained unprovable in empirical terms and so depended crucially on the disputed substantive a priori intuition.)"

First, I note my agreement that the statements of the empirical sciences are synthetic – I use the word artificial instead. Next I note that the paragraph quoted here is not a simple expression of Meaning. In Fact, Meaning there is not clear at all to the Uninitiated, of whom I am one. But I can Calculate Meaning from any Representation. And, the direction in which Meaning drifts here can be seen. What we see is rejection after rejection in an endless stream. I have highlighted a relatively simple statement within the arrangement. The criterion of falsification enabled by "testability in principle" became associated with Popper and did not originate with him. That criterion is of

Ancient origin as you will see in due course. The Meaning that can be taken from the part I highlighted is: - statements that do not meet the requirement of testability can be ignored. A pseudo-problem is not a real problem. Ignoring Meaning is a Strategy in Artificial logic. That strategy is one of Denial – problems are defined as ignorable but if they are actually Real they will not go away and will grow. We also glimpse here the threads of their Personal Bias reaching into the Moral Domain. Their Bias took hold in that domain and I will present Evidence that the potential of that Bias to Cause harm has been realised. We recall that the Positivist Approach is governed by an "empiricist criterion of meaning and a broadly logistic conception of mathematics". According to Uebel (2012):

> "Following the logicism of Frege and Russell, the Circle considered logic and mathematics to be analytic in nature. Extending Wittgenstein's insight about logical truths to mathematical ones as well, the Circle considered both to be tautological. **Like true statements of logic, mathematical statements did not express factual truths**: devoid of empirical content they only concerned ways of representing the world, spelling out implication relations between statements. **The knowledge claims of logic and mathematics gained their justification on purely formal grounds, by proof of their derivability by stated rules from stated axioms and premises.** (Depending on the standing of these axioms and premises, justification was conditional or unconditional.) Thus defanged of appeals to rational intuition, the contribution of pure reason to human knowledge (in the form of logic and mathematics) was thought easily integrated into the empiricist framework."

Thus Truth was cast aside while the word true continued in use. The words knowledge, truth and justification are taken in vain and abused. It is indisputable that Reasoning was cast aside. Specialists in philosophy openly make statements that are Illogical. They say that "true statements of logic" are like mathematical statements in that they do not express "factual truths". That actually means: None of the statements in logic and mathematics are True. Statements that are True are statements that are Verified. Truth does have fangs and bites hard. The Power of Truth to bite them was removed by peer agreement that Verification is impossible. Truth is accessed through Reasoning. Mathematics and other Formal logics are defanged of Truth and of Reasoning, but Specialists in Formal logic do not admit to that. Their own fangs are Artificial and Unstable but their Appearance is formidable. When we Sense Danger we do Not Approach. Being devoid of empirical content statements of logic in general and mathematical statements in particular spell out implication relations between statements. And relations determine the Meaning that is perceived in empirical content.

Now Hume preceded the Vienna Circle in time and yet we see these clever fellows laying False Claim to Justification for Axioms and Premises. The word justification has been redefined as knowledge claims from successful tests. And testing is defined in Terms of Formal logic and frequently expressed Quantitatively. The definitions are Not Justifiable and not acceptable. Claims must be Proved and Proof is Verification. The tests of science are Inductive. No claim can be proved by Induction. Where there is No Justification there cannot ever be Verification or Falsification and there is No Pure Science. But when the Process of Pure Science is engaged, that which is Unjustifiable is exposed by type. In the term empirical sciences the word empirical is attached as a description of science. Before exploring the Meaning represented in the word empirical it is useful to consider the Political Context within which the discussions of the Vienna Circle occurred. As we reflect we should bear in mind that the problem of Justification is Insoluble for Induction. And that the Assertion of Beliefs as Truths is part of Ideologies.

The Ideology of the Nazi Party was Active in Germany and becoming increasingly powerful during 1924 to 1936. Uebel (2012) states that the Vienna Circle:

> "…. often associated their attempt to overcome metaphysics with their public engagement for scientific Enlightenment reason in the ever-darkening political situation of 1920s and 1930s central Europe."

The violence and oppression of the times directly affected their lives. The Association of Metaphysics in general with the Nazi regime in particular gave the Appearance of Justification for their attack on all forms of Metaphysics. They attacked all Ideologies with one exception. They were unable to Justify their own position. That was based on a different set of Beliefs and was an Ideology. The type of philosophy did not change in response to their experiences in the ever-darkening political situation of the 1920s and 1930s. It is clear from recorded Observations that Hitler had studied the literature of Specialist philosophy even though he was Not Marked by a Qualification in that. A rigid Formal Structure was characteristic of the Nazi Ideology and the Vienna Circle members did not pick up the relation of that type of Structure to the effects clearly visible at that time in Social Dynamics. The type of philosophy could only have changed if they had changed the Point of Reference. The signs that ignoring Truth and Justification lead to disaster were all around them. But it is Reasonable to presume that they Knew Not What They Did. From Ubel's description it Appears that they all looked hard to find Justification for a different position and wanted to find that. It is a Fact that they looked in the wrong place.

The word Ideology represents theories that are believed to be true and/or are asserted as being true. Theories are composed of many different parts with Meaning that is co-ordinated. Each different theory has particular Meaning as a Whole that distinguishes it from other theories. The type of Meaning in the theory makes a difference to the Actions that follow. Formal philosophy took a Quantitative turn long ago. Probability Theory developed within that Quantitative Approach to the World. It will emerge from Evidence during the course of this narrative and others that the widespread application of Probability Theory has led to damage that is less visible and less clearly attributable than that of some other Ideologies. The harm is different in some ways and the same in others, but cannot be described as less or more damage. Damage can only be understood in Qualitative Terms. I note that harm does not necessarily come from an Ideology. The difference between those that do damage and those that do not is determined by differences in the type of Ideology. Differences in type are Qualitative and are revealed through Reactions. Ideologies are Theories and no more than that until they gain the False Certainty of Belief and are Asserted. Then Power is gained through them. Beliefs are rooted in Emotions. Emotions are a Source of Power.

THE TURN TO REPRESENTATION

Uebel tells us that:

> "In line with the Tractatus claim that all philosophy is really a critique of language, the Vienna Circle took the so-called linguistic turn, the turn to representation as the proper subject matter of philosophy."

A critique is commonly defined as a critical analysis and a critical analysis is defined as fault-finding. Their critique is an assembly of negative views in relation to Natural Language. There is nothing positive in a critique. But an analysis is Not Biased against something at the start and is an impartial assembly of pros and cons from which a conclusion emerges. The Selection of parts for the assembly in an Analysis within the Boundaries of Pure Science is Natural and governed by Testing Meaning for Justification. An Analysis is Comparative. It is a detailed examination of elements and structure. Within the Boundaries of Pure Science the conclusion of an Analysis is not determined at the outset. The conjunction of the words critical and analysis is peer agreed and misleading. Language is a Natural tool for the Representation of Meaning. The Vienna Circle turned against that particular type of Representation. The "proper" subject matter of philosophy is said to be other forms of Representation. They defined philosophy as Formal logic and turned Attention to Artificial tools and away from what they were doing with the tools they used. They still used Natural Language as a tool and they turned that tool against itself.

A tool must be fit for the purpose it serves. Formal logics are Artificial languages. Within that type of logic the members of the Vienna Circle had a Bias towards the Quantitative form. In order to give Power to one Bias it is necessary to Act against an opposing Bias. That is a Simple Representation of a particular Natural Dynamic. They found Fault in Natural Language where Fault had been put and denied that it was them who put it there. Nature was blamed for the misdeeds of philosophers.

It is a Fact that Natural Language is a type of Representation in which Meaning can be accurately portrayed. That tool is fit for the purpose of Representing Meaning. Is Formal logic as good as or relatively better than, Natural Language for the purpose of Representing Meaning? Specialists in philosophy have not asked that question and yet they have given us the answer as a Premise. They Assert that Formal logics are a superior tool for working with Meaning. But they do not make clear the Fact that Meaning is the subject matter of philosophy. In order to Verify the fitness of Natural Language to Represent Meaning I need to conduct an examination of their Formal Structures and Test the fit of both types of Structure to the Structure of Meaning. Tests of fit are analytic and analysis is comparative. That is a Simple and Natural Principle of Meaning. In order to Test whether an item of clothing fits us we try it on and Observe the alignment at many different points. Specialists in Formal logic do not do that. We compare what we see to a vision of what we want to see. Specialists in Formal logic do that. They tell us what to wear and what not to wear, irrespective of its fit. They know the effect they want to produce. They have Acted against the use of Natural Language in Representing Meaning.

I repeat a Fact: Specialists in Formal logic misuse the tool of Natural Language and point to the misuse as proof that the tool is defective. That is a Clinical Indicator of the presence of an excessive desire to Control other things. An excessive desire to Control other things is symptomatic of Psychological Disorder. We need to understand the Causal Process that leads to the State of Disorder. We do not have to look far for part of that. These particular specialists in Inductive science were surrounded by traumatic events and the risk to their personal survival was ever increasing. But that is only part and not the Whole of the Causal Process.

Having failed to gain admission to the Vienna Circle, Popper conducted his argument from outside the group. Essentially, he expressed the same Meaning in different ways in Natural Language. The term 'official opposition' was conferred but he did not oppose their views. Giving the Appearance that he did oppose them was useful in drawing Attention to them all. Popper was excluded from the group for personal reasons but he was not a threat to their ideas. The Action to marginalise Metaphysics was co-ordinated. Theories are Representations of Meaning and Representation is said to be the proper subject matter of philosophy. Popper described a fundamental problem with some of the theories popular in the 1920's by pointing to Marx's theory of history, the psychoanalytic theory of Freud and the individual psychology of Adler. Popper picked up the criterion of refutation that Mach had identified as a definitive part of Science and concluded that:

> "There was no conceivable human behaviour which could contradict them."

In other words, the particular theories he chose provided an explanation for every conceivable thing. A reason could be given for every happening. The theories were Inductive. Inductive theories are Not Justified and are Not Justifiable. The Structure of these Theories could deal with any input without being disturbed by it. The theories Popper used as examples were not designed through engagement with Pure Science. There is an Active Premise that things that do not fit the category science fall into the category Metaphysics. Pure Science does not fit the category science for the definition of that is restricted to Inductive science. So is Pure Science Metaphysics? Well it depends what you mean by that word. Metaphysics is a word that covers different types of Meaning and Action. Pure Science is a particular and distinctive type of Qualitative Action that is Not Inductive.

Having given examples of Inductive theories within the category Metaphysics and having stated that they could not be refuted, Popper asked:

> "When should a theory be ranked as scientific?" or "Is there a criterion for the scientific character or status of a theory?"

He already knew the answer he could give to that question. There must be some conceivable human behaviour which could contradict them. The criterion of Verification had been abandoned. There was now a Single Criterion. Popper focussed Attention on Falsification. I will return to that later.

Is Meaning secondary to its Representation? According to MOM, there is no order of precedence in Dynamic Systems. Meaning and its Representation are Intrinsically Related even when they do not have that Appearance. Where one Meaning does not fit a Representation, a different Meaning will fit. But there is the special case where all Meanings fit a Representation. Then the Representation has no particular Meaning and can deal with the particular Meaning of any part that it meets. Then the Representation is Universal in its Operation. That which is Universal catches all Meanings and Acts upon them in the same way. Theories are Propositional in relation to Truth. The problem with a Theory lays in the transformation of its Appearance from Proposition to Fact leading to the acceptance by an audience that Truth is there when it is not. Popper clearly saw that Falsification is Intrinsically Related to Testing and that in order for Testing to take place it is necessary to have particular Meaning to Test. Thus, particular Meaning must be drawn from a general Theory. Since it is the particular Meaning that is Testable, it does Appear that General Meaning in theories is Not Testable. But Appearances are Deceptive. Representations are Appearances of Meaning. Focussing Attention only on Appearances leads us into Errors of Judgment.

Some things are necessary in order for a tool to Operate. There must be a Source of Energy and there must be an understanding of how the tool works. Humans directly or indirectly provide the Energy that activates the tools they use. Set aside for now the Fact that humans are not the Source of their own Energy for that is not relevant to the matter in hand. In order to use a tool to Represent Meaning it is necessary to know what Meaning is to be Represented and how the tool can do that. I note that Natural Languages are not just used to Represent Meaning, and are also used in the Calculation of Meaning. But whether we are Representing or Calculating Meaning, tools must be fit for purpose and can only be used if we know how to Operate them. It is a Fact that all humans are pre-programmed to use Natural Languages. The Manual for Natural Languages is built into the human system. It is a Fact that we must learn to use Artificial languages. That Manual is not pre-installed and is only provided to the few. I repeat that unless we know how to Operate tools, we cannot use them effectively. And we cannot Test how they Function either.

So, Natural Languages are Universal tools for delivering Meaning and Artificial languages are Specialist tools that few can use. In Quantitative terms in relation to the number of users, Artificial languages are not better than Natural Languages. Yet those who use Artificial languages have Asserted that Artificial codes are better than Natural Language. So, it should be emerging that numbers matter in unexpected ways. I will gather together a few Facts: Meaning that is Not Represented cannot be Tested. Focussing Attention on a Representation is necessary and not sufficient to find Meaning. Meaning is hidden from those who cannot translate the Representation into terms they can understand.

Meaning was not taken as "the proper subject matter of philosophy". It does not make sense to Focus Attention on a tool and not on the use to which it is put. And it is a Fact that where there is Deception the Attention of those who Deceive is not focussed in the direction it Appears to be. Deception is enabled by Misdirecting Attention. That Fact has been known for Centuries. I propose that Specialists in philosophy pay Attention to Meaning and to the Manipulation of that, while they turn the Attention of others towards their Artificial Representations where Meaning and the direction of its movement cannot be seen. Intention is set aside in this narrative. Attention can be Unintentionally Misdirected. I am happy to accept the absence of Intention to Deceive as being within the range of possibilities. Intention requires there to be Causal Knowledge and I accept that Specialists in philosophy do not have that. Deception is a Natural phenomenon and Specialists in philosophy are not concerned with that which is Natural, so it is possible that they do not understand the Process of Deception as a Whole. Deception is a Psychological Process. Intention can be set aside for now, but not for ever.

Popper cited Einstein's Theory of Relativity in order to provide a Contrast to Metaphysics. The topics with which Einstein was occupied were different to those that occupied Marx, Freud and Adler. Einstein was not dealing with the topic of Human Social Dynamics. The objects to which Einstein paid Attention were not human. Differences in subject matter are recognised by philosophers and said to be irrelevant to the Method of dealing with them and with that particular statement I am in agreement. But the type of Method is relevant. Pure Science is a Process with the Potential for Universal Application. Pure Science has a particular type of Meaning and a distinctive type of Action. The parts of Pure Science are Unified as a Whole Process and that Process eventually Unifies the Perception of Meaning. While Einstein's Theory of Relativity is described in mathematical terms it was not discovered using Artificial Induction.

Look again at the use of the term "unified science" which was common to members of the Vienna Circle and Popper. I want to embed in you the Meaning of that term and showing you the same Meaning from different angles of view will have that effect. So, what does that term actually Represent? We have another term, "inductive sciences". The word science is common and is the noun, which is the subject of the terms. So, the subject is described in two different ways. The adjective inductive represents a particular type of science. The adjective unified suggests there was more than one type of science to begin with and that there is not more than one type now. That is a False suggestion. There has only ever been one type of Science. I take the traditional view. Pure Science is Not Inductive given the Artificial sense of that word. Popper seems to have understood that the traditional view was correct. He replaced the word inductive with the word deductive but did not change the type of Action to that of Natural Deduction. That is an Act of Deception. Specialists in philosophy call that an Act of Definition. Definitions can be misused to deliver False Meaning to an audience and when they are used in that way the Action is Deception. There is no Evidence of Malign Intent but that does not change the type of Action that it was. Dropping the difference in type from view does not actually remove the difference in type that is there.

Deception can occur in the absence of Intention. We can even deceive ourselves. Popper focussed on Representation as the proper subject matter of philosophy and directed Attention away from differences in Meaning. He did what the Circle did. He used Natural Language to attach the Meaning of the word science to something that is not even part of the Action of Science. They used the word Unification for the Action of Dis-unification simply in order to gain

Power. They created Disorder for the purpose of establishing their own Order. They did not Test the differences in Functioning of different States of Being. But remember, these people have been trained to be Clever people and they are Not Acting as Intelligent people do. They do not use Reasoning to Calculate Meaning. Thus:

Chorus: They Know Not What They Do.

So, Popper destabilised the traditional view that science can be distinguished from non-science on the basis of its Inductive methodology by arguing that the Baconian/Newtonian insistence on the primacy of 'pure' observation, as the initial step in the formation of theories, is completely misguided. According to Popper all observation is selective and theory-laden—there are no pure or theory-free observations. Thus all Observations are Uncertain in relation to Truth. In other words, we can never be Certain of the Meaning of that which we Detect with our Senses. That removes the Power of the Senses to guide us. Let us look closer at that detail. And as we do please bear in mind a Fact: Nature designed the Senses to guide Active agents in their environment. Pojman (2011) describes differences between Mach and Planck on the topic of Observations and on a unified approach to discovery. As described by Pojman (2011):

> "At issue were differences on how to make science free of human subjectivity, and how to achieve a unified science. Planck [1909] argues that while we once defined heat according to sensations, the study of heat has gone beyond this, now being under the purview of electrodynamics and kinetic theory; similarly, tones and colour are now understood as frequencies or wavelengths. Although physics had its beginnings in the analysis of sense impressions, its current success is due to removing these anthropomorphic elements. Planck formulates a vision of a human-independent science in reaction to Mach's claim that science is human-dependent."

Anthropomorphism is of course the attribution of human characteristics to a thing. Pure Science and science are different types of human Action and thus cannot be independent of human Factors. But that which is subjective is related to the Capacity for Emotion and humans also have the Capacity for Reason. I argue that within the Process of Pure Science, human Actions are controlled by a search for the True Meaning of sensations and thus sensations do not govern that Process of Analysis. That view is different to the views of both Planck and Mach but draws them into alignment. The problem is not how to make Judgments of Meaning free from human Factors and is how to make Judgments of Meaning free from human Error. The demarcation of Pure Science has been complicated by arguments about the role of sensory experience in finding knowledge. That argument is one of many Active Distractions. We pay Attention to the World through our Senses and there are different Worlds and different parts to Worlds. When what happens is Observed the Observer is involved but only distantly. There is a difference between Observing and Experiencing that is hard to tell. We Sense things that are external to us in the World that surrounds us as single individuals and we sense things in the World that is internal to us. We relate the two Sources of Sensations. Meaning Emerges from Relations. Now set aside the notion that crept in there of our being something different to our body for I do not want to talk about souls and the Meaning of that word, though I can see that the Ancient idea cannot be dismissed. The fly in the ointment is that I argue there can be no Whole without parts. I think the fly can be taken out but it would take too much time to do that, so it must remain where it is for now.

Wholes are different to the parts of which they are composed and there is considerable Evidence for that from Pure Science. According to MOM, it is not possible to know what part Sensory Experience plays in finding Knowledge without knowing how the Senses might work together to produce Knowledge. That takes us beyond the analysis of Sensory Impressions and into the search for Meaning and Causation, but does not take us away from them. Going beyond Sense Data does not involve Dissociation from the information received by our Senses. Nor does it involve an Assumption that Differences in Sense Data have Trivial Meaning. When we Observe a World, we are paying Attention to Stimuli through our Senses and we are searching for their Meaning Relatively. Pure Sensations have no Meaning. The Meaning of Sensations comes through relating them. When we actually know the Meaning of a Sensation, we do not need a theory. But when we do not know the Meaning of a Stimulus, a theory is useful and having more than one theory is even more useful for then we can engage in Comparative Analysis. Popper saw disadvantage where there is advantage.

Specialist philosophers wanted to be perceived as being the same as those who were acknowledged to be Scientists. They aimed to remove the Perception that there was a difference in type. Dropping differences in Meaning is what Specialist philosophers commonly do. Uebel (2012) informs us that:

> "Given their empiricism, **all of the members of the Vienna Circle** also called into question the principled separation of the natural and the human sciences. They were happy enough to admit to differences in their object domains, but **denied the categorical difference in both their overarching methodologies and ultimate goals in inquiry**, which the historicist tradition in the still only emerging social sciences and the idealist tradition in philosophy insisted on. **The Circle's own methodologically monist position was sometimes represented under the heading of "unified science".** Precisely how such a unification of the sciences was to be effected or understood remained a matter for further discussion."

Here Ubel seems to recognise that the representation of Inductive science as unified with Science was False. A "methodologically monist position" stands alone. These Specialists in philosophy had to draw Power from somewhere else in order to dominate the field of social sciences. They misrepresented their Intention. I can think of no other way of putting that Meaning clearly. Except perhaps to say that their Actions did not fit the Meaning they Associated with those. Associations are Powerful and can affect all who see them. So, I note that people are not always aware of what motivates them and thus I will enter Uncertainty in relation to Intention and repeat the Mantra:

<u>Chorus</u>: They Know Not What They Do.

People act on the basis of Beliefs and Good Intentions can have bad effects. It was not the topic that divided "the natural and the human sciences" and it was the difference in Method. All of the members of the Vienna Circle drew Attention to the Trivial Difference of topic that could be easily dismissed and thus drew Attention away from the Significant Difference which was the difference in type of Action. Having decided they wanted to be seen as being the same as those to whom the word science was applied with Justification, these clever fellows discussed how to convince others to see them in that way. Effort was applied to the Creation of an Illusion that their Method was the same in type as the Method of Pure Science. That is not Unification. Popper was also in Denial. He took the same view with the Appearance of a difference. He simply denied there was a single Method that could be applied. Thus, any Method becomes science. The differences between Pure Science and science vanish either way. The Mechanism of Denial is a Psychological Mechanism. It is a very effective Mechanism in the toolbox of the Arts of Persuasion. Thornton (2011) states:

> "Popper stresses in particular that there is no unique way, no single method such as induction, which functions as the route to scientific theory, a view which Einstein personally endorsed with his affirmation that 'There is no logical path leading to the highly universal laws of science'. They can only be reached by intuition, based upon something like an intellectual love of the objects of experience'."

Acceptance of vague statements leads us into Errors of Judgment. Here the Authority of Einstein is Associated with a statement that has no particular Meaning and his Authority was used to support Popper's position. It is a Fact that we can use a Natural Tool without effort and without grasping the detail of its Operation. That which is Innate does not require us to be Conscious for its Operation or require us to Consciously Operate it. According to MOM, there is a Natural Logical path leading to the Universal Laws of Nature. The Natural Laws cannot be reached using Artificial logic. The "highly universal laws of science" are Artificial having been derived by a Formal Inductive Method and those products of Induction are Not Universal

Laws of Nature. And Einstein took the view that the success of physics did involve sense impressions, but he could not specify how that was done. The term "intellectual love of the objects of experience" indicates the necessity of immersion in sensory experience as part of the search for True Meaning and thus as part of the Process of Pure Science.

Einstein was a Physicist; that was his Special-ism. He said forget the –isms and that was good advice. Physics has long had a Quantitative Bias, and that is Not a Natural Bias. There is no Formal logical path that leads to the Laws of Nature and their Meaning. But there are Laws of Nature and a Natural Method of Processing Data that leads to abstracting and understanding them. The particular statement quoted from Einstein lacked the Point of Reference that would show the difference in type of logic. Perhaps Einstein did not see the difference between that which is Natural and that which is Artificial. Einstein used the tool of mathematics to describe happenings that are Natural. That tool is not sufficient to describe Meaning and the use of mathematics can interfere with the Perception of Meaning. Psychological Processes are involved in the Perception of Meaning. Einstein understood that, but was not engaged in their study. Psychological processes were not the Focus of his Attention. Intuitions and Insights are products of Intrinsically Related Natural Psychological Processes. We do need to understand those.

I propose that there is a path that leads to the Universal Laws of Nature and that it is the Natural Logical path of Pure Science. Einstein was a Physicist, but he was also an Innovator and he was not restricted to the custom and practice of his discipline. If he had been we would not have the Theory of Relativity. It is recorded that Einstein disagreed with the views of members of the Vienna Circle and with logical Positivism as a Whole. According to Howard (2017):

> "Einstein's philosophy of science is an original synthesis drawing upon many philosophical resources, from neo-Kantianism to Machian empiricism and Duhemian conventionalism. Other thinkers and movements, most notably the logical empiricists, drew upon the same resources. But Einstein put the pieces together in a manner importantly different from Moritz Schlick,

> Hans Reichenbach, and Rudolf Carnap, and he argued with them for decades about who was right (however much they obscured these differences in representing Einstein publicly as a friend of logical empiricism and scientific philosophy)."

The Real Opposition was not described as Opposition. The Vienna Circle carefully drew on the Power of Association. Einstein was an Authority in his field. In Physics, Artificial logic is brought to bear on happenings in the Natural World in the absence of Knowledge of the Meaning of the Natural World. But Einstein was looking for Meaning and his discoveries emerged through doing so. The Structure of Meaning is Natural. Einstein used Inductive logic but he also engaged in the Process of Pure Science. While he did not Mark the differences between them in Action or Meaning he did leave clues.

PART TWO of Chapter two

EMPIRICISM

Paying Attention to differences is of crucial importance in Tracking Meaning. What are the differences between logical Positivism and Empiricism? It is hard to see any. The word empirical is widely used as a synonym for science and science is Inductive logic which is empirical. If there is a difference in Meaning and Action it is Trivial. It is easy to see things in common between logical Positivism and Empiricism. The possibility of Verification was removed, and science cannot Verify. The possibility of Falsification remained in place for a while but Nothing is Certain when Induction is in Operation and the Method of science is Induction in all parts and as a Whole.

Positivism is at the foundations of Empiricism. Empiricism is constructed in a particular way from a particular Theory which was Abstracted by Induction without paying Attention to Meaning in the Real World. Empiricism is based on Probability Theory and is constructed in Quantitative Terms. All happenings in all Contexts including the Action of experiment have been Represented and interpreted in Quantitative terms governed by Probability Theory which is based on Uncertainty. Uncertainty was given the Appearance of being Measurable and Quantifiable. Empiricism is Inductive science and Inductive science is empirical. Empiricism dominates Decision Making in Social Contexts. The Evidence threaded through this narrative Verifies that dominance of Formal logics is a Fault leading to Errors of Inference.

The Fact that words can be misused does not detract from the Fact that the same words can be used well to Calculate Meaning once the Process of Pure Science is engaged. Natural Language provides the building blocks and the Structure within which words can be placed to Represent Meaning precisely. What we do with words makes a Significant Difference to the Perception of Meaning. We have Observed the Action that led to the Illusion that Inductive sciences are Pure Science and the Illusion that Pure Science is Inductive science. The word empirical is part of the maintenance of that

Illusion. The common definition of empirical in The Pocket Oxford Dictionary is:

empirical: based on observation, experience or experiment, not on theory

That definition is a Premise and its Meaning relies on the definitions of each of the words that are parts within it. Psychology is one of the social sciences. The Dictionary of Psychology gives us a Specialist view of empiricism:

empirical: "An extremely common term in psychological parlance. It enjoys (or perhaps suffers from) a number of diverse and rather specialised meanings"

The common thread that binds the diverse, specialised Meanings is stated in The Dictionary of Psychology as a note:

> "...........all usages are based on or relate directly to data, to their collection, analysis or evaluation." An empirical investigation "involves the collection and evaluation of data, and then a process of induction from observation to theoretical propositions which are tested through further observation with an emphasis on experiments."

The Bias of Specialist philosophy is observable in that definition which is a description of Inductive logic with its deductive part. It is not a co-incidence that many definitions suffer from the problem of a number of diverse and rather specialised Meanings. Definitions are designed and are Active as an Authority. There is Authority here to cover any type of Inductive Action by the word empirical and thus by the word science. In practice it is the Quantitative Method to which those words refer.

Within the Boundaries of Pure Science, the Method of Natural Deduction is applied to Observations - different propositions in relation to particular Observations are Tested. Propositions are small theories and Testing is Experimental. Data is produced by observing what happens under particular conditions. But the Action on data is

Not Formal Induction. Formal Induction is Associative and Pure Science does not draw its Power from the Mechanism of Association. It is assumed we know what data is and the dictionaries are there to guide us should we feel Uncertain in relation to that. There are different types of data. According to the Dictionary of Psychology, data is "the body of evidence or facts gathered in an experiment or study". That aligns with the common use of the word. The common definition in the Pocket Oxford Dictionary is:

data: n. pl. (also treated as sing., though the singular form is strictly datum) 1. known facts used for inference or in reckoning. 2. quantities of characters operated on by a computer etc.

Before Facts are known, data must be processed in a particular way paying Attention to their Meaning and its Truth. Quantities of characters operated on by a computer are not known facts. The word data is used to cover input to a computer and in order to know which type of data we are dealing with it is necessary to specify the type. Sense data is not mentioned in this definition. Sense data is not known facts, is used for inference or in the Calculation of Meaning and is not quantities of characters operated on by Formal logic. Sense data is the input to human Senses. Pure Science has been engaged in investigations of how that is gathered and what happens during the processing of that type of data. Thus, we know many Facts in relation to sense data. Facts are not produced by Induction. Specialists in philosophy have agreed that is the case. Because that is so and is indisputably so, they have agreed there are No Facts and that is an Error. They also agreed to continue using the word fact and to include in the definition of that word a part to suit the Uncertainty in relation to Truth that characterises the Artificial Inductive Method. That which is Not Fact bears the name fact. That happened long ago.

Natural Deduction is a particular type of Action and can be engaged in the collection and evaluation of all types of data. Yet Natural Deduction is not mentioned in the definition of empirical. Within the Process of Pure Science, Natural Deduction is used to Process that which is detected by observation, experience and experiment. And theory produced by Natural Deduction guides further detection. Interestingly and curiously, Natural Deduction is not defined in the dictionary of psychology that I have. Only deduction as part of Induction is mentioned there. Natural Deduction was not taught as part of my training in Psychology. The contrast with my earlier experience of Action in the Physical Sciences was visible. There I learned the Natural Philosophy of Pure Science informally through experience. Come to think of it, Induction and Formal logics were not taught during my training in Psychology either. I searched for particular Meaning as the peer approved gifts were passed to me during my studies in psychology and most of my questions went unanswered and were actually discouraged. Yet, I had previously observed Pure Science in Action and questioning is an essential part of Pure Science. The discouragement puzzled me given that psychology is categorised as science.

The other essential part of the Process of Pure Science is ensuring that the particular questions asked are the ones that are answered. As time passed, I did learn many Facts about the Functioning of Active Agents. Facts are only produced by Natural Deduction and Experiment within the Boundaries of Pure Science. Natural Deduction may not be defined or taught in Psychology or generally in social studies, but that type of Action is Naturally Occurring. More would occur if that type of Action was not suppressed. Actions and Reactions are determined by something. Where something is Acted against there is an Active particular purpose giving direction. We will observe later that the differences in Meaning that distinguish Induction from Deduction have been blurred. A merger of Meaning as it is Represented in those words has long been underway.

There is an interesting note in the definition of the word empirical in the Dictionary of Psychology that aroused my Curiosity:

> "Note that such usage places empirical in opposition (more or less) to RATIONAL."

So, I turned to their definition of the word rational and found:

<u>rational</u>: 1. relating to or suggestive of the use of reason, the process of reasoning, or the property of being reasonable. 2. pertaining to that which is correct, or at least justifiable by reason, e.g. a rational decision. 3. descriptive of an organism capable of high mental

functioning: our own species is often called (seriously as well as in jest) the rational animal 4. connoting sanity or lucidity 5. primarily cognitive in nature as opposed to emotional

So, empirical investigation is data collection and its evaluation using Induction followed by further data collection. And that which is empirical is NOT RATIONAL. Thus, briefly and according to those definitions we can pick up the following Meaning. Formal empirical investigation is not sane or lucid, is primarily emotional in nature and is opposed to that which is cognitive, is in opposition to the use of Reason and is Not Justifiable. The word empirical is descriptive of an organism incapable of high mental functioning. I have never heard that particular Meaning expressed, yet it is there to be found when we follow the clues along a Track. That Meaning is not peer agreed and thus has been hidden. Perhaps the designers of these definitions had a sense of humour and used irony - perhaps True Meaning was expressed in secret and left for others to find and to use. Or perhaps they failed to relate the definitions in the dictionary as they put them there and thus failed to pick up some particular Meaning. When we look at the possible relations between the Meaning put into Single words we can find Meaning that is hidden, which may be True or False and must be Tested for Truth if Meaning is to be Verified.

When Meaning is well hidden it is easily missed. All I have done is draw into view particular Meaning from an accessible place. Are those who construct our dictionaries inattentive to the relations between the Meaning they put into one word and that in another? Are they expressing in secret that which they think they know while feeling that what they know cannot be expressed in public? Or is there one Meaning for a public audience and different Meaning for a private audience? Are some Specialists using Natural Language to pass secret messages to each other? To do that would require a high level of precision in design. We are told that Meaning cannot be delivered precisely in Natural Language. Is that telling True or is it is False? It is possible to Test for Truth using the Process of Pure Science.

True Meaning must be shown if we are to find Knowledge. I do not accept deliveries of Secret Meaning that is Not Tested for Truth. And once I find True Meaning through Independent Testing, I tend to deliver it to others. I was brought up to tell the Truth, while Observing that others did not do that. Those who placed that requirement on me did not place that requirement on themselves. Curiouser and Curiouser said I at a very young age. Seeing that I had no choice but to do that which I was told to do, I developed a habit of telling the truth or keeping quiet. I saw there was no choice for I saw that punishment follows disobedience. I was extremely quiet for a long time. Children are not clever at lying for that is a skill and skills must be developed. But it is not necessary to do that which we are skilled at doing. It is a Fact that Meaning is in the relations between things and that it is difficult to strike the Balance that must be struck in Representing Meaning in Social Dynamics. Habits are hard to break. But, as I advanced in years the habit of being quiet left me, for there was a strong competing Bias - it was necessary to Function in social situations. I learned how to tell the Truth without being punished for doing so. And while I was not punished, I found I was not rewarded for telling the Truth. So I remained quiet in relation to particular things.

Experience is stimulation acquired through the Senses and traditionally there are five of those. The five different types of sensory receptor gather data through their stimulation and the signals detected are managed as they travel onwards through a Complex System. Empiricism does not give Observation precedence in the formation of theories. Empiricism gives precedence to Observation coupled with Induction. The definition of empiricism in the Dictionary of Psychology states that data plus Induction leads to theoretical propositions which are tested through further observation with an emphasis on experiments. Thus, theory is not disembodied from sensory experience but sense data is subject to processing by Induction before theoretical propositions are formed. Thus theory is disconnected from some parts of the Meaning in Sense Data. That would introduce Error into observation statements. Unjustifiable Beliefs are hidden as long as there is a sprinkling of empirical content in the Appearance and it is thus that they survive. That which is hidden escapes attack. The word Belief is not mentioned in the definition of empirical in the Dictionary of Psychology and theory is excluded from the definition of empirical. Thus the word empirical is Dissociated from Beliefs and Propositional Meaning. And the definition emphasises that "theoretical propositions …. are tested through further observation with an emphasis on experiments". Psychology lacks good theories and is full of unrelated propositions. Probability Theory is Active in psychology and that theory was gifted by Specialists in philosophy. That Theory guides empirical testing and was Not Tested for Justification prior to its acceptance. Testing is possible.

Uebel notes:

> "……… that Vienna Circle philosophy was by no means identical with the post-World War II logical empiricism that has come to be known as the "received view" of scientific theories, even though it would be hard to imagine the latter without the former."

Now, recall that:

> "…… the word 'positivism' seems to be used ….. to represent that branch of the enlightenment tradition that failed; it has become the 'other' which we all know is wrong without having to know what it is."

And reflect! It follows that it is hard to imagine empiricism without the fore-running logical Empiricism referred to as Positivism, which is known to be wrong in some unspecified way. That is hugely convenient in Social Dynamics. If we do not know what it is that fails, then we do not know what it is that will succeed. Particular pronouns point Meaning in particular directions, and I will demonstrate that immediately by changing the pronoun. If you do not know what it is that fails, then you do not know what it is that will succeed. And to take that Meaning further: If you do not know what it is that succeeds and I do, then I will succeed and you will fail. It is a Fact that if precisely what is wrong cannot be seen then its occurrence cannot be prevented. Logical Positivism/Empiricism is a philosophy that failed as a Whole, but many parts survive with a different Appearance. The type of Meaning and Action remained constant. Quite simply, the philosophy of Positivism became less positive. Uncertainty rose to Power. But Uncertainty rose to Power on the back of Certainty. And as I will Verify, Uncertainty in one position gives Certainty in another. But it is peer agreed that Verification is impossible and if that is True then I cannot Verify any proposition no matter how much Testing I do, and nothing I say can be Fact. Acceptance of the Premise that Verification is impossible renders the notion of Testability Meaningless.

Popper succeeded in unsettling the idea of proof and truth, while seeming to retain in some form the idea that Empiricism is an alternative to Metaphysics and different in type. On the distinction between science and pseudo-science, Popper (1963) stated that he:

> "..knew, of course, the most widely accepted answer to my problem: that science is distinguished from pseudoscience —or from "metaphysics"—by its empirical method, which is essentially inductive, proceeding from observation or experiment. But this did not satisfy me. On the contrary, I often formulated my problem as one of distinguishing between a genuinely empirical method and a non-empirical or even pseudo-empirical method — that is to say, a method which, although it appeals to observation and experiment, nevertheless does not come up to scientific standards. The latter method may be exemplified by astrology, with its stupendous mass of empirical evidence based on observation — on horoscopes and on biographies."

Popper diverted Attention. Popper defines pseudoscience as metaphysics and science as Inductive. That is not just in Error. That is a Reversal of Meaning. Popper was part of the Inductive "analytical" philosophy of science which is a reflection of Natural Science. The Action in reflections and its Meaning is reversed. The most widely accepted view at that time was not what he claimed it to be. It seems from Popper's description that metaphysics does not proceed from observation and experiment. That is in Error. He states that some methods use observation and experiment but do not reach scientific standards. But look at the example he gave. Horoscopes and biographies go far beyond anything that is Observable. The "stupendous mass" of data in astrology is not Evidence of anything, is not based on Sense Data or Observation and does not involve Experiments that Test Meaning for Truth. Astrology is pseudo-science and is Inductive. The suggestion here is that Pure Science is not different in type to pseudo-science and is a matter of degree. That suggestion is part of the effort to remove the competition and the suggestion is False. Popper's version of Unification is to remove all the opposition. That is an Ancient policy in warfare. It is a Fact that all forms of Inductive science are pseudo-science. And although a genuinely empirical method is said to proceed from observation, Popper was Biased against Observation. According to Popper, no Observation can possibly lead to the discovery of True Meaning. The Power of Experiments was also removed by the Quantitative Method.

Definitions are constructed using the Mechanism of Association. Once definitions are set, they are taken as being true. Outside the boundaries of Pure Science, you can play fast and loose with definitions and nobody will be disturbed by that, for that allowance is gifted by Specialist philosophers and the gift has been accepted by us all. Where do we think definitions come from? It is a Fact that they are gifts of Meaning by human beings that are sent through time. It is a Fact that Meaning moves even when there is an attempt to fix it in one position. But changes in type of Meaning are Discontinuities and those happen infrequently under particular conditions in Natural Systems. Changes in type of Meaning have a high frequency of occurrence in Formal logic and that is Not Natural. Acceptance that Meaning in Definitions has the Quality of Truth leads us into Errors of Judgment. It is necessary to Calculate Meaning and to distinguish Meaning that is True from that which is False. Only then will we have a Stable basis for Judgments.

The common definition of metaphysics in the Oxford Pocket Dictionary is different in part to that given by Popper:

metaphysics: 1. Branch of philosophy dealing with the nature of existence, truth and knowledge. 2. Colloq. Abstract talk: mere theory……

Popper dropped one type of metaphysics from view. The Dictionary of Psychology gives us:

metaphysics: A branch of philosophy that seeks out first principles and, of necessity, must go beyond what can be learned by mechanical or physical analyses. Often, because of its intellectual heritage, the term is used as a label for any philosophy that is abstruse. Indeed, William James characterised metaphysics as 'nothing more than an unusually obstinate effort to think clearly'.

The definitions we have seen are puzzling. Popper defined metaphysics as pseudo-science and that is inconsistent with the term "dealing with the nature of existence, truth and knowledge" and with the seeking out of first principles. Pseudoscience is commonly defined as:

pseudoscience: Generally the improper use of the scientific method, a fake or false science. Most pseudoscience has the superficial trappings of real science but is contaminated by such factors as poor control over experimental conditions, improper analysis of data, illogical deductive reasoning and, occasionally, outright fraud.

According to MOM and the traditional view, Formal logic is pseudoscience. Pure Science is Not Inductive and Formal logic is. The empirical scientific Method is fake or false science. While we do not acknowledge what Pure Science is we cannot distinguish between fake science and the Real thing. Meaning is Relative and is seen through comparison. Unification simply dropped Pure Science from view where it could be compared with the Inductive Method. That which is hidden can be brought into view for Comparative Analysis.

The tools of the Art of Persuasion weave Truth and Untruth together. Having set the practice of observing and the statements that record observations to be perceived as Unstable, Popper moved to develop his requirement that theory be falsifiable by linking that criterion to new observations and experiments. He copied parts from Aristotle and built upon them. Popper demanded that theories have empirical content. He set the amount of empirical content of a theory to be "the number of states of the world that are forbidden" by the theory. The value of a theory was thus Quantified. In a book titled "The Logic of Scientific Discovery" (1972), which I have read, he stated:

> "…. the more they prohibit the more they say" (p 41) …. : "I define the empirical content of a statement p as the class of its potential falsifiers" (p 120).

Hasok (2009) informs us that:

> "Similarly but staying away from the strict falsificationist idiom, Imre Lakatos understood empirical content as the number of empirically testable predictions."

Something that is similar is also different. We recall that effort is put into minimising the Significance of differences when they are expressed. I will pick up some particular differences between Popper and Lakatos in due course. Hasok continues:

> "It is difficult to craft an exact quantitative measure, but we can at least say that the amount of empirical content depends on the number of empirically testable relations that a theory specifies. That, in turn, depends on the number of independently measurable parameters. Increasing the latter, or at least maintaining it, was something that Bridgman sought to achieve with his operationalism."

I beg to differ. It is impossible to "craft an exact quantitative measure" that is accurate and counting gives the False Appearance that can be done. And, that is not my perception of what Bridgman sought to achieve. Operationalism was not Bridgman's creation

OPERATIONALISM

Bridgman was an American Physicist born in 1882 who died in 1961. His work on the physics of high pressures was original. His philosophical reflections were vaguely represented and also influential. Green (2002) tells us that:

> "In 1927 Percy Bridgman, an eminent high-pressure physicist who would win the Nobel prize in 1946, published The Logic of Modern Physics. In it he proposed operational analysis as a

check against the kinds of mistakes which, it seemed, had led to the collapse of Newtonian physics."

Particular aspects of Operationalism (commonly shortened to Operationism) relate to empiricism, logical positivism, and much else besides. Bridgman published some ideas which were picked up by others and their interpretations lie at the core of the development of Operationalism. Later, Bridgman voiced concern at the way in which his ideas had been interpreted and developed, particularly in the Context of psychology. Disagreements had ensued but they were not heated. I infer from the descriptions I have seen that Bridgman was a gentle man. Hasok (2009) includes a quote from Bridgman at the American Association for the Advancement of Science symposium in 1953. Bridgman commented:

> "There would seem to be no reason why I am better fitted than anyone else to open this discussion. As I listened to the papers I felt that I have only a historical connection with this thing called "operationalism." In short, I feel that I have created a Frankenstein, which has certainly got away from me. I abhor the word *operationalism* or *operationism*, which seems to imply a dogma, or at least a thesis of some kind. The thing I have envisaged is too simple to be dignified by so pretentious a name."

Bridgman felt he had created a monster. Bridgman is not responsible for the way his ideas were developed. It is difficult to predict all the effects and the outcome that will follow when ideas are released into the public domain. An idea is a stimulus to further thought and if Meaning is left free then it can be taken and used to suit any purpose. Of course Specialist philosophers found Bridgman's ideas easy to Unsettle. Hasok comments:

> "When subjected to the scrutiny of professional philosophers, however, Bridgman's ideas were soon exposed as unsystematic and undeveloped, as he freely admitted himself. Moreover, it became evident that his ideas did not help logical positivists solve the key problems that they were struggling with. After the

initial fascination, the standard positivist (and post-positivist) reaction to operationalism was disappointment, and operationalism was often seen as a failed philosophy that did not live up to its promises."

Little survives the scrutiny of professional philosophers. The Action of Unsettling Meaning and its Representation is the Core Operation of their specialism. Vague statements and generalisations feed their practice. We do know what this "failed" philosophy that succeeded in social studies is. Operationalism is the 'other' which we all know is wrong. This "dogma" was allowed to survive, for it could have been defeated. But what is this dogma for which Bridgman is not responsible? I can answer that question. It bears the names empiricism and science. But I did not "of course know the answer" to that question before I Tested Meaning for Truth and Justification. I will present at length Evidence of the Truth of my answer. Bridgman (1927) suggested that we do not know the Meaning of a concept unless we have a method of measurement for it. He stated:

"We mean by any concept nothing more than a set of operations; the concept is synonymous with the corresponding set of operations".

That arrangement is open to misinterpretation as Bridgman found. He also found it impossible to correct the misunderstanding once it took flight. A focus on Operations which are visible and tangible had the Appearance of being a viable alternative for those who could be accused of Metaphysical speculations. Good Concepts are difficult to develop and in a sense they are nebulous but there are two types of Meaning Represented in that word. Only one of those Meanings is a possible analogy for a Quality of a Concept.

nebulous: adj. 1, cloudlike. 2. indistinct, vague Latin: related to NEBULA

Concepts might be cloudlike but they are not indistinct, vague. It is difficult to grasp this difference in Meaning so I will specify a

Significant Difference. Clouds have particular Meaning and that which is vague does not.

Measurements are Operations. It seemed to follow from Bridgman's statement that when you produce a measure, you create a "concept" but in this quotation Bridgman actually said that a concept is synonymous with a "set of operations". Measurement is usually Quantitative and even a set of Quantitative Measures would not create a Concept. But the interpretation that took hold is that Concepts are measurable and measures are Quantifications. Illusion developed from that idea which was set as a Premise. If the set of measurement operations contained different types of Operation then that might make a Significant Difference to the idea, but we have to understand what a Concept is before we can create one. If a Quantitative measure, or measures, is taken as validating a hypothetical "concept", which is assumed to exist, then an Illusion that it does exist is constructed. A measure is a Construct. Something Falsely Appears to be there because we have what Appears to be a precise measure of it in terms of numbers. Given physical form, that which is metaphysical and speculative is brought into existence. You can see how appealing that is. That which is said to be Impossible to do is seen to be done. But that is a Magic Trick involving sleight of hand. Bridgman was not the Magician.

Formal philosophy with its focus on Representations is Operationalism. The Representation is the Appearance of Meaning and that is mistaken for the Meaning of the thing that has the Appearance. Yet, we are told "operationalism was seen as a failed philosophy that did not live up to its promises", which suggests that Formal philosophy is different to that. That suggestion is Testable.

According to Hasok (2009):

> "One lesson we can take from Bridgman's troubles is that meaning is unruly and promiscuous. The kind of absolute control on the meaning of scientific concepts that Bridgman wished for is not possible. The most control that can be achieved is for the scientific community to agree on an explicit definition and to respect it. But even firm definitions can only constrain the uses of a concept. The entire world can agree to define length by the

> standard meter in Paris (or by the wavelength of a certain atomic radiation), and that still comes nowhere near exhausting all that we mean by length. Bridgman himself later specifically admitted that his statement that meanings were synonymous with operations was "obviously going too far when taken out of context" (1938, 117).

Philosophers are used to Asserting Assumptions as truths. We do not know that Bridgman wished for absolute control on the Meaning of Concepts. There is no Evidence for such a Premise. But anyway, his wishes are not relevant for they were disregarded. I must also register my particular disagreement with the first sentence of that paragraph taken from Hasok. Meaning is Active but it has no Will of its own. It is not Meaning that is unruly and promiscuous. It is human agents that are unruly and promiscuous with Meaning and it is some human agents that act in that way, not all of them. Such unruly and promiscuous Action with Meaning is common, for it is encouraged by Specialists in philosophy. In Fact the only part of that particular Opinion quoted from Hasok on which I fully agree is that Bridgman's Meaning was taken out of Context and its Meaning changed then. Bridgman's attempts to develop his Meaning and to correct the errors in its Perception by others were ignored, while they profited by using his reputation to give Authority to their own Actions. The use of an Authority to give Power to your own Opinions is an Ancient Magic Trick. I would add that Bridgman's Meaning was not Tested at all in other Contexts and was not Tested for Truth in the original Context. He was feeling his way towards understanding the things he saw. That type of Action is Justifiable. Bridgman was indulging in Speculation. Speculation is part of Pure Science. Green (1992) describes recognition of an Error:

> "Bridgman himself was troubled by the question regarding the nature of operations and admitted late in his life that he had not really provided "an analysis of what it is that makes an operation suitable", or "in what terms can operations be specified" (Bridgman in Frank 1956, 77)."

Green adds the comment that:

> "Not all instrumental operations are measurement operations as such (for instance, we may know how to make iron melt without thereby obtaining any precise idea of the temperature at which that happens)."

Bridgman knew that. Green tells us that:

> "In his last general philosophical treatise, The Way Things Are, Bridgman came back to this theme and stated that an operational analysis was only "a particular case of an analysis in terms of activities — doings or happenings", instead of analysis "in terms of objects or static abstractions", or "in terms of things or static elements" (1959a, 3; also 1959b, 522)."

MOM is in line with his last statement on the theme. I will return to Bridgman and differences in Methods of Analysis in the next Chapter.

FALSIFIABILITY

At last we can move away from the "-isms" but we do not move far. Popper was part of the Action to remove the Power of Natural Language to Represent Meaning. He used the Mechanism of Vagueness. At p18 of "The Logic of Scientific Discovery", Popper adds a note that falsifiability is "a criterion of demarcation, but not of meaning". I cannot claim to understand that statement, except perhaps as part of an attack on the views of the Vienna Circle. Popper states that: "Falsifiability separates two kinds of perfectly meaningful statements; the falsifiable and the non-falsifiable. It draws a line inside meaningful language, not around it." I cannot claim to understand that statement either. Of course for him, meaningful language is Not Natural Language. Popper fails to say much that is Meaningful when he uses Natural Language but he did deliver large Quantities of statements using that tool and he did deliver Meaning. He used Natural Language in an Artificial way. The Vienna Circle had set "an empiricist criterion of meaning" and declared metaphysics to be not simply false, but to be "cognitively empty and meaningless." Natural Language was not meaningful for them either. But they all recognised the Power of Natural Language to deliver Meaning, and to do so irrespective of its Truth. These Specialists in the Quantitative Method aimed to take away the Power of Metaphysics by attacking Natural Language. Surely that was the wrong target. It is the failure to deliver Meaning that is Justified and aligned to Truth that should be challenged. They could not do that without attacking their own position. So they attacked the tool that enables Meaning to be Calculated and Communicated.

Having Unsettled the idea that Verification is possible using Inductive logics, Popper Asserted that Falsification is the single criterion that demarks science and that was coupled with the idea that science is one type of thing. To dispense with the notion of Verification leaves Uncertainty everywhere, for there is no proof of any statement. However, if a proposition is Falsified then you have the Certainty that it is wrong. So, while according to Popper, Certainty of Truth is not possible with the Quantitative Approach, he left the idea that there can be Certainty of things that are Not True. That contradicts the idea that there can only be Uncertainty. The Contradiction was not picked up by him. Other Specialists in philosophy dispatched any idea that there can be Certainty at all. They were Certain about that!

Certainty can be produced through Qualitative Analysis, and Testing through Experiment is part of that. Yet Active agents have been moving towards Universal application of the Quantitative Method in testing and experiments. There Appears to be no purpose to testing if you can never arrive at a Fact. Facts can be found and they cannot be found using Induction. So, let us consider the proposition that purposes are served by preventing access to Facts. That is a Reasonable proposition. It is Natural for active agents to search for the True Meaning of things. Knowledge of the Meaning of things is related to Survival. In order to enable a search for Facts, the Method of Analysis must fit the purpose of finding True Meaning. In order to Disable a Naturally Occurring Search for True Meaning it is necessary to produce an Illusion of Testing. Caught within that Illusion, agents believe they are searching for Facts and that Facts can be found, yet they are simply kept busy and can never find anything for Certain. False Meaning must be supported in some way. Popper used the notion of Falsification in the absence of Verification to support the Illusion of Testing. A proposition can be wrong but can never be right. Thus the search party is directed towards Unsettling Meaning and towards the production of Uncertainty. Popper was a clever fellow.

If it is True that propositions can never be proved and can only be disproved, then you can test things to your heart's content but you will never produce Facts or Knowledge in Social Dynamics. There will just be a temporary Appearance of Knowing that is easily Unsettled when there is a desire to do that. For now, I propose that the Complex composition of Observation Statements is an aspect of Human Action on the World that underscores the need to engage the Action of Pure Science in Social Dynamics. True Meaning must be found if Functioning is to be stabilised.

In science, the principle of Falsifiability is underpinned by the criterion of Testability. The Action in testing using the Quantitative Method in the Context of Social Studies has long been ruled by Probability Theory. Recall that according to Popper,

> "Every "good" scientific theory is a prohibition: it forbids certain things to happen. The more a theory forbids, the better it is."

Does the type of thing a theory forbids make a difference to our Judgment of the worth of that theory? Certainly that must make a difference. So, what does Probability Theory prohibit? Well, the main thing that Probability Theory prohibits is Certainty. It prohibits all Certainties and that is a great deal to prohibit. Does that make it a "better" theory? Well the Theory is Not Testable if every Single Possible Certainty must be specified and Falsified. It is impossible to do that. And if Certainty is forbidden to happen, then testing any single possibility can lead only to results that are Unstable. The effects of peer agreement are Significant on the Action in Social Dynamics. Under Formal Rule, something can be accepted as being true only if peers agree that it is and while the agreement continues. And the purpose of testing is Not Specified. But it is clear peers have agreed that Meaning that is Testable will Not be Tested for Truth and Justification. Surely that defeats the Principle of Falsifiability. And if that which peers have agreed will not happen actually does happen, then defeat to that happening can be delivered in different ways.

Falsification is still standing as the Mark of science outside the Context of Specialist philosophy, but within Specialist philosophy that criterion was Unsettled. Nothing replaced that. Thornton (2011) notes:

> "While falsifiability is simple as a logical principle, in practice it is exceedingly complicated—no single observation can ever be taken to falsify a theory, for there is always the possibility (a) that the observation itself is mistaken, or (b) that the assumed background knowledge is faulty or defective."

The same applied to Verification. To Verify is to demonstrate that something is True. To Falsify is to demonstrate that something is Not True. They are opposite sides of the same coin. The coin is Truth. That coin is not in circulation in Social Dynamics.

Lakatos (1973, 1976) was critical of Popper's position. He stated:

> "...... in 1934, Karl Popper, one of the most influential philosophers of our time, argued that the mathematical probability of all theories, scientific or pseudoscientific, given any amount of evidence is zero. If Popper is right, scientific theories are not only equally unprovable but also equally improbable."

If that is what Popper argued and if Popper is right there is no difference between scientific theories and theories classified as metaphysical. To fail to account for anything is as useless as seeming to account for everything. Popper was wrong in suggesting there was a difference in type between science and metaphysics. And so were the members of the Vienna Circle. Whether they were right or wrong is Independently Testable. If both science and metaphysics are Inductive, they are of the same type. According to MOM, Pure Science is a type of Metaphysics that is Not Inductive.

Since Popper's stated view was that all observations and the statements related to them are theory laden, he must have recognised that his own Opinions were theory laden and improbable. Popper did not Test for True Meaning. He was able to persuade others to accept his Unjustifiable Opinions through the Force of his Will and use of the tools of the Arts of Persuasion. That was his Method of Acting upon the World. That is the Formal philosophical Method of Acting upon the World. That type of Action has peer agreement. I note it is reported that the philosophers of the Vienna Circle had experienced discomfort when interacting directly with Popper. That Method of Acting upon the World creates Friction. Yet they did not question the Method. Perhaps they did not understand the nature of the Formal philosophical position in Qualitative Terms.

Lakatos commented that the criterion of falsifiability is:

> "...... a rather stunning one. A theory may be scientific even if there is not a shred of evidence in its favour, and it may be pseudoscientific even if all the available evidence is in its favour. That is, the scientific or non-scientific character of a theory can be determined independently of the facts".

That is an accurate statement. And Lakatos referred to the human factor:

> "Is, then, Popper's falsifiability criterion the solution to the problem of demarcating science from pseudoscience? No. For Popper's criterion ignores the remarkable tenacity of scientific theories. Scientists have thick skins. They do not abandon a theory [merely] because facts contradict it."

In reflecting on alternative possibilities for demarcation Lakatos said:

> "....... all the research programmes I admire have one characteristic in common. They all predict novel facts, facts which had been either undreamt of, or have indeed been contradicted by previous or rival programmes."

That statement positions specific predictions at the forefront of a search for Knowledge and that does have the Appearance of Meaning in common with Popper. Appearances are Deceptive. A prediction is a proposition about an event or happening. But a predicted novel

Fact is a prediction that came True. Lakatos points to Verification as a distinctive mark of Science and that is the traditional view. That is a Significant Difference in Opinion from Popper. Lakatos drew Attention to the importance of distinguishing Pure Science from pseudo-science as follows:

> "The problem of demarcation between science and pseudoscience has grave implications also for the institutionalization of criticism. Copernicus's theory was banned by the Catholic Church in 1616 because it was said to be pseudoscientific. It was taken off the index in 1820 because by that time the Church deemed that facts had proved it and therefore it became scientific............the problem of demarcation between science and pseudoscience is not a pseudo-problem of armchair philosophers: it has grave ethical and political implications."

The problem of demarcation between "science and pseudo-science" is a Significant Problem and is not a problem that armchair philosophers can deal with. The problem has not been solved. The Construct of Ethics stands where a Concept of Moral should be, so I will use the word moral. In order to understand the Moral and Political implications of Meaning and Actions it is necessary to experience happenings in the Real World and engage in Pure Science to Process Sense Data. We need to solve the Problem of Demarcation and that can be done. But can we solve the related problem of groups of people with interests in common suppressing ideas that are Speculative in relation to Truth and thus suppressing the Knowledge to which such Speculation can lead? For two Centuries those people who aligned to the Facts that Emerged to Copernicus from his Observations of the sky were treated as Heretics. They held an Unorthodox Opinion. Heretics are excluded from their communities and have sometimes been executed. It is difficult to change peer agreed Beliefs. But that which is difficult to do is not impossible to do. We do not have Centuries left to do that. Fortunately, Discontinuities in type of Cognition can happen fast under the right conditions. But that which is fortunate can also be unfortunate and the difference depends on the type of State that is reached.

The criterion of Falsification did not withstand philosophical scrutiny. Thornton (2011) sums up as follows:

> "Popper's final position is that he acknowledges that it is impossible to discriminate science from non-science on the basis of the falsifiability of the scientific statements alone; he recognizes that scientific theories are predictive, and consequently prohibitive, only when taken in conjunction with auxiliary hypotheses, and he also recognizes that readjustment or modification of the latter is an integral part of scientific practice. Hence his final concern is to outline conditions which indicate when such modification is genuinely scientific, and when it is merely ad hoc."............. "..the shift in Popper's own basic position is taken by some critics as an indicator that falsificationism, for all its apparent merits, fares no better in the final analysis than verificationism."

The Fact that is never mentioned is that the difficulties of science are not difficulties of Pure Science. Dropping the difference between Pure Science and Inductive science is a Source of Confusion. In all Contexts Popper is still regarded by some as one of the greatest philosophers of the 20th Century. The shift in his position seems to have done little to dislodge the effect of his earlier work in Contexts beyond Specialist philosophy. It was not much of a shift either. Popper did not abandon the criterion of Falsification. Popper had a thick skin. Thornton continues:

> "...... in what seems to be his last statement of his position, Popper declared that falsifiability is both a necessary and a sufficient criterion. "A sentence (or a theory) is empirical-scientific if and only if it is falsifiable". Furthermore, he emphasized that the falsifiability referred to here "only has to do with the logical structure of sentences and classes of sentences" (Popper [1989] 1994, 82). A (theoretical) sentence, he says, is falsifiable if and only if it logically contradicts some (empirical) sentence that describes a logically possible event that it would be logically possible to observe (Popper [1989] 1994, 83)."

Popper's last statement is tortuous. Popper clung tenaciously to his Beliefs. Popper was Certain of his own views. He did not clearly express Meaning so it is not surprising that people do not understand the criterion. That criterion still stands for there is no other in view to take its place and the necessity of making a distinction is there. Popper's Actions in Natural Language were in line with those of Wittgenstein. It is easy to understand that these two philosophers did not get on if neither of them would say what they meant and perhaps did not even know what they meant. When there is no particular Meaning the audience must guess what statements Mean and the guess can be rejected. Popper did not respond well to rejection. But it is clear from the "last statement" that Falsifiability has nothing to do with Meaning and its Truth.

Chief Justice Rehnquist made the following comment in the Daubert ruling:

> "The Court speaks of its confidence that federal judges can make a "preliminary assessment of whether the reasoning or methodology underlying the testimony is scientifically valid, and of whether that reasoning or methodology properly can be applied to the facts in issue." ..... The Court then states that a "key question" to be answered in deciding whether something is "scientific knowledge" "will be whether it can be (and has been) tested."

Following that sentence are three quotations from treatises, and I pick up a comment of the Judge in relation to one of those:

> He quotes from Popper the phrase the 'criterion of the scientific status of a theory is its falsifiability, or refutability, or testability' And he comments " I defer to no one in my confidence in federal judges; but I am at a loss to know what is meant when it is said that the scientific status of a theory depends on its "falsifiability," and I suspect some of them will be, too."

The Judge indicates that he is looking for Meaning and is not looking only at the Representation. Falsifiability cannot be a necessary and a sufficient criterion if nobody understands the

Meaning of the word and the Action entailed. Inductive science interferes with the Perception of the type of thing that Pure Science is. Definitions are a Significant Variable and all Specialists in philosophy know that Fact. Vagueness is a Quality of Representation that prevents access to particular Meaning and is a variable that is manipulated to produce particular effects. Where Meaning is not clear it is difficult to know what Response to emit. Vagueness leads to Uncertainty in the audience. Vagueness in Representation increases the possibility of non-response or the wrong response to the Stimulus. Thus, the Action of the Stimulus can remain undisturbed. Vagueness leads to Certainty for the designer under the condition that the audience is not a designer using the same strategy. If both use vagueness, then you have Discord and Disharmony. For then the need for Meaning to be perceived remains unsatisfied.

PART THREE of Chapter two

POPPER DEFINED TESTABILITY TO SUIT INDUCTIVE SCIENCE

In "The Logic of Scientific Discovery", Popper defined testability in a way that defies Testing. He Asserted:

> "There is only one way to be sure of the validity of a chain of logical reasoning. This is to put it in the form in which it is most easily testable: we break it up into many small steps, each easy to check by anybody who has learnt the mathematical or logical technique of transforming sentences. If after this anybody still raises doubts then we can only beg him to point out an error in the steps of the proof, or to think the matter over again."

That passage has the False Appearance of being Reasonable. Definitions Act together in Defence of particular purpose. The word valid is the key to a Deception. Validity is a matter of peer agreement. Validity is a Construct of Truth that is Dissociated from a Concept of Truth and stands in its place. Truth is whatever peers agree to be true irrespective of Truth. I do not challenge the claim that logical reasoning is valid. I challenge the claim that it is Reasoning.

Artificial logics are rigid Structures. They are puzzles. Each part has a particular place within the puzzle and only fits snugly in one place within the Whole arrangement. If you break it up into many small steps you disassemble the puzzle. What you then test is that the parts fit together and produce the image that is pre-determined. There is no way of making the many small steps fit a different conclusion than the one they were designed to Represent. Some puzzles are designed to Represent more than one Meaning. Those gift Instability. Instability and Uncertainty are Intrinsically Related and are also Related to Certainty though they are Not Intrinsically Related to that Quality. The Principle is the same whether there is one or more Meaning represented. If you agree that the procedure has been followed then, since the conclusion necessarily followed from those steps, you are forced to agree to the conclusion to which it leads. You then have either Certainty or Uncertainty in terms of Meaning. What you do not have is the Certainty of Truth.

A mathematical technique is a logical technique of transforming sentences expressed in the code of numbers. The rules within Formal logics are deductive. The logical rules for building a chain of steps within the Context of Induction gift the False Appearance of Reasoning. They bear a resemblance to the Rules of Natural Deduction but their Meaning is different. Popper knew there is No Justification for conclusions derived by Induction and that the Appearance of deductive rules did not change that. It is the deductive part to which he is directing Attention. Testability was defined in that part of his narrative as checking that the procedure rules had been followed. In order to do that it is necessary to know what the Formal rules are. They are complicated and it is not possible to work them out by following Meaning using Natural Deduction so it is necessary to learn the rules by rote or look at the instructions to check the rules have been followed. That Necessity arises when there is No Logical Relation and thus No Meaning that is Justifiable. Popper misused the word test and he enabled that happening by failing to make the Meaning of the Action of Testing clear. I can do that. That is simply done by specifying what it is that is tested and what it is tested for. When Attention is focussed on checking that the procedure followed fits the rules for procedure, Attention is not focussed on Testing the Justification for Premises and the Justification for changes in Meaning at each step or on Testing the Qualities of the conclusion to which the procedure leads.

Popper Asserts that effective criticism is not possible without having learnt the mathematical or logical technique of transforming sentences. But that Action does not enable effective criticism. That prevents criticism from having any effect. That Action enables the Perception that the peer agreed rules have been followed as peers agreed they should be. Premises begin chains of Inductive logic and are threaded through proofs. Assumptions determine the Qualities of the conclusion. Everybody should have doubts while Assumptions remain Untested. It is not relevant to check Assumptions for their fit to the procedure and the conclusion. They are carefully designed to do that. Assumptions should be Tested for Justification in relation to different things.

The requirement to learn the mathematical or logical technique of transforming sentences is a Defence against Attack. That is a cleverly designed Strategy and an Ancient one. I repeat that checking a Method has been followed correctly is not a Test of the Method or the Quality of its products. A particular Bias Acts against the audience doing what they presume they are doing and want to do. It is a Fact that any person who has not learnt the mathematical or logical technique of transforming sentences cannot argue within those webs of complexity. Or at least cannot do so easily and cannot succeed for much is Asserted. But it is not necessary to argue in that way. In Fact arguing within those webs of complexity is contraindicated. Independent Testing of a chain of Formal logic is not possible by design. A rule is fixed and a challenge on rules is doomed since the rules have been agreed and are valid. But when rules are Not Reasonable that is their weakness and the weakness can be exposed through Testing Premises for Justification. Premises are the foundations on which Formal logic is built.

So, following the path set by Popper is no use at all in Testing Meaning. But to send others up that path is very useful indeed. If the challenger complies with Popper's instructions they cannot succeed. Popper arranged conditions that gave Certainty to designers that their conclusions cannot be Unsettled. To direct the Attention of the enemy away from the weak spot in your defences is an Ancient Strategy in warfare. Assumptions are the weak spot and the Action of Assertion is part of the defence of that weakness. The Analogy with warfare will be dismissed. But there is Evidence of a sustained campaign against the Process of Pure Science. That Evidence should not be dismissed.

Popper reached far and wide, drawing on similarities:

> "In the case of the empirical sciences, the situation is much the same. Any empirical scientific statement can be presented (by describing experimental arrangements, etc.) in such a way that anyone who has learnt the relevant technique can test it." .... "If, as a result, he rejects the statement, then it will not satisfy us if he tells us all about his feelings of doubt or about his feelings of conviction as to his perceptions...."

In science retesting amounts to repeating the steps in the experiment precisely. It is the procedure that is checked. Repeating an experimental arrangement may or may not reproduce a particular effect but repetition does not Test the Meaning of that which is repeated. Also, it is a Fact that nothing can be precisely repeated in every detail. The failure to replicate can always be attributed to a difference that is inevitably there. Popper knew that Fact. Popper stated that Fact. Single Cases are characterised by the presence of differences and the requirement to replicate experiments in precisely the same way is impossible to fulfil. He did not state the Meaning of that Fact. Only one conclusion is possible. The conditions that have been agreed in science are the conditions that lead to the Maintenance of Uncertainty. That is not due to Chance. Uncertainty serves particular purposes.

Recall that if the experimental arrangements etc. as described by the designer have been repeated and the critic still rejects the conclusion then the rejection will not be accepted. According to Popper there is only one course left open to those who are Doubters. Popper continues thus:

> "What he must do is to formulate an assertion which contradicts our own, and give us his instructions for testing it." .... "If he fails to do this we can only ask him to take another and perhaps a more careful look at our experiment, and think again."

The experiments in Inductive science by philosophers are experiments with statements – they are described as thought experiments using Formal logic. They are experiments in reasoning that are Not Experiments in Reasoning. It is possible to distinguish between those different types of Action using Natural Language. Men of science have sought to determine that distinguishing different types of Meaning and Action will not be possible using Natural Language and that Meaning will not be Verifiable by any means. But that which it is not possible to do without Pure Science is possible to do when that Process has been engaged.

When differences in the Focus of Attention are not seen, differences in Meaning are not seen. Different Points of Reference are not dropped accidentally. Chance does Not Determine the

Perception of Meaning. The Action in science is a vague reflection of the Action in Pure Science that is good enough to deceive the audience. Popper may have accepted that Falsification is possible, but he made it very difficult to do. Inductive reasoning based on Competing Premises and complying step by step with the rules of Formal logic requires skill in that type of logic and the time to construct the Structures. Those without that type of skill and the time required are disabled by the Call to Action that Popper Asserts as a necessity. Even if a contradictory proposition is formulated and instructions given for testing it, there is no possibility it can be Verified. Compliance with Assertions is contraindicated. We need to Act differently. Note the term "formulate an assertion". Popper seems to have got the idea in his mind that Assertion is Justifiable. That is what he did. Assertion is difficult to deal with and he knew that Fact. Within the Process of Pure Science the Meaning in propositions is formulated precisely as statements to make them Operational, and thus their Truth becomes Testable. If you want to use the word assumption to cover that you can but you must recognise that the Action of assumption has different Meaning in the Context of Pure Science. And Assertion of Truth has no place at all in Pure Science.

Popper neatly removed the possibility of Falsifying an arrangement by his insistence on producing different "assertions" for testing coupled with the Formal definition of what testing is. But when we substitute the word proposition for the word assertion then the requirement does Appear to be Reasonable. Setting up different propositions for Testing is an integral part of Pure Science. It is the Action of these clever fellows against Testing that scuppers the opposition. They brought Formal logic to bear on Natural Logic. And it is a Fact that the absence of an alternative proposition does not make the existing proposition more likely to be True or to Function well. Popper was Will-Full and he misunderstood the Meaning of many things. Or at least I will presume that he did not perceive the True Meaning of his Actions. He managed to pass on his misunderstandings to many people.

Within the Boundaries of Pure Science, a Testable proposition is one that is expressed as a Certainty, but it is the Meaning that is made Certain and not the Truth of it. It is necessary to represent Meaning precisely in order to Test it and the statements within Pure Science are designed to enable their Testing. I repeat: There is no Action of Assertion within the Boundaries of Pure Science. It is the Contrast of Certainties that makes Meaning propositional and enables Testing by Experiment. It might be this particular Meaning or it might be that particular Meaning that has the Quality of Truth. Those who engage in the Process of Pure Science are not open to any conclusion at all and are open to any conclusion that is Justifiable, and they always keep looking for other possible Meanings. There are No Premises in Pure Science. Outside the Boundaries of Pure Science the contradictory statement is absent. Premises Assert that one statement is True when it is Not True for Certain and might not be True at all.

Pure Science is first and foremost a Qualitative Method of Testing. It is Meaning that is Tested and it is Tested for Truth. It is a Fact that Quantitative Approaches cannot Analyse Qualitative Differences. Popper was a Master of Distraction. We recall that Popper described a fundamental problem with some theories popular in the 1920's as: "There was no conceivable human behaviour which could contradict them." It was inconceivable that those who do not Know the mathematical and other logical techniques of transforming sentences can contradict those who are armed with them. But, the mathematical and logical techniques of transforming sentences are Universal Statements that have No Justification and are not Verified. It has only been assumed that they cannot be Falsified. It may have been the case in the past that there was no conceivable human behaviour which could contradict them. But that is not the case now or in the future, given that you read my narratives. Meaning can be Independently Tested given a Method that is fit for that purpose. Pure Science is a tool that can defeat Unjustifiable Meaning and the Action that is aligned to that. Pure Science is the only type of Active Process that can do that. The type of Action that is necessary to access True Meaning is Natural. There will be continued Resistance to the Process of Pure Science for there is Opposition to the Emergence of Truth. Resistance is Natural when a position is challenged. And it is clear that the use of Formal logic involves Confrontation. Pure Science is in need of Defenders. But it has its own Defences.

SO, WHAT BECAME OF PROOF AND TRUTH?

Those things simply disappeared. Where there is Proof there is Verification and that which has been Verified is True. According to Popper there is never Proof and from that it follows that there can be No Truth. But in Induction there are reasons with the False Appearance of Facts. Popper did not tell the Truth about the relation of Inductive science to Pure Science, which is Natural Science. I presume he did not know the Truth but that does not excuse the Assertion of False Meaning. Nobody could prove him wrong, or so it Appears. Under the guidance of Probability Theory the Concept of Truth was replaced by fractions of possibilities of truth. Popper and the members of the Circle all moved towards accepting the idea of corroboration as a weak form of proof. The substitution of Constructs for Concepts is Operationism. Proof was set at one end of a dimension and disproof at the other end, and neither extreme could be reached by Inductive evidence which is Qualitatively weak. So, the ends of the scale of fractions merely enable sectioning of the space between them and are never realised. Thus, the impossibility of proving anything at all by Induction was reflected while the value of gathering evidence was maintained and transferred to their Inductive science where numbers are used in the weighing of evidence. The differences in Meaning and Action between Pure Science and pseudo-science are Significant - evidence is different in type, is gathered in a different way, represented in a different way and put together in a different way. The only thing they have in common is the words that Represent Actions. The words are the same but the Actions and their Meaning are different. The resemblance is superficial.

Once the difference in type between the True and the False Appearance of Meaning was dropped and the idea of truth in fractions introduced, Truth vanished. That is a Trick of the light. The enlightenment movement plays Tricks with lighting. Stage lighting in Artistic productions is a Model of that. Changes in the strength and direction of the lighting shift the Attention of the audience to particular areas of the stage. The audience does not see what happens in the shadows. Pure Art Represents Meaning that emerges from Natural Deduction more often than is commonly supposed.

Popper did not deny that there can be overwhelming Evidence for particular propositions. But he did deny that there are Facts. He pointed out the limits of experience, the unreliability of observation and the subjective quality of processing. Thus he showed how to remove Certainty even in the face of overwhelming Evidence. That is a Fault. Popper was Not Uncertain in his theory laden opinions and statements. They all led Meaning in one particular direction. Popper moved Meaning for those in his audience in the direction of Uncertainty. But he did not move it in that direction for all people. He had read the teachings of the fathers of Formal logic. His own teachings were aligned to them. I will show you some of the relevant teachings in due course in this narrative.

Formal logics provide the reasons to support conclusions and there are a multitude of those. Selection of different reasons gives the Appearance of having derived the reasons yourself, and sometimes we do. But in Formal Systems the reasons are there for the taking and the Appearance that you have produced by personally Calculating Meaning the reasons you give for the conclusions you have accepted is an Illusion supported by the profusion of choice and the existence of Inductive formulae to deal with new data. When you have accepted a conclusion from those who use Formal logic, the conclusion can be backed up in diverse ways by showing some of the steps that have also been pre-prepared for delivery. Formal logics do not ignore the difference between True and False Representations of Meaning, they make the difference disappear. That is not an accidental happening. Chance is not involved in that. And there is no such thing as Magic but there are Magic Tricks with that Appearance.

The perception that there is a Single type of science took hold and Popper became an Authority on science. Thus, his Authority was generalised. There is No Justification for the Power he acquired. It is possible to Verify the Absence of Justification through Testing. Testing involves taking alternative views for Comparative Analysis. The criterion of Verification was dismissed by peer agreement. The criterion of Falsification was retained. In "The Logic of Scientific Discovery" Popper stated:

> "We must clearly distinguish between falsifiability and falsification. We have introduced falsifiability solely as a

criterion for the empirical character of a system of statements. As to falsification, special rules must be introduced which will determine under what conditions a system is to be regarded as falsified. We say that a theory is falsified only if we have accepted basic statements which contradict it – a necessary but not sufficient condition; for we have seen that **non-reproducible single occurrences are of no use to science**."

I added the highlight in order to draw Attention to a Significant Difference between science and Pure Science. Non-reproducible single occurrences are of use to Pure Science. It is a Fact that Single Cases are of no use to Inductive science. A single event can be coded by number, its presence and absence can be recorded, and there are simple statistics that can deal with nominal data. But interval data are required for most Statistical Operations. And there are different approaches to doing that. Statistics relies on counting the frequency of occurrence of Selected Variables. Characteristics of Single Cases are variables that can be counted and single variables can be split into fractions. Counting in numbers is a Simple Operation that is Independent of Meaning. Numbers have rank order and mathematical equations change the numbers. Thus statistical equations change the weight of that which has been coded. Single Cases have particular Meaning and are not rank ordered in Natural Systems.

Here Popper is referring to falsifiability and falsification of theories, not of specific propositions. But whatever the words are applied to, surely the distinction between falsifiability and falsification is a Simple one. Falsifiability refers to the Potential for falsification and falsification happens when the Potential for Meaning to be proved wrong is realised. The Potential for Falsification is not different in type to the Potential for Verification. When a proposition is falsified then it is verified that the proposition is False. We are told that "special rules" can be Asserted to determine whether the Potential is realised or not. And one of those rules limits the weight of Meaning to be drawn from Single Cases. That is a Fault.

The basic statements which contradict a theory would be particular propositions that have been tested using science and shown to be False. Popper states that unless propositions fail testing on many occasions they will not be accepted as being False. He stated that:

> "We shall take it as falsified only if we discover a reproducible effect which refutes the theory."

He is setting up a basis for peer agreement. He is indicating that peers should agree to demand something that others will find impossible to do. That is a Fact – he shows in his narratives that theories are irrefutable by science. And men of science do not accept opposing views. A single occurrence is not a reproducible effect. It is a Fact that no single occurrence is precisely reproducible. Non-reproducible single occurrences are the happenings that Pure Science deals with and it does so by Calculating their Meaning. Meaning is reproducible. Inductive science deals with non-reproducible single occurrences by abstracting characteristics in common. That which is common to Single Cases is necessarily repeated across groups of them and can be counted. To Focus Attention on that which is common introduces Error when drawing inferences. Reproducible characteristics are not indicative of differences in the Meaning of Single Cases. It is the differences in Single Cases that make the differences in their Meaning. A difference makes a difference. Specialists in philosophy point to differences in procedure in order to Unsettle conclusions from Single Cases. That is Unjustifiable.

Now, mathematics is a useful tool within the Process of Pure Science, but Inductive science is a different type of thing and the Meaning of that tool is different in that Context. Those who mimic the type of Action in Pure Science rely on the audience being unable to tell the difference in type and thus associating the Qualities that give strength to the Process of Pure Science with that which resembles it in Appearance. The Quantitative Method in Pure Science serves a different purpose to the Quantitative Method in Inductive science. And note that serve is the word I used. The Quantitative Method in Pure Science is used to serve the particular purpose of discovering Truths. Mathematics is merely one tool among many in the Process of Pure Science that serve the purpose of discovering True Meaning in relation to the Cosmos. Mathematics is not dominant in the Context of Pure Science. Mathematics is just one form of Representation. Numbers are used to describe the characteristics of things. While descriptions can lead to Discoveries, the Action of Describing is Not The Action of Discovering. I will demonstrate the difference and begin to show how the different types of Action can work together within the Process of Pure Science. To illustrate his position regarding prediction of novel facts, Lakatos gave a well-known example of a prediction in Astronomy enabled by a theory:

> "Halley, working in Newton's programme, calculated on the basis of observing a brief stretch of a comet's path that it would return in seventy-two years' time; he calculated to the minute when it would be seen again at a well-defined point of the sky. This was incredible. But seventy-two years later, [when both Newton and Halley were long dead,] Halley's comet returned exactly as Halley predicted. Similarly, Newtonian scientists predicted the existence and exact motion of small planets which had never been observed before."

Something that is novel is of a new kind or nature that has not occurred before. Something that is novel is different in type and has particular Meaning. The word novel fits the prediction and does not fit the happenings. There is no suggestion that it does, but it is important to note the difference. The clue is in the word return. The common definition of the word novel is: - something that has no prior frequency of occurrence. Halley predicted a future happening that came to pass and the prediction was Verified as a Fact when the time came. Halley's prediction relied on the proposition that the sighting was not a novel event and was systematic a particular Comet in Orbit reappeared periodically. Given that the happening was regular and systematic, its Motion could be represented in mathematical terms. That which is regular and systematic repeatedly occurs in a pattern that is determined by the particular Forces Acting upon it. If all those Forces are known, are represented in some form and are accurately related to each other, then the course of an object can be charted. The Quantitative Method is one way of doing that and it only suits particular things. Mathematical formulae relate numbers in particular ways. Different relations have different Meaning. Mathematical signs are used to Represent the particular relation and the direction in which movement is occurring.

Halley's prediction was an Experiment in Observation and description for if the Comet reappeared around the time in the area he predicted then it was confirmed that it was a Single Comet in orbit and that it was a particular Comet that had been sighted in his time and before then. He took measurements of happenings during his time on earth and related those to each other, and in his mathematical calculations he took account of particular happenings in the past. Then the mathematical calculations were projected into the future thus providing a future measure of their accuracy in the form of a prediction. But we should not accept the Association of mathematics with the Quality of exactness and precision. And it is a Fact that the prediction was enabled by Meaning taken from elsewhere. Mathematics is not a measure of Meaning. Something that recurs is not due to chance and the Fact that it is repeated does not tell us anything about the determinants of the doing or happening. The determinants of the doing or happening do tell us about what has occurred, will occur and whether it will recur. And what you mean by a single occurrence depends on the definition of that term. Each Appearance of the Comet was a Single happening and there was more than one of those, but the single occurrences were not themselves reproducible or precisely predictable in time and space. But narrowing predictions to a range of possible values on particular variables has value under the condition we recognise that is what has been done.

The small planets which had never been observed before did exist and were not novel in that sense. The prediction that they existed was a novel prediction enabled by Newton's Theory. The prediction of their existence would have accounted for Error in other calculations. In other words, it was noticed that something that had influence was missing and the type of thing that was missing was proposed. Newton's Theory led to some predictions that were Testable and Verified but the Theory was not itself Verified and was eventually Falsified as a Whole. But it did produce some Facts. Theories are not abandoned because Facts contradict it and they should not be accepted because some Facts fit the theory. The Qualities of Meaning as a Whole are different to the Qualities of Meaning of the parts from which the Whole is composed. Theory is Speculative. And while the Quantitative Method can lead to novel predictions based on systematic regularities and the predictions can lead to Facts when Tested, that Method cannot predict novel doings or happenings. Nor can it predict when doings or happenings that are reproducible will occur when the Forces that are in Operation cannot be described in numbers. Where there is Life, there is a Force of Will that Comets and the like do not have; mathematics cannot describe that type of Force. And novel events are not Continuities, they are Discontinuities. That is a Significant Fact. Once a novel event has occurred it may be repeated providing the conditions for its survival are right. And that which is repeated may be rare or frequent but either way it is systematic. There are Causal Processes in Nature that lead to Discontinuities and to repeated occurrences. We need to discover what those are.

Acceptance of the Premises that there is only one type of science and that "non-reproducible single occurrences are of no use to science", draws the Power to pass Meaning on out of Single Cases. Single things, their single doings and the happenings related to them are not perfectly reproducible in Appearance for each one is unique in its details. But Meaning is reproducible and its reproduction is variable in form. Meaning is passed on from Single Case to Single Case. That is another Significant Fact. I presume these Facts were known to the Ancients. But they have not been Related in a Theory for Speculation. There is no Theory of Meaning. Or at least there has not been one so far.

Pure Science can lead through Verification of what is True and what is False to Causal Knowledge. Inductive science does not lead to Verification and thus does not lead to Causal Knowledge. Popper neatly dropped Causation from the "unified science" that is Inductive science and provided the Appearance of Justification for doing that. At p 39 of "The Logic of Scientific Discovery" Popper stated:

> "In physics the use of the expression 'causal explanation' is restricted as a rule to the special case in which the universal laws have the form of laws of 'action by contact'........This restriction will not be assumed here. Furthermore, I shall not make any general assertions as to the universal applicability of this deductive method of theoretical explanation. Thus I shall not assert any 'principle of causality' (or principle of universal

> causation). The 'principle of causality' is the assertion that any event whatsoever can be causally explained – that it can be deductively predicted.".....''I shall, therefore, neither adopt nor reject the 'principle of causality'; I shall be content simply to exclude it, as 'metaphysical', from the sphere of science."

The Meaning we pick up from that depends in part on the definition of contact. And here Popper does not tell us the Meaning in use or what he is and is not assuming. Specialist philosophers and their disciples play mind games with large and small audiences. Having stated he was excluding Causation from science Popper develops the Appearance of not dropping Causation. He manipulates the audience as players do with static pieces arranged on a board and with moving pieces that have No Will of their own. The audience is Relatively Passive when Rhetoric is delivered through speech or written narratives. The player of the game is Active. Popper states:

> "I shall, however, propose a methodological rule which corresponds so closely to the 'principle of causality' that the latter might be regarded as its metaphysical version. It is the simple rule that we are not to abandon the search for universal laws and for a coherent theoretical system, nor ever give up our attempts to explain causally any kind of event we can describe."

That is Operationism. The Construct of Causality was to replace the Concept of Causation. Concepts are Qualitative and thus Metaphysical. The Construct of Causality has the properties of the Quantitative Method: - Causation was set at one end of a dimension and its absence at the other end, enabling sectioning of the space between them while Verification of Guilt or Innocence in terms of Causation can never be realised. Then Popper expresses an allowance that the Construct enables. They can be allowed to search for Causal Knowledge for it is clear that they will never find what they are looking for. There is no such thing as Proof of Meaning in this Ideology. And thus there is No Particular Meaning. The need for Meaning that people have is never satisfied by this Ideology and thus they look elsewhere.

It is Certain that those who seek Causal Knowledge will never find that if they are persuaded to follow Popper's directions. Under the condition that pseudo-science is accepted as Science nobody will ever be able to prove the True Meaning of any kind of event. That is hugely popular. But, that flies against the Natural Driving Force of Nature in individual active agents, which is exerted in the direction of Survival. So I can only presume that if they do know that Causal Knowledge is Intrinsically Related to Survival they think their own Survival can be protected by their Construct. Either way:

Chorus: They Know Not What They Do.

Now the Chorus is a Mantra that is a statement designed to Act in a particular way and I will tell you the way in which I use it. It is not a Premise. The Truth of that statement is not Assumed or Asserted and is presumed prior to Testing. I do not know if that Mantra is True in any particular case but the Bias is in place to prevent overreaction to the Meaning that emerges from the Evidence. We tend to blame people. I propose not only that blame does not lie as a Whole with Single Individuals but also that apportioning blame distracts us from finding the solutions we need. The Evidence I present is not in relation to any particular person and anyway no crime has been committed. That which is defined as Criminal is a matter of peer agreement and nobody has agreed to categorise Formal logic as a Crime against Humanity and the Natural World. The Evidence I present is Meaning that Emerges from a Comparative Analysis of two different types of Structure for Cognition and the type of Functioning they determine. It is custom and practice to name the particular place from which Meaning has been taken, but it should be clear by now that the Source of the Meaning Represented by us all lies elsewhere.

There is No Certainty when you cannot prove what is True or what is False. From some angles of view that is a benefit. That benefit disappears when Pure Science is engaged and Meaning from all angles of view is seen, for that Action enables Truth to be taken into account. I will phrase that differently in an effort to make Meaning clear. All Meaning is Relative and it is Relative to many different things. If you cannot see all possible Meanings then you might miss seeing the one Meaning that is True and you cannot Judge Meaning that is hidden from view by any Standard. I propose that in the long run Universal Engagement with the Process of Pure Science would benefit all of us and not just the few. Pure Science has a distinctive type of Action that enables Verification of what is True and what is False in the Representation of Meaning. The Action against the Process of Pure Science was in essence Simple and was directed against the type of Point of Reference that gave it particular Meaning.

So do not despair for Pure Science can be Activated in the Social Domain. There can be properly formulated propositions and there are Tests that can distinguish the Representation of True Meaning from the Representation of Meaning that is False. There is such a thing as the Certainty of Truth. Certainty of that type comes through Calculating Meaning and Testing it for Justification. We need to know how to find that type of Meaning and Mark that difference in Method. Then we need to know how to Act on the differences in Meaning that we find in Single Cases. Uncertainty can be Marked and while it cannot be measured in Quantitative Terms, it can be expressed in Qualitative Terms and we can deal with Uncertainty in a way that is Justifiable. The limits of personal experience need not limit us in terms of Knowledge. Knowledge can be shared once it is found through Testing for True Meaning. The Error in Observations can be gauged and corrected. The Method of Processing data can change to Natural Deduction in search of Truth. I am not going to tell you it is easy to achieve all that but I will say it is possible to do that, with considerable effort. I aim to provide Evidence that the outcome would be well worth the effort that is required.

THEORY AND EVIDENCE

The word theory is used in Inductive science so let us see what it means in that Context. The position of theory in the Social sciences did not change after the condemnation of metaphysics and catch-all theories. According to the Dictionary of Psychology, in the social sciences the term theory is:

> "……. awarded to almost any honest attempt to provide an explanation of some body of fact or data."

Observation confirms that is still True. That doing is a systematic regularity. An explanation is not a Fact or a proposition. An explanation is an attempt to account for a happening by providing details to support a particular conclusion. Explanations are Inductive. Inclusion of the word honest suggests that the word theory is not awarded to a dishonest account of some happening. A dishonest account is an account that is Not True and where there is Intention to Deceive. Both those Qualities must be present for the word dishonest to apply. Intention is difficult to Verify even when Pure Science is engaged. Verification is impossible using Inductive science and Pure Science is not systematically engaged. Thus the word honest suggests we can tell a difference that we cannot tell.

Data is a word that is used to cover different types of information and I would include Facts as a type of data. I define Facts as Representations that are Verified as True in relation to the Meaning they describe. Data must be collected from a Source. Hempel [(1966); Source Kelly (2008)] concluded that:

> "In sum, the maxim that data should be gathered without guidance by antecedent hypotheses about the connections among the facts under study is self-defeating, and it is certainly not followed in scientific inquiry. On the contrary, tentative hypotheses are needed to give direction to a scientific investigation. Such hypotheses determine, among other things, what data should be collected at a given point in a scientific investigation."

Disconnected parts defeat a search for Meaning as a Whole. Meaning requires there to be a relation between parts. Pure Science Tests particular propositions about the relations between Facts and/or propositions, and it does so in small units. That is easy to understand.

Hypotheses are proposed answers to questions. The Selection of data should not depend on the answer. The word tentative is in line with that. In a Natural search for answers that are True, the selection of data to be collected depends on the questions that are asked and the range of potential answers is relevant to the search. Hempel used the plural "hypotheses". I will make clear that the gathering of data should not be guided by any particular hypothesis. That Miss-take of

Meaning is made in Inductive sciences. Hempel did not make that mistake in the part I quoted. But he did not draw Attention to the role of questions in that part.

Tentative hypotheses do give direction to enquiries within the Process of Pure Science. But the Significant Issue is one of Guidance. Pure Science is guided by a search for answers that are True. When Pure Science is engaged, the hypotheses that are possible answers do not give direction to the enquiry as a Whole. The questions do that. Questions are probes. Probes penetrate a thing that is under investigation. It is True Meaning that is under investigation when the Process of Pure Science is engaged. In general, probes pick up and transmit information to a Central Processor. Probes can be sharp or blunt and can be sharpened during use. Causal Meaning lies in the Intrinsic Relations between things and the type of relation can be found by probing in the right place with a precision instrument designed for that purpose. Outside the Boundaries of Pure Science, Premises short-circuit the asking of questions. Premises are hypotheses that are asserted as being True. Where truth is assumed and asserted, Meaning is not questioned. Where the Action of Probing by Questions is absent, nothing can be discovered, until and unless Meaning becomes Manifest. At some Qualitative point during Action in the Real World, True Meaning does become Manifest and the point at which that happens is not predictable or precisely measurable in terms of numbers.

In Inductive Systems, Premises determine, among other things, what data should be collected at a given point in an investigation and that has particular effects on the conclusions from the inquiry. Premises are answers to questions that have not been shown and for which Evidence has not been sought through Testing. Premises are not tentative and are not hypotheses yet they give direction to Formal logic. Premises precede hypotheses in Artificial Systems and determine the answers that are possible. Formal logic is not a search for Knowledge. But we all need Knowledge in order to Function, so that is a valuable commodity. Those who use Formal logics must look elsewhere for Facts and they look to Pure Science to produce them. As they look they interfere with that Process and they do so Selectively. The Significance of Questions is recognised and their formulation is controlled. In Formal Systems deviations from questions that have been designed for a particular purpose are not permitted. That is a difference between an Inductive System and the System of Natural Deduction. In a search for True Meaning questions are flexible for they Relate to the answers to previous questions and those are not known in advance. Questions change in Response to the answers. That Fact is known. It follows by Natural Deduction that the effect of restricting questions is known. And it is a Fact that in the light of that Fact, questions are restricted. But we must set aside Intention for people Act automatically and the Ancients set the core of the programme that is followed in Formal Systems.

Russell (2004) stated:

> "So far, no method has been found which would make it possible to invent hypotheses by rule. Usually some hypothesis is a necessary preliminary to the collection of facts, since the selection of facts demands some way of determining relevance. Without something of this kind, the mere multiplicity of facts is baffling."

It is clear that Russell picked up the same Meaning as Hempel. The Method for deriving hypotheses is clear and it is the Natural Method of Deduction. So, it is puzzling they should introduce Uncertainty as they did. Hypotheses are possible answers to questions. Questions determine the answers that are possible. The answer that follows a question should relate to the question. The particular answer that is given in response to a question is information. An arrangement of questions is the way to determine the relevance of information that is elicited. Relevant information can be Evidence for a particular hypothesis or against it. Thus the number and type of potential answers progressively reduces. It is information that is Selected not Facts. In short, questions guide the Selection of information from the multiplicity of information available. Then the information must be Tested for Truth before being accepted as Fact. I am puzzled that these Specialists in philosophy did not find that Reasoning. But the puzzle is solved by the realisation that they do not use Reasoning and that we are subject to the Illusion that they do.

Theories are bundles of Facts and tentative hypotheses in hypothetical relations that are also tentative until the units of which the theory is composed have been Tested for Truth and Meaning has been Verified or Falsified. Theory goes beyond what we Observe and beyond what we Know is True into the Realm of Uncertainty. As Einstein saw and told us: Within the Process of Pure Science, Theory is Speculative. If we take the view that the Ultimate Goal of Processing Sensory Stimulation is to Predict Meaning in relation to doings and happenings in the environment and that those occur in relation to particular things that are there, then Theories are Mediating Variables in the processing of Sense Data. A theory takes Meaning from Single doings and happenings that have actually occurred or are imagined to have occurred in order to predict Single doings and happenings that could occur and have yet to occur. The value of a theory can be judged by the Truth of the predictions. But the Truth of a theory as a Whole must be judged independently for the Meaning of a Whole is different to the Meaning of the parts of which it is composed. Theory is a Speculative Mechanism within the Process of Pure Science.

Uncertainty is dealt with by the Mechanism of Experiment. When that which has been Verified and is True is mixed in an arrangement with that which has not been Verified and might be False without Marking that difference in Quality of Meaning, we fail to show Significant Differences. Losing sight of that which is True and that which is Not True leads us into Errors of Judgment. That is a Simple Fact. The relations between parts of a theory can also be True or False. Relations are Significant in terms of Meaning. Different types of Relation draw different types of Meaning from parts. We must speculate with all parts of a theory and with relations between them if we are to find Facts and Knowledge. We must also speculate with different theories and the relations between those.

<u>In short</u> - Within the Boundaries of Pure Science, Theories are Speculative. The word Theory conveys the Meaning that the arrangement is only one possibility among others. If there is only one Theory in use, then it is not being used as a Whole for Speculation. Attention is necessarily focussed on a single theory while the parts within the Theory are in the Process of being Tested for Truth and

Justification. That is also Speculative. Otherwise, Attention is focussed on comparing one Theory with another.

When the part Assumptions and the Whole Assumption that are Fundamental to the theory are removed from Testing, as the Premises do, the Theory is protected and cannot collapse. Falsification of a Theory is not possible if it is agreed that non-reproducible single occurrences are of no use beyond the single occurrence. If a statement with particular Meaning is Falsified then we have Verification that the statement is Not True. Verification is possible in the Single Case. Popper seems to have accepted that. Particular statements can be Falsified and the Premise is that the Theory, which is a Whole composed of many different statements, cannot be Falsified by doing that irrespective of the number of statements that are demonstrated to be False. But I will specify a condition under which that Premise is False. When the proposition that is Falsified is the Point of Reference that governs Meaning throughout the Theory, then the Theory is Falsified. It is a Simple Fact that when the Foundations of a Structure are Destabilised then the Structure is liable to fall and if the Foundations are removed then the Structure cannot survive.

Many people across the Centuries have had Intuitions about the Structure of Meaning. Howard (2017) quotes Einstein as follows:

> "Furthermore this conceptual system that is univocally coordinated with the world of experience is reducible to a few basic laws from which the whole system can be developed logically. With every new important advance the researcher here sees his expectations surpassed, in that those basic laws are more and more simplified under the press of experience. With astonishment he sees apparent chaos resolved into a sublime order that is to be attributed not to the rule of the individual mind, but to the constitution of the world of experience; this is what Leibniz so happily characterized as "pre-established harmony." Physicists strenuously reproach many epistemologists for their insufficient appreciation of this circumstance."

Einstein used Natural Language to communicate Meaning to other people and here he used it well. That small narrative is consistent with an Approach for which the word Holistic is used. Taking that view, Theory has Meaning as a Whole and the Whole is different to the parts from which it emerges. That perspective will be picked up in due course. Howard picked up a Significant Difference between Einstein and Specialists in Formal logic and I will highlight part of that:

> "Missing from the empiricists' picture was what Einstein thought most important in creative theoretical physics, namely "free inventions" by the human intellect. Not that the theorist was free to make up any picture whatsoever. Theorising was constrained by the requirements of fit with experience. But **Einstein's own experience had taught him that creative theorising could not be replaced by an algorithm for building and testing theories.** However, this cannot be taken to mean that the "free inventions" once created have any scientific credentials at that point. This creative element is an essential generative phase but it is pre-scientific. Adding a fit of inventions with experience does not bestow scientific credence that is explanation. As Popper clearly pointed out broad explanatory scope in a theory contraindicates science."

I see value in some parts of that arrangement. But I would not mention Popper in the same breath as Einstein without showing the differences between them. They were talking about different types of thing using the same words. The muddling of Pure Science with science is systematic and harmful. Howard seems to accept the Premise that science is one type of thing. Popper and other Specialists in philosophy Acted against Qualitative Analysis and that is the Approach Einstein is highlighting here as important in Pure Science. The members of the Vienna Circle held themselves apart from Popper and suggested there were differences between them that either did not exist or were Trivial. But Specialists in philosophy held themselves close to Einstein and ignored the Significant Differences between them. Quantitative Measurement dominates the Specialist discipline of Physics. Algorithms are Quantitative Formulae. Though Einstein used the Quantitative Method and did so effectively, it is clear that Einstein also perceived the importance and value of a Qualitative Approach to Physics. The creative element is an essential generative phase and it is part of Pure Science.

Specialists in Formal logics did not ignore the notion of "free inventions" for that idea is well suited to Induction. What those self-styled scientists did was to separate inventions from experience at one end of a spectrum and tie it to the other end. Freedom from "the requirements of fit with experience" enabled experience to be fitted to "inventions". The World has not gained from that Action. Once again, I will draw on Einstein's words and adapt them to suit my purpose:

> With astonishment we have seen a sublime order that is to be attributed not to the rule of the individual mind but to the constitution of the Natural World in conjunction with its allowance for Natural adaptation to experience, resolve into chaos in the hands of human beings. The pre-established harmony is disappearing at an increasing pace. Epistemologists should be strenuously reproached for their insufficient appreciation and outright denial of this circumstance and for their further denial of the part their Artificial logics play in maintaining disorder. The pre-established harmony of Nature has been thrown into disharmony by human beings.

Those who are in Denial have been strenuously reproached by a few people for their insufficient appreciation of this circumstance. That makes No Difference. Perhaps if the many reproach them, that will make a Difference. But those who gain Power cling on to Power regardless of criticism and the Political Structure enables them to do that. They do not abandon Power [merely] because Facts contradict them or because nobody Believes what they say any more. Politics and Inductive science are Intrinsically Related. People who Appear to be working in different Contexts are working in the same Context. The Point of Reference that governs politics and Inductive science is the same. Formal logic bears the name science and is the Method they use in politics and in many other Contexts as well. The Social Systems set up by the Ancients are still in existence and those Formal Structures enable those who gain Power through creating the Appearance of Justification, to remain in Power once their Actions have dispelled that Appearance. That is a Fault. If Nature thought it could take a nap or could work away creating other Universes while those in whom it had Installed Reasoning did the work for it in this World, then it is in for a surprise when it perceives what has been done. And whatever it is, it will awaken. Or perhaps it is just Testing us and Observing what happens prior to designing something that fits its purpose better. Centuries on earth pass in the blink of an eye relative to time in the Universe. And all Active agents learn from Errors. Nature is an Active Force.

That which is Natural by Design of a type of Force that we do not understand has become Artificial by design of man. And it is a very small number of people who did that. I propose that humans are a Force we can understand. And an interesting Fact to which I am drawing Attention is that some men have determined that we will not understand the type of Force they apply within their Sphere of Influence. That which men have determined should be hidden can be made visible and cast aside. Perhaps as Tolkien suggested in the Lord of the Rings, no man can destroy a Dark Lord or those protected by the Power of the Ring. His character revealed her identity and spoke the words "I am no man. I am a woman." It takes a difference to make a difference. But gender is a Trivial characteristic among many others. All human beings are capable of Reasoning. Let us hope that Common-Sense can prevail. Common-Sense is Natural and Formal logic is Artificial. But hope is not sufficient to change direction. Tolkien's character thrust a sword into the core of the dark matter that threatened to destroy her and those around her. That is of course a Metaphor. Natural Language is the weapon of choice and its use must be Justified.

Popper was not inclined to clearly express any particular Meaning. He told his audience to watch how he used words and find his Meaning. Popper learned from the legacy of Plato and Aristotle. An Enthymeme is a Structure in which parts are arranged in particular relations to lead the audience to a seemingly inevitable conclusion. That Structure is characteristic of Induction. The designer is not perceived to be responsible for the conclusion and Appears merely to have drawn Attention to a fact. But the facts of Induction are not Facts, nor are they propositions. They are Assertions of a Will that has acquired the Power to Act through the Quantitative Construct of Peer Approval.

Einstein perceived that creative theorising could not be replaced by an algorithm for building and testing theories. And Evidence Verifies that is True. Algorithms are products of Formal logic in general and the Quantitative Method in particular, and lack the flexibility required for Speculation. Einstein perceived that "free inventions" are not entirely free and Evidence Verifies that is True. The phrase "free inventions" represents the products of human imagination and gives the impression of ideas unconstrained by experience. Yet there can be no total separation of ideas from experience and Einstein saw that. Freedom allows Attention to wander without limits and in any direction, but the lead is retained so the wanderer is not lost and returns to base. Freedom from constraints can produce novel arrangements and new Meaning. But it cannot do that without parts to work on. New Meaning is found when the relations between parts change. Einstein searched for Meaning. Specialists in Formal logics do not search for Meaning for they are held by the Illusion that they already know what things Mean.

Long ago it was Verified that the mind is Not a Blank Slate and the Truth of that finding was agreed when Justification was shown – though some people are still in Denial of that Fact. Experience has Meaning and Meaning is written in our memory. While imagination can go beyond experience, imagination cannot fully detach from experience. There can be a lack of fit of ideas to experience, but that is a different difference. Knowledge is surely a matter of a good fit of ideas and experience to Truth and the Natural World. When Sense Data and the ideas about the way things relate to each other are in alignment to Meaning in the Natural World then accurate Prediction is enabled. But alignments must Emerge from parts. Imaginings that are free are ideas that are in disorder and disorder can lead to a sublime order and novel ideas. It is a Principle of Meaning that Order does not tend to Disorder as Specialists in Formal logic Assert. The Natural World is Not Biased towards Disorder and is a Dynamic System. Beyond Order there is New Order which is reached only through Disorder.

A focus of Attention on Operations does not necessarily interfere with finding Meaning. Different types of Operation enable and disable different things. Theory is a particular type of Operation on Meaning. Hasok [(2009), citing Frank (1956) as his source] informs us that:

> "Bridgman ...... acknowledged from early on that there were good theoretical concepts that were not amenable to direct operationalization, ..."there need be no qualms that the operational point of view will ever place the slightest restriction on the freedom of the theoretical physicist to explore the consequences of any free mental construction". All that was required was that the theoretical system touched the operational ground somewhere, eventually. "

Bridgman recognised that theoretical Concepts are Speculative in terms of Meaning and go beyond sensory experience by relating experiences of different types in different ways. But the Operational point of view was used by others to restrict Speculation in terms of Meaning. That misuse of an idea troubled Bridgman. Misuse is abuse and is disturbing. Concepts have Meaning and are Active. Action and Meaning are Intrinsically Related. Can we have good theoretical Concepts without being able to materialise them? Actions are material and are physical but are not concrete, so the answer depends on definitions. I propose that good theoretical Concepts are Concepts that are Justifiable in terms of the Actions to which they lead. Some Concepts are Not Justifiable. The presence or absence of Justification can be detected by foreseeing the consequences of Actions through Abstracting Meaning and Reasoning with that. Induction does not abstract Meaning and that is a Significant Difference.

Within the Process of Pure Science, Testing is guided by the search for True Meaning and provides Evidence in relation to Truth. In the Context of Specialist philosophy, the word evidence has been endowed with the vagueness that enables frequent misuse. Narratives in science are sprinkled with the phrase evidence based. The term evidence based is accepted as an indicator that something is confirmed to some degree in the absence of the showing of the details of the evidence. Once shown, details can be Tested for their Qualities. A failure to produce details coupled with a failure to Test their accuracy in terms of Meaning leads to Error in Judgments. Kelly (2008) discusses the Formal philosophical position on evidence. I will pick up a few parts from that paper in order to demonstrate that the Appearance of Certainty given to the word evidence in the Context of pseudo-science is an Illusion. There is no agreement on the type of thing that evidence is. At the beginning of his article Kelly uses the following quotations as examples:

> "For my own part, I think that if one were looking for a single phrase to capture the stage to which philosophy has progressed, 'the study of evidence' would be a better choice than 'the study of language'. — A.J. Ayer, *Philosophy in the Twentieth Century*
>
> And when we try to define 'evidence' … we find it very difficult. — R.G. Collingwood, *The Idea of History*"

When we look at those statements and relate them to Facts we already know, we perceive that the study of evidence is the study of representations. And, a failure to define evidence is a failure to pin Meaning down to any Point of Reference. It is very difficult indeed to grasp the Meaning of anything that Specialist philosophers have touched. They remove anything that enables differences in Meaning to be perceived. That is a Strange thing to do and it is Not a Natural thing to do unless you are a Predator. Surely Specialists in philosophy are not predatory! Well politics and philosophy are Intrinsically Related and psychology makes up a trio. Those are Powerful Factors and they are Active in conjunction. The question is: Are those factors used to enable Functioning as a Whole or only to enable the Functioning of some parts in a Dynamic System? Kelly gives examples of differences in the Meaning represented by those who use the word evidence:

> "Russell, the greatest empiricist of the first half of the twentieth century, tended to think of evidence as sense data, mental items of one's present consciousness with which one is immediately acquainted. Quine, the greatest empiricist of the second half of the century, maintained throughout his career that evidence consisted of the stimulation of one's sensory receptors."

You would think that an empiricist would know what they Mean when they use the word evidence. They rely on evidence to support their Opinions. It seems that They Do Not Know what evidence is. There is no agreement among peers in philosophy other than an agreement to differ in their definitions and to be vague. Kelly also describes the Opinions of some philosophers during the twentieth and into the 21st century, who were presumably less "great" by agreement between peers. I will just quote a small part of that:

> "According to one recent and influential study, one's evidence consists of the totality of propositions that one knows (Williamson 2000). According to another, one's evidence consists of the occurrent thoughts that one is having at a given time (Conee and Feldman 2004). Within contemporary confirmation theory, a prominent version of Bayesianism is naturally understood as identifying one's evidence with those beliefs of which one is psychologically certain."

Note and remember that Bayesian notion of evidence. And also remember that it is a Fact that one is psychologically Certain of one's Beliefs which are not Tested for Truth and cannot be Justified for they are born of Induction. Beliefs are Not Evidence. It is a Fact that the word evidence is misused in the prominent version of Bayesianism and that it is Universally misused outside the boundaries of Pure Science. Kelly quotes Crupi (2014) as stating that:

> "…. being capable of receiving genuine support from empirical evidence is itself a very distinctive trait of scientific hypotheses as compared to other kinds of statements."

Curiouser and Curiouser said I. There is no consensus on what constitutes evidence. There can be no genuine support for anything if we do not know the type of thing that evidence actually is. Then we remember Operationalism. If you do look further, a Quantitative Construct of evidence stands in place of a Concept of Evidence. Quantitative Constructs have the Appearance of Evidence and the Appearance of Certainty but their philosophical foundations are weak. The Appearances are Deceptive. So, is the support of empirical evidence in Quantitative Terms genuine? Collins dictionary defines genuine as:

genuine: 1. Of the original stock; purebred 2. Really being what it is said to be; true; authentic 3. Sincere and frank; honest

What more can I say? You will immediately grasp the point I wish to make. The word genuine is misused to associate particular Meaning with the term empirical evidence. Of course, there is always more that can be said beyond any particular point. Kelly quotes the following from Hume's "An Enquiry Concerning Human Understanding":

> "It is characteristic of rational thinkers to respect their evidence. Rationality is no guarantee of correctness. Indeed, in a given case one might be led astray by following one's evidence, as when one's evidence is misleading. But being mistaken is not the same as being unreasonable. To the extent that one respects one's evidence, one is not unreasonable even when one is wrong."

We have seen suggestions of that type before and will see them again. But, surely, if evidence leads rational thinkers astray in their conclusions, then it is not Evidence and they are not being Rational. In order to be Rational it is necessary to use Reasoning. Under the condition that "evidence" misleads us there is only the Appearance of Evidence, an Illusion of its Truth and a Discontinuity between the data and the conclusion. Within the Process of Pure Science, Evidence is particular Meaning drawn from Sense Data through Relations and the Whole arrangement is governed by a search for Truth in relation to the Real World. Meaning is Tested by Experiment as it is drawn from the Senses. That which is Not Verified by Testing is Marked as Unverified, or False if it has been Falsified. That which is Verified is Marked as Fact. That is a Simple Principle under the condition that the word Testing is accurately defined as the type of Action that occurs within the Boundaries of Pure Science.

When Induction is in Action differences are usually ignored and dropped from sight. There is an alternative way in Formal logic of reducing the Power of Differences to affect the Perception of Meaning, which has the same effect as dropping them from sight. Simply allow all Opinions to enter the arena with Equal Force. That Operation is enabled by the removal of Certainty through the failure to Test Meaning for Truth and Justification. That alternative Strategy is Significant and has been used in the Quantitative Ideology to deal with the Problem of Justification that is Insoluble for the Inductive Method. The Illusion of Equality of Opinion creates a Significant Problem in Social Dynamics. That which is Not True has been given the Power that it needs to Act against that which is True. Inductive logic is the Source of that problem.

It Appears that Confusion reigns in the Perception of Meaning. But Confusion does not reign in all minds. Active agents who have Strong Will are Not Confused and they reign in Social Systems armed with their Beliefs. The population they Rule is made Subject to their whims and fancies. That is a Simple Truth. And Truth is a type of Meaning and Meaning is a Complex Material composed of many different threads. In other words, Truth as a Whole is composed of many parts. Agents with Strong Will are threaded throughout Social Systems and the Actions of them all are supported by the philosophy of Formal logic. Meaning travels through a System. That which has the Appearance of being a Simple Principle as a Whole is actually Complex in terms of Meaning and Action. Active agents with Strong Will are Rulers in whichever territory, large and small, they happen to be in. These Will-Full Rulers produce Disorder in their Subjects for they are Not Constrained by Reasoning and are not prepared to be constrained. That is a difficult problem to solve, but it is Not an Insoluble Problem. The key to the solution is to change the type of philosophy in Operation in Social Dynamics.

A NOTE IN MOVING FORWARD

Uebel 2012, states that:

> "For most critics (even within the ranks of logical empiricism), the problem of ruling out metaphysical statements while retaining the terms of high theory remained unsolved. By 1950 ….. Hempel conceded that it was "useless to continue to search

> for an adequate criterion of testability in terms of deductive relationships to observation sentences" (1950 [1959, 116])."

Theory is metaphysical. As far as I can see, the term "high theory" covers the Artificial terms of Formal logic and arrangements of those - those terms are not couched as propositional, so the word theory is misused here. The second-order reflections of first-order sciences are reversed in order in the looking glass. Theory goes beyond that which is Observable. But Theory is geared to understanding what we Observe through our Senses. The sentences that represent Observations are the result of Action on data from the World and are part of the Action that is to come. Representations must be Tested for their fit to Meaning in whatever form they take. Specialists in philosophy agree they cannot solve the problem of Verification for metaphysical statements. They agree they cannot solve the problem of Testability or Falsification either. They agree they cannot provide Verification. That Appears to leave us all in a State of Uncertainty. In that State there are No Facts. In that State there should be no Premises but there are. Where there is Assertion of Assumptions, there is Deception in Relation to Truth and peer agreement to Deception as a Method of Constructing Meaning. But there is No Evidence of Intention to Deceive, so this is Not Pure Deception. This is a type of Deception based on Beliefs. The Ideology that has been Installed and guides their Actions comes from a Source elsewhere. Appearances can be Tested for Meaning but Specialists in philosophy have agreed they cannot or will not do that. It necessarily falls to others to do what they cannot or will not do. They must step aside and accept the lead of Pure Science through those who will engage in that Process. They will not do that unless Force is applied. Truth is a Powerful Force and that should be sufficient. The Discovery and the Disclosure of True Meaning is feared.

Observations are a necessary part of Theory. Deductive relations necessarily fail to produce True Meaning and Causal Knowledge when they are governed by Induction. The type of Action as a Whole in any Structure is determined by that which is in the position of Power and determines Meaning. Deductive relationships are deductive in name only and are accurately described as inductive. There is merely some resemblance to Natural Deduction in its

Appearance. I will look in detail at the differences in Action covered by the words deduction and induction in the next part of this narrative. But note to carry forward that differences cannot Act when they are suppressed or not drawn into Action. Specialists in philosophy have not Tested Meaning at any level. Justification depends on Testing Action for its Meaning. The Power of Testing lies in its ability to discover Meaning through disclosing differences. Truth cannot be brought to bear if the Action of Testing is not engaged. The Illusion of Testing gives rise to the Illusion of Justification. I have seen it said that Illusions cannot be dispelled. That statement is False. It is puzzling that we are told there can never be Proof of anything. It is puzzling that the tools of the Arts of Persuasion have been applied to disable Natural Deduction and thus to disable the operation of Pure Science.

When we ask the right questions puzzles can be solved. The asking of Questions is Restricted in Formal Systems. Only peer agreed questions are to be asked and given an answer, which is also subject to peer agreement. Curiouser and Curiouser said I. That blocks access to True Meaning. And sometimes no answers need be given to any questions, but that does not necessarily block access to True Meaning and can block the possibility of Miss-taken interpretations. Dynamic Systems are Complex Systems and care must be taken in Calculating Meaning. Rules give direction to Meaning and Action; they hold parts together systematically. The rules of Induction are determined on a personal basis. In other words, agents who use Induction make up their own rules. That means there are no general rules for Induction. The deductive part of Induction suggests otherwise. When the Appearance of Meaning does not fit the Action that has occurred, then the audience falls into Errors of Prediction for it also does not fit the Action that will occur. Prediction is a necessary part of Judgments. Poor Predictions lead to bad Judgments. Remember a Fact to take forward as you read the rest of my narrative and carry it forward to the narratives soon to come:

ERROR LEADS TO UNCERTAINTY AND UNCERTAINTY LEADS TO ERROR

CHAPTER THREE: A SELECTION OF TOOLS WITH A PHILOSOPHICAL EDGE

PART ONE of Chapter three

PHILOSOPHERS FIGHT BATTLES ON THE GROUND OF DEFINITION

In order to gain Knowledge of Social Dynamics it is necessary to gain Knowledge of Psychological Processes. Philosophers are human beings and subject to Normal Psychological Processes and so are psychologists. Now, all my adult life I have been under the Spell of an Illusion that the Discipline of Psychology was genuinely focussed on a Search for Knowledge of the Processes that underlie Normal and Abnormal Human Behaviour. That is only one of many Illusions that held me under a spell. All my life I was also under the Illusion that Truth is an Issue and that True Meaning is valued when it is found. I had succumbed to the suggestion that the only problem is the difficulty of finding True Meaning and of understanding Human Actions. I do not succumb easily to Persuasion, but all the evidence pointed to that one conclusion. The problem is that the Evidence of my Senses did not fit the Meaning I was systematically given. In other words, I repeatedly Observed that the Meaning presented to me did not fit the Actions I saw. Finally the difference between Evidence and evidence emerged.

Verification is only an Insoluble Problem until Meaning is Tested for Truth. I accept that it is difficult to Verify True Meaning but it is Not Reasonable to accept that it is impossible to do that. It is Not Reasonable to Ignore True Meaning once it is found. Truth must be dealt with. But, given there is no requirement to be Reasonable, Truth can be ignored once discovered. And if Truth cannot be found there is no need to deal with it at all. Removing the Method by which data patterns can be Tested for True Meaning, gives the False Appearance that the Discovery of True Meaning is an Insoluble Problem.

Illusions can be dispelled by carefully designed Experiments directed at finding True Meaning. Reasoning is threaded through Experiments within the Process of Pure Science. Definitions are

Gifts of Meaning. Gifts of Meaning should be Tested for their Qualities and accepted or rejected on the basis of the findings. It is stated in the Penguin Dictionary of Psychology that it is futile to attempt to define psychology. In relation to the word psychology it is stated that:

> "When a formal definition is provided, it is usually done to satisfy some particular practical and/or legal issue."

That need not be so but it does seem to be so. We recall that in Court the definition of expertise in psychology is restricted to Abnormal Psychological Processes. That satisfies some particular practical and legal issue. Universally definitions serve particular purposes and there are different types of purpose. In setting definitions, Meaning is set in relation to the issue that is to be satisfied. In Formal Systems, Truth is not one of the issues to be satisfied and definitions are not set for its satisfaction. Practical and legal issues have been separated from Truth in Reality. We have already Observed some of the Evidence that Verifies that statement is a Fact.

The giving of Opinion in all Formal Contexts is either restricted to Specialists or weighted in favour of Specialist Opinion. In general, the movement of Meaning from one territory to another is restricted. The territory in which a Specialist can give Opinion is Marked. Sharp divisions in Meaning and Action are Not Natural and are man-made to serve a purpose. The discussion in the Specialist definition of psychology that fails to define the area covered by the specialism goes on to show the futility of various possible criteria that could be set to distinguish the type of thing to which psychologists devote their Attention. I note the following part:

> "The … requirement that a psychologist study psychological processes is equally futile, for each new development in psychology implicates processes once thought to lie in other domains............terms derived from sociology, anthropology, biology, physiology, medicine, philosophy, computer sciences, linguistics, mathematics, chemistry and physics; the scientists

> who studied these processes often did not presume to be studying psychology but they were. What shall we call them?"

I suggest that we call them all human beings, including those to whom the word psychologist is applied. I would not add psychology as part of the definition of the area of study of other specialists. Nor would I add those other areas to the definition of psychology. A definition of an area specifies where Attention is focussed not where it naturally shifts to and from as relevance dictates. And a topic does not define the type of Meaning and Action involved in a field.

It Appears that the problem of defining psychology is Insoluble, yet there are Specialists in the study of psychology. So what are they studying? Perhaps that is the wrong question. We already have an answer to a different question that is related to that. What do psychologists have No Special Knowledge of? Whatever Specialists in psychology are studying, it is practically and legally accepted by Formal ruling in Court that they do not have Special Knowledge of Normal Human Psychological Processes. The Turner Rule satisfies the practical issue of ruling out Opinion testimony on Normal Human Psychological Processes in the Legal Context and does so without defining that term. It is presumed that we all know what those processes are. And that presumption is an expression of a Premise that is Active in all Contexts. Normal Psychological Processes cannot be ruled out of any Context that involves human beings. The problem of mere Opinion evidence in relation to those Processes is solved by Denial that there can be "a special sort of witness" to those. That need not rule out witness testimony in relation to Normal Psychological Processes for the rule that is applied to other witnesses could fill the gap. That Rule rules out hearsay and thus rules out the notion of peer agreed dogmas. Thus, a witness to Normal Psychological Processes "must have personal knowledge, must state facts, not opinions". That could be a viable solution to the presentation of Evidence on Psychological Processes under the condition that there are Facts in relation to those Processes and under the condition that those Facts are positioned in their Intrinsic Relations. Facts and the relations of Facts to each other and to propositions in Single Cases must be Tested for Truth and thus Verified. So, now we have returned to the Problem of the Cosmos as the Judges succinctly described it. All Facts have been transformed into Uncertainties by philosophical talk. The Formal philosophical position is that there are No Facts. Denial is hard to deal with. Denial is impossible to deal with if the possibility of Verification and Falsification is erased and the Error of removing that possibility is peer approved. I have shown you part of the doings leading to that happening.

It is not stated in the Penguin Dictionary of Psychology that it is futile to attempt to define philosophy. The definition given there is clear and all-encompassing. I will quote the main part of the definition of philosophy from the Dictionary of Psychology:

> "Philosophy is 'in that most hackneyed of phrases, the search for the truth'."

I challenge the general definition of philosophy as the search for truth. There are different types of philosophy and the word is misused to cover the different types. The definition of Natural Philosophy as the search for the truth is True. That definition is Not True of Artificial philosophy. It is peer agreed that Specialist philosophy is not a search for the truth and that particular agreement is actually True for they do not do that. But they give the Appearance of Justification for that failure. The word hackneyed indicates that Meaning in the term 'search for the truth' has been rendered Trivial by misuse. There has definitely been Action against the Concept of Truth. Specialists in Artificial philosophy have undermined the search area. The licence to define and use words without restrictions should be withdrawn. The definition of both types of philosophy as the search for truth does not merely satisfy some particular practical and/or legal issue. That False definition gives Universal Power to Specialist philosophers to determine the perception of Meaning. Those who use Formal logic become accepted as experts in truth. Surely there should be agreement that definitions are Justifiable and that Reasoning is shown – it makes Sense to put that requirement in place.

Specialists in philosophy are Acting on the human mind and do so through activating Psychological Processes. Their Opinion on Normal Psychological Processes is no better than that of any other human being and yet they are accepted as Authorities to guide human Actions. That Authority is based on definitions. The practical issue served by the Formal definition of philosophy is clear. That definition is the foundation on which their Authority is built. Specialists in philosophy have transferred the Power of Truth to their discipline through the Mechanism of Association. The Vehicle of Definition carried particular Meaning to the place in which they want it to be seen. Before Truth can be told it must be Calculated and they do not do that. The Illusion that Philosophers Search for True Meaning is aligned to the Illusion that Philosophers Tell the Truth. That is clever. That is diabolical. The Spirit (or spirits) that Created those Illusions is Ancient and is not now in this World but the presence is still felt. Meaning flows through time and space. And it will become clear in due course that more is known about Psychological Processes than is commonly acknowledged, and that much has been known for Centuries.

RATIONAL AND IRRATIONAL

It is very easy to produce Confusion in the Perception of Meaning. A failure to clearly define the terms in which Meaning is Represented will do that, and a failure to Test given definitions for Truth and Justification will take the perception of Meaning into Error. The effects and the outcome of Meaning and Action in every arrangement depend on purpose. Where things are determined by a human agent there is always a purpose that is the Focus of Attention. The Action of asking questions can produce Confusion. But that same Action can lead to Clarification of True Meaning. Action that has the same Appearance may not have the same Meaning. A difference in purpose leads to a difference in the type of questioning. Differences in type are determined by Points of Reference. Points of Reference govern Meaning in every arrangement. And differences in type are disclosed in the details of an arrangement. So note and carry forward a Fact: The questions that confuse us are different in type to the questions that clarify Meaning, but they have in common their being questions. The commonality Masks the difference in type.

Is it Rational to Fail to Test Meaning for Truth, or is that Irrational? And is it Normal to be Rational or Normal to be Irrational? The common definition of rational is:

rational: adj. 1. of or based on reason. 2. sensible. 3. endowed with reason. 4. rejecting what is unreasonable or cannot be tested by reason in religion or custom. 5. (of a quantity or ratio) expressible as a ratio of whole numbers [Latin: related to RATION]

And ration is defined as a fixed allowance. The definition of the word rational reflects the length of time that the Quantitative Method has been with us. The word rational has been defined as the Quantitative Method and the Quantitative Method has been defined as rational. That which is Not Rational is Irrational. That which is not the Quantitative Method is Irrational by definition. Metaphysics has been defined as Irrational by those who use Formal logic. Definitions can be Tested for Truth. For Centuries the Quantitative Method has been Associated with the use of Reason, and with being sensible. In part 4 you see Meaning that fits something Hume said in relation to volumes of divinity and metaphysics. He went further and urged the destruction of all documents that do not contain Formal logic. That is actually Irrational. This definition of rational is in Error. The Fact that Meaning has long moved in one direction does not make that the direction in which movement should have taken place. Parts 1, 2 and 3 of the definition of rational would fit into a Qualitative Concept of Rational under the condition that Natural Reasoning is specified as the type of reason in Operation and the word sensible is defined as that which can be detected by the Senses and to which Reasoning is applied. That is not how sensible is defined in the common dictionary:

sensible: adj. 1. having or showing wisdom or common sense 2 a. perceptible by the senses b. great enough to be perceived. 3. practical. 4. (foll by *of*) aware.

The Error in part 1 of that definition is the failure to distinguish between wisdom and common sense. The Premise that those words have one type of Meaning will be Tested. And that Error is compounded by the association with sensory experience in the other parts. The definition of sensibility is interesting and relevant so I note that:

sensibility: n. 1. capacity to feel. 2 a. sensitiveness. b. exceptional degree of this. 3. (in *pl*) tendency to feel offended etc. Note Usage *Sensibility* should not be used in standard English to mean 'possession of good sense'.

Different agents have different Beliefs. It is a Fact that Beliefs are not determined by the Capacity to Reason. Beliefs are governed by the Capacity for Emotion. Beliefs are Not Rational but it may be Rational to have them. Beliefs and Opinions fill Gaps in Meaning. But they should not replace Knowledge when we have that. And a definition of rational that separates Reason from Emotion is inaccurate and misleading. I would define rational as the Operation of Reason and Emotion in Conjunction governed by a Search for True Meaning using Natural Deduction. So it is clear that I would align the definition of rational with that of Pure Science. Whether it is rational to have a particular Belief depends on the type of Belief you have and how it relates to Functioning. The Method for producing change in Social Dynamics has for Centuries involved fighting for Beliefs and that is Not Rational. Emotion takes charge of us when we fight and when we flee. The shift in Point of Reference from Reason to Emotion under certain conditions is pre-programmed by Nature. The setting is adjustable by environmental conditions. Human beings can change the conditions that affect other things. An exceptional degree of feeling is not incompatible with the possession of good sense, but does interfere with the exercise of that.

It is characteristic of an Ideology that those who adhere to it Act against different Ideologies. It is clear that Inductive science has that characteristic. Ideologies are arrangements of Beliefs. Emotion is the Dominant Operational Factor in Inductive Ideologies. Pure Science is not an Ideology and does not Universally Act against Beliefs and Opinions. Beliefs and Opinions are a guide where there are Gaps in Knowledge. The mind is not and cannot be a Blank Space in part or as a Whole. There is a need for Meaning and there cannot be none of that in the mind. Where a gap in Meaning Appears, it Acts as a vacuum and sucks Meaning in. While Beliefs and Opinions are Not True by definition, we can safely accept those that are Justifiable. Justification involves Testing Meaning for its effects and for the outcome of those effects in terms of Action. More than one Belief can be Justified and co-exist. If there is No Justifiable Meaning then humans fight over their Beliefs. While only one Meaning can be Justified as a Whole and can be True in the end, that one Meaning can be Represented in different ways that all have Justification. In other words, there are different forms of one type of thing and dropping that idea leads to Confusion. It is a Fact that the same type of Meaning becomes Manifest in different forms. That which is Justifiable is not necessarily True, but it is Functional and does not lead to Malfunction. Thus that which is Justifiable aligns to that which is Natural. A Concept of Truth and Justification can be developed from those parts. It is Rational to search for Truth and Justification and to represent that in material form.

I will repeat and develop Meaning from this part in a different form. In Social Dynamics as it is currently constructed, one Belief is traditionally pitted against another Belief. Words are used to Summon Emotion. Emotion leads to Action. Both Capacities, those of Reason and Emotion, lead to Action and the Balance between them can be disturbed. The production of Stability in Social Dynamics requires that words are used to Summon Reasoning and Emotion in conjunction. Perhaps that should be done in equal measure and perhaps not. That remains to be Calculated. The type of Emotion is a Factor that must be taken into account in the Calculation. I propose that a Stable Outcome can only emerge from exercise of the Capacity to Reason in conjunction with the Capacity for Emotion. And I propose that Pure Science involves Emotion. Is it possible that Pure Science can produce the necessary Causal Knowledge to enable us to resolve apparent chaos into sublime harmonious order? That would astonish us. I think it could have happened if Meaning had been Calculated just a few Centuries ago. The task is more daunting now.

THE APPEARANCE OF REASONING GIVES THE APPEARANCE OF JUSTIFICATION

Different types of thing have different Meaning. In order to identify the Meaning of a particular thing it is necessary to Verify the type of thing that it is. That is possible through Observing Actions and Reactions. Though I would add that questions and propositions come from Observation under conditions that are Not Controlled and from those, the setting of Controlled Conditions can lead to Observations that Verify Meaning. Specialists in philosophy Assert that Verification is impossible and they remove the Marks that enable the perception of Differences in Type of Meaning. The Formal philosophical position is that Attention should be focussed on the Representation of Meaning and not on Meaning. In other words philosophers tell us we should accept the Meaning shown in the Appearance of something. That instruction is extraordinary. It is a Fact that Appearances are frequently Deceptive in relation to Meaning. Why would they tell us to do something that potentially leads to our being Deceived? Intention is not an issue in this narrative. So, though the question of why they would do that springs to mind and is a puzzle, we do not need an answer to that question. It is Reasonable to change that particular question and thus Divert Attention from Intention at this point in time. Many other questions spring to mind and must be answered first.

Meaning can be Tracked using Reasoning. Do we know what Reasoning actually is? Formal logic claims to be science and claims to be reasoning using deduction. Those claims are False and the False claims are supported in many ways. The Meaning of reasoning does not Appear to be Testable. The word reasoning is variously defined and always the Meaning assigned is represented vaguely. For example, the online encyclopaedia Wikipedia defines reasoning broadly as:

> "the capacity for consciously making sense of things, applying logic, for establishing and verifying facts, and changing or justifying practices, institutions, and beliefs based on new or existing information."

The Method by which sense is made of things is here associated with "applying logic" and that is associated with the Action of justifying. The type of logic is not specified. Definitions in different sources are co-ordinated. In Collins Dictionary, logic is in part defined as the "science of reasoning" and the word reasoning is defined as:

reasoning: 1. the drawing of inferences or conclusions from known or assumed facts. 2. the reasons, proofs, etc. used in this process.

That is a description of Formal logic. The common definition identifies reasoning as Inductive and as you have clearly seen, science is also defined as Inductive. An assumed fact is a Premise, and it is truth that is assumed. Assumptions are Not Facts and Facts are Not Assumptions. The word assumed is misused when positioned as an adjective to describe facts. Definitions fix Meaning in a term for use and this definition is agreed by Specialists in Formal logics. They assume that statements are facts when they are Not Verified and they produce reasons for doing so. They have defined the word reasoning to fit their Actions. And the Construct of justification by peer agreement has been positioned where a Concept of Justification should be. According to MOM, the definition of reasoning as Induction is an Error of Agreement and there is a possibility that might be an Agreement that there will be Error. The second part of the definition backs up the Error by including "proofs" for the drawing of inferences from assumed facts as being reasoning. Premises are Not Tested for Truth and thus are Not Proved. If the proofs actually Verified that an assumption was True then that would no longer be an assumption and would be a Fact. Evidence of Fact is only produced by Testing for True Meaning in Reality, and is not produced by giving reasons to support inferences. The definition of fact from Collins dictionary is:

fact: 1. thing that is known to exist or to be true 2. item of verified information 3. truth, reality 4. thing assumed as the basis for argument.......

Once again, different types of Meaning are Confused in that definition. Facts are a category of things that are known to be True. Are things that exist True? They are there and have form in Reality. A thing assumed as the basis for argument exists in some form and is Not Fact. Written narratives exist. Measures exist. Analysis of Operationism demonstrated that not everything that exists is True. Parts 2 and 3 are Justified Meanings. The Power of Association is in harness. That which is Justified is used to support that which is Not Justified. The definition of reasoning in the Penguin Dictionary of Psychology is:

reasoning: 1. In general, thinking, with the implication that the process is logical and coherent. 2. More specifically, problem-solving, whereby well-formed hypotheses are tested systematically and solutions are logically deduced. Note that the term is used with the sense so that it is the cognitive processes that are of concern, not whether the correct outcome is achieved. Perfectly logical reasoning can lead to the wrong solution if one's initial assumptions are at fault.

I will pick up just one part Meaning from that, which is an Observation given form as a statement: "it is the cognitive processes that are of concern, not whether the correct outcome is achieved". That prompts a particular question: Why isn't finding the right solution and achieving a correct outcome of concern in Formal logic? I will allow the answer to that question to emerge in due course. And of course, there is another Significant Observation encapsulated in that Simple Statement. It is cognitive processes that are of concern to Specialists in philosophy. Cognitive Processes are Psychological Processes. Cognitive Processes are of concern in Specialist psychology. But the Focus of Attention is not what it Appears to be in either of those Special-isms.

The Illusion of Justification for Induction is in part created by presenting reasons and proofs. Assumptions are threaded through Inductive reasoning. According to Richardson (2013):

> "Explicit reasoning is responsibly conducted thinking, in which the reasoner, guided by her assessments of her reasons ….. and

> of any applicable requirements of rationality ....., attempts to reach a well-supported answer to a well-defined question."

Applying Natural Deduction leads to the conclusion that the "reasoner" who cannot Justify their inferences is not responsibly conducting their assessments, which do not merit the title "reasoned" that is applied to them. Induction is Unjustifiable.

DIFFERENCES DIFFER IN SIGNIFICANCE

Differences are more frequent than commonalities, but the difference in frequency is not relevant. The Fact that is relevant is that the Meaning held in differences outweighs that held in commonalities. And differences in Meaning are less frequent than differences in form. I note that Specialist Philosophers do not ignore all differences. Selections are threaded through Dynamic Systems and are a Causal Factor in the Perception of Meaning and in the assignment of Authority. Specialist philosophers drop many differences and Mark a few for Attention. It is a Fact that they drop or pick up differences without Justification. A Mark indicates that something is important relative to other things. A Mark provides Contrast and draws something particular into view. Marks Operate on the Mechanism of Attention. Those who use Induction highlight differences in form. It is the type of Meaning to which we should pay Attention.

True Meaning must be expressed if it is to have any effect. It is a Fact that a Representation is the Appearance of Meaning and it is a Fact that Meaning often does not fit the Appearance. It is a Fact that Specialist Philosophers do not Mark Meaning that is Real or Meaning that is True. If different things were to be Marked and drawn to Attention, the effects and the outcome would be different. Their Intention really does not matter. I do not need to prove that agents Know they are not telling the Truth in order to Verify that what they say is Not True. They tell us that they do not know what is Real and what is Not Real and that they do not know what is True and what is Not True. I am inclined to take their word for that, not because they say that they do not know and have agreed on the truth of that statement, but because I perceive that the Method they use cannot lead to Facts or Knowledge. They rely on picking up Meaning from those who engage in the Process of Pure Science and they apply the Method of Induction to the Facts they pick up. That is a Fault. Only Pure Science can draw Facts into their Intrinsic Relations and thus produce Knowledge.

The Actions of Philosophers with words are Confusing. The philosophical consensus warrants freedom in the use of words and gives the freedom to define words as you wish them to be used and to use them as you wish. But Freedom is Relative, all Meaning is determined by a Point of Reference and the Real Focus of Attention can be hidden from view. There are exceptions to the Appearance that any type of Meaning can be represented in any way you wish. The exceptions are conditions, and one condition is Simple. True Meaning can be expressed under the condition that there is no Certainty of its Truth. When the audience will not Believe Meaning is True, it is not considered necessary to Block the Representation of True Meaning. In Fact, allowing its Representation can be more helpful to those who do not wish True Meaning to be perceived than blocking its Representation altogether. I will give you the Reasoning for that conclusion. Since Deduction is Natural people can Calculate True Meaning, and an Unquantifiable number of them do. Drawing Meaning out for them and ensuring it is Associated with the Quality of Being False defuses the effect of its Perception. True Meaning is Marked as False Meaning and vice versa. The difference in type of Meaning is the Significant Difference and that difference is dropped.

Beliefs are powerful and difficult to dislodge. True Meaning becomes an Issue when it is represented as Fact and Believed to be True. Belief is a Powerful Factor and that Factor is misused. As I have said before and will say again, some Beliefs are Justifiable and some are not. And when a Belief is Verified as having the Quality of Truth it is a Fact, so we change the word that we use to Mark its type. When we accept Meaning that is misrepresented by type, we are misled.

It is a Fact that Meaning cannot have its effect until and unless it can be seen. While Attention is drawn to one particular thing it is not drawn to a different thing. The Attention of audiences is drawn by Specialists in Formal logic to differences with Trivial Meaning. Meaning that is Significant is not attended to while we are absorbed by Trivia. And the effect of an arrangement on an audience is Selective, for it depends on the angle of view of each person. But angles of view can be fixed by Designers of Illusions. In Fact restricting Points of Reference is a Causal Factor for Illusion. When the angles of view are restricted, people are drawn to see Meaning in the same way.

Natural Deduction is not defined. A search through our dictionaries shows only the Formal Construct of deduction. The difference between that which is Natural and that which is Artificial is a Significant Difference in type of Meaning and Action. The difference between Induction and Natural Deduction is Not Marked. In the Specialist philosophical literature differences are marked between induction, abduction and deduction. I propose that the Marked differences are not differences in type and are Trivial in terms of Meaning. In order to Test that proposition, the Meaning Represented by those words must be identified.

PART TWO of Chapter three

IS THERE A DEFINITION OF INDUCTION?

The short answer is No. That question and the answer are the beginning of a puzzle. Differences between Induction and deduction have been described in the narratives of Specialist philosophers. For example, it is clearly specified that Induction starts with sense data and that deduction does not. My Curiosity was aroused by the Action of Marking a sharp distinction when so much Meaning in Specialist philosophy is vague. Specialist Philosophers are Certain of that distinction. They are Certain of that because they designed and agreed that part of a definition. Many definitions are Premises. In other words, it is Assumed and Asserted that Definitions are true. Yet the Act of defining does not change Meaning in the Real World and can only change its Appearance. Meaning is sometimes what it Appears to be and sometimes it is not. Definitions should not be accepted without Testing their fit to True Meaning in Reality. And we should not accept a gap in Meaning or vagueness in a definition. To define is to clearly specify particular Meaning.

The Attention of Specialists in philosophy is focussed on the Representation of Meaning. That is the Appearance they give and it might or might not be so. The Representation is how Meaning Appears to be. A Representation is Meaning Operationalised and there is a necessity for that to happen if Meaning is to have an effect. Thus Operationism has the False Appearance of Justification. We detect the Appearance of things with our Senses. If the Fit of Sense Data to Meaning in Reality is Not Tested then True Meaning cannot be found until and unless the Action that is Intrinsically Related to True Meaning overtakes you. Then you are exposed to the consequences of the failure to Test Meaning. But what determines the failure to Test Meaning for Truth which leads us to experience consequences that we would have prevented had we perceived True Meaning? That is an important question. A particular answer is Asserted as a Premise by those who use Formal logic: - The failure to Test Meaning for Truth is the Fault of those who did not Test. Thus the consequences are also the Fault of those who did not Test Meaning in the Representation for its Truth. Is that position

Justifiable or Unjustifiable? The Formal philosophical position sets the conditions that produce a Bias against the Testing of Appearances for Truth. Natural Language is a tool on which we rely for the Calculation and Representation of Meaning. Specialist philosophers created the Appearance that they turned against Natural Language for their Actions on the World due to its deficiencies.

The Natural Philosophical Position sets the conditions that Bias us towards Testing Appearances for Truth. One type of Bias is Active against a different type of Bias. It does not make Sense to accept Sense Data on its face value, yet it Appears that is the course that Specialist philosophers took. Or perhaps they only encouraged others to take that course, thereby enabling the acceptance by their audience of gifts of Meaning as they Represent it. It is not necessary to Test those two alternatives for Truth in terms of Intention. It is necessary to Test the effects and outcome of Actions and then Judge whether or not it makes Sense to take a particular course. In order to understand the type of Method that Induction is, we need more than the Premise that it starts with Sense Data. And definitions are subject to change. When we look closely we find that even that small Mark cannot be relied upon and is rubbed out. Certainty Appears and then Disappears in Wonderland, and it does so without warning. There is the Appearance of Certainty when there is a material form, but:

THE ONE THING YOU CAN BE CERTAIN OF IN WONDERLAND

IS EXPERIENCING UNCERTAINTY

Vickers (2012) informs us that:

> "In recent times inductive methods have fissioned and multiplied, to an extent that attempting to define induction would be more difficult than rewarding".

That is a familiar theme and describes well the position in which the word science is currently placed. That is only to be expected since it is a Fact that the term Inductive Method and the word science have the same Meaning, though Specialists in philosophy will not tell us what that Meaning actually is.

GUESSING IS THE NAME OF THE GAME IN WONDERLAND

Or it is if you are prepared to play that game and I am not prepared to do that. I am prepared to Speculate and Experiment. The Action of Speculating is a different type of Action to that of guessing. Pure Science has particular Meaning and is different in type of Action to Inductive science. The Qualities of Pure Science are Distinctive and have not been Marked. I will specify the Qualities of Pure Science in due course as they emerge from the Evidence. In order to perceive the difference between Pure Science and Inductive science we need to find the Qualities of Induction as well, for then Comparative Analysis is possible. We are told that Induction Appears in many different forms and that there is no limit to the Meaning represented by that word. Is it possible to Mark the boundary between Pure Science when that has not been defined and something else that also has the Appearance of being nothing in particular? It is Certainly difficult to do that and it might be Assumed that cannot be done, but it is necessary to Test Assumptions and to reject Assertions of their truth. It is a Fact that Human Actions are determined by particular things and Induction is a Human Action. It Appears that Induction is nothing in particular and yet the assignment of a name suggests that it is something particular. There Appears to be a Contradiction, but is there? I propose that Induction is anything and everything and something particular. Once you understand the Structure of Meaning you will understand the Sense of that proposition. That proposition can be Tested through Observation coupled with the Calculation of True Meaning. Even given vagueness without limits it is possible to Track particular Meaning through close inspection of details. That is a Fact you should note.

So far this Calculation of Meaning has reached the stage where we see that Induction is a particular Method of Acting on the World and the Action Appears to begin with Sense Data. Vickers (2012) goes on to describe "Carnap's taxonomy of the varieties of inductive inference" and suggests that the classification "may help to appreciate the complexity of the contemporary concept". We can use all the help we can get.

Carnap's Taxonomy of Inductive Inference

1) Direct inference typically infers the relative frequency of a trait in a sample from its relative frequency in the population from which the sample is drawn.

2) Predictive inference is inference from one sample to another sample not overlapping the first. This, according to Carnap, is "the most important and fundamental kind of inductive inference"..... It includes the special case, known as singular predictive inference, in which the second sample consists of just one individual."

3) Inference by analogy is inference from the traits of one individual to those of another on the basis of traits that they share.

4) Inverse inference infers something about a population on the basis of premises about a sample from that population.

5) Universal inference is inference from a sample to a hypothesis of universal form."

Like all general statements these are vague. The word frequency is only mentioned in part 1 of this definition but all the parts involve frequencies. Traits are imagined to exist before and after being constructed from patterns of frequencies using equations of Formal logic, which are frequently mathematical equations. The word Meaning is not used in any part of Carnap's Taxonomy, but it is Meaning that Inductive logic aims to deal with, and the word inference Represents that Fact. This taxonomy is part of a catch-all theory that deals with Meaning by counting things. There is No Justification for any part of it. Specialists in philosophy agree there is No Justification for conclusions reached by Inductive logic. That is stated and ignored. Empiricism is Inductive inference and this taxonomy is a description that fits the Action in Inductive science. That taxonomy and the Probability Theory to which it is related are alien to Pure Science. There is not even a resemblance. Pure Science is Not Artificial logic and is Not Inductive. At part 2 it is stated that "singular predictive inference" is one of "the most important and fundamental kind(s) of inductive inference", and it is specified that "the second sample consists of just one individual". I note that the first sample, from which Meaning is transferred to the Single Case, is not specified by type.

Vickers (2012) describes the Problem of Justification for Induction thus:

> "The original problem of induction can be simply put. It concerns the support or justification of inductive methods; methods that predict or infer, in Hume's words, that "instances of which we have had no experience resemble those of which we have had experience"

Induction has many problems and the lack of a definition is one of them. Vickers observes that:

> ".... much of what contemporary epistemology, logic, and the philosophy of science count as induction infers neither from observation nor particulars and does not lead to general laws or principles".

So, it was clear that Induction is limited by "instances of which we have had experience". Now, it is not clear that Induction is limited by previous experience and it is clear that it is often not directly related to experience. The Marks of Induction were pencilled in and rubbed out. The suggestion that the starting point for Induction is Sense Data was Unsettled. It seems to be agreed that is a criterion and that is not a criterion that is distinctive of Induction. Specialist philosophers Appear to be unable to settle on any particular Meaning. They agree to be in constant disagreement. Curiouser and Curiouser said I. And if Induction does not lead to Natural Laws and Principles, where does Induction lead? I propose that Induction leads to and from Opinions or Beliefs that do not have the Quality of Truth. It seems that since we are not told by Inductive scientists what the word Induction Represents, we are left to guess its Meaning in the absence of a definition. There is an alternative course of Action. The alternative Method is to Calculate the True Meaning of Induction.

The categories of Inductive inference described by Carnap are all Actions using resemblances. Numbers are one form in which similarities are expressed for the drawing of inferences. The reliance on resemblances is characteristic of all forms of Induction. Resemblances are similarities in Appearance and reliance on them leads to Error in inferences. Apart from the use of similarities, Specialist philosophers do not tell us what Induction is but there are some things it Certainly is Not and that Meaning has been specified. Inferences from Induction are Not Justifiable and Truth is not sought or told when that Method is in operation. Those particular deficiencies are a Mark of Induction. So, I have a small list of Marks to enable detection of that Method:

MOM's Marks of Induction

1. Induction deals Meaning through resemblances.

2. Truth is not the Point of Reference for Inductive inference.

3. Inductive inferences cannot be Justified.

As far as I can see, those Marks are peer agreed by Specialists in philosophy. They do not Deny them, but they have failed to make the Meaning of those Marks clear. That is a Simple and expressive list. All the parts are Intrinsically Related. These are Facts that are distinctive of Induction and the list does not need to be longer. But I need to clarify at length the Meaning that each of those Marks Represents and the Meaning they Represent as a Whole.

MOVING TOWARDS A CONCEPT OF TRUTH AND JUSTIFICATION

A Real Difference between Inductive science and Pure Science in the direction in which Meaning is moved can be Marked by the words Truth and Justification. Natural Induction and Natural Deduction are part of the Action in the Process of Pure Science, and Control is not surrendered to Induction. Truth is the Point of Reference governing Processing of Sense Data when Pure Science is engaged. The Point of Reference determines the Quality of Meaning that is drawn from all parts and aligned as a Whole Meaning. The Point of Reference distinguishes Pure Science from Inductive science. True Meaning may or may not be Justifiable. Justification is aligned to the search for Truth as a Whole. Justification is a Quality of Meaning that is dependent on a Natural Standard for Judgment. And that which is Natural is Not Artificial.

We have seen the suggestion that support can Function in the same way as Justification. The suggestion was delivered through the term "support or justification". Should you accept the suggestion that support is a Reasonable replacement for Justification? According to MOM, things that Function in the same way have the same type of Meaning. If support is different in type to Justification, then support has different Meaning and Functions in a different way. The type of Meaning is determined by the Point of Reference. Induction is the Context in which things are supported and thus support is different in type of Meaning and Action to Justification. The Point of Reference for Justification is Truth and Truth is not the Point of Reference for Induction where the Action of support takes place. Different types of Point of Reference lead to different types of Meaning. Premises are a form of support and those are put in place in Inductive logic to enable acceptance of whatever follows from them.

When we compare the common definitions of the words justification and support their different types of Meaning and Action are visible.

justify: [....<L. Justus, just + facere, to make] 1. to show to be just, right or reasonable................

support: 1. carry all or part of the weight of; keep from falling, sinking, or failing.....[Latin *porto* carry]

The difference between the Action of to support and to justify can be clearly seen, but the Meaning represented in the word Justify is not precisely the Meaning I want to represent using that word. The word reasonable covers different types of Processing and the type is not specified in the definition. It is difficult to see the problem in Meaning of an arrangement unless you look at the detail, so in order to show you a problem here I will take part of another definition. The word just is an adjective and is part of the word justify, which is a verb. The Action of Justifying makes Meaning Just and that word as I use it Means:

just: 1. Morally right or fair ....

Immediately the importance of the Meaning of Justification emerges. But its Significance is lost if the Meaning of the words moral, right and fair is specified in a way that is Not Justifiable, and I will come to that in due course. The other parts of the definition of the word just set its Meaning free and I will not pick those up. But I will show how that freedom is used. The small word just is also part of the word justice.

justice: 1. justness, fairness 2. authority exercised in the maintenance of right. 3. judicial proceedings.........

As we have seen, a Construct of Justice by Agreement stands where a Concept of Justice by Discovery of Truth should be. The Law in general terms is decided by agreement among a small group of peers in Power outside the Courts. And Justice is administered by peer agreement among a small group of peers in Power inside the Courts. The Judges do not decide what the Law will be, but they can determine the detail of the Law as they apply it in Single Cases. And it is a Fact that changing a small detail can change Meaning as a Whole. As we have also seen, when the Judges search for Justification in evidence it is hard to find. When cases come to Court, and Truth cannot be determined, the Judges set the conditions that Force agreement to be reached. And an Unquantifiable number of people in every Context search for Truth and Justification despite the difficulties of doing that. We need to be able to distinguish those who use Reasoning in a search for Truth in relation to happenings in the Real World from those who do not. Now, I will take parts of the common definition of the words true and real:

true: 1. adj. In accordance with fact or reality ....

real: adj. 1. actually existing or occurring.................

The other parts of the definitions of those words are not relevant here. That which actually exists or occurs in the past, present and future, and its Meaning is the Focus of Attention when Pure Science is engaged. Meaning that is actually True aligns with what actually exists or occurs. And the word true and the word justified have Significantly Different Meaning. That which is True is not necessarily that which is morally right or fair. Common Meaning in the word justify is represented in the following way:

justify: 1. show the justness or correctness of (a person, act, assertion etc.). 2. (esp in passive) cite or constitute adequate grounds for (conduct, a claim, etc.); vindicate. 3. (as justified adj.) just, right (justified in assuming) 4. *Printing* adjust (a line of type) to give even margins.

That definition gives plenty of scope for misuse of that word. Dictionaries are worth reading for they tell stories of Action on Natural Language over the Centuries. Yet while we look at them often, we do not read them and do not grasp that they can tell us stories. The Concept of Truth and Justification awaits full development but I have made a start on that task in this part of my narrative and will add to that Meaning as I travel along the course set for me by the Evidence. Definitions are just a beginning in developing Meaning and the Qualities of the starting point for the Calculation are Significant for the possibilities in later development.

Agents search for Truth when they have a need to do so, and then find the path is Blocked. I was puzzled when I recognised that type of Action is Systematic in all Contexts. Those who enter Specialist Contexts may not notice that the path to True Meaning is Blocked as they follow the leaders there along a different path. Those who undergo training in Specialist Contexts that have been invaded by Inductive logic are not directly told that there is no need for Truth in that Context but many of them will necessarily reach that conclusion. The standard justificatory procedure may lead them to believe it is so difficult to find True Meaning that its very existence is doubtful. But whether they Believe that or not, it is a Fact that where the discovery of True Meaning is not a Focus of Attention, they will not perceive or understand what is actually existing and occurring in the Natural World.

Gaps in Meaning are filled in one way or another. The notion of support filled the gap in Meaning that was left by the disappearance of the Concept of Truth and Justification. The Concept was replaced by the Construct of Support, which was Quantified and that Construct is Not Justifiable. Many examples can be found of a concerted effort to fill the gaps left by the removal of Concepts with fractions of things that never actually amount to any Whole thing but retain the Appearance that they can amount to that. There is a word that is used to cover that position and that word is probability. The conclusion of an Inductive argument does not have the Quality of Truth and is said to follow with some degree of probability. Numbers have rank order and the limits of the Quantitative Scale of Probability are fixed. Proximity to the limits guarantees that the Quality of the limit will be transferred to the conclusion by Association. And none of this Action is novel or recent. I will provide the Evidence for that statement in due course.

The Uncertainty in Premises guarantees that there will be Uncertainty in the conclusion and that the Quantitative Construct of Probability can be used to produce an Illusion of Truth for Meaning that is actually determined by human beings and has the Appearance of being determined independently by mathematical equations. Degrees of probability are a form of Temptation to accept conclusions that are Unjustifiable. There is the Appearance that these Constructs might provide Knowledge for we can see the fixed limits that indicate where Certainty lies and we see in the mind's eye the fractions that hang suspended as steps between them. It looks as though there is a ladder that can be climbed. There is always movement one way or another and movement suggests we are getting somewhere. Induction is not governed by rules. Mathematical equations are a form of Inductive deduction. They create the Illusion of Deduction by Number.

Recall that Popper won the argument with Carnap and the others in the Circle. Peer agreement on a Formal philosophy of No Possibility of Confirmation was reached. That has the Appearance of peer agreement that there is no such thing as Certainty, but that Appearance is Deceptive. There is never any guarantee of the truth of a conclusion using their Method. But Certainty was not eliminated. The Fact that Natural Deduction can lead to the discovery of True Meaning is known. It was found that including a form of deduction within Induction does not solve the problem of Justification for Induction but it does solve the problem of having conclusions accepted by an audience. But they could not take the credit for that if credit were due. Thus they cannot take the blame for that if blame were due. And anyway, the Construct of Blame has no place within the Boundaries of Pure Science.

If the difference between Induction and Natural Deduction was to be clearly Marked then Induction would lose its Appearance of Justification and much of its Power. Or at least that is a Testable Proposition. But the Fact that there is an alternative to Induction is not acknowledged. It is Assumed and Asserted that there is no alternative. The conclusion that all we can have is some degree of Quantitative Probability is an Illusion. It is said that there is a difference between Induction and Abduction. The difference in Meaning suggested by their different names is easily shown to be Trivial and yet that which is Trivial in Meaning can have a Significant effect once Power is assigned to it. Induction begins with Premises and those are the Assertion of truth of the start of an arrangement.

ABDUCTION ENDS INDUCTION WITH THE ASSERTION OF TRUTH OF A CONCLUSION

Abduction is Induction coupled with the Selection of a particular conclusion. Abduction is described as Inductive "reasoning" to the "best" explanation. Abduction is a common form of Induction that begins with a Premise and ends with a Premise. A Single Conclusion is Selected, usually from others previously selected as possible, but no Justifiable rules are specified for the Final Selection. There is the Appearance that testing governs the Selection. Where Probability Theory governs testing there is Error of unknown Quantities in the Judgment and the Illusion that the Quantity of Error is known. Error is measured using mathematical equations. The weight to be given to inferences is determined by mathematical equations that take Error Variance into account. The "best" explanation is usually the one considered to be most probable as Judged in Quantitative terms. But even Qualitative terms would not make a difference to the Quality of a Judgment derived by Induction. Justified conclusions are only produced through Calculating Meaning by Natural Deduction in a search for True Meaning.

Douven (2011) states that:

> "Many ….. examples of scientific uses of abduction have been discussed in the literature;…. Abduction is also said to be the predominant mode of reasoning in medical diagnosis: physicians tend to go for the hypothesis that best explains the patient's symptoms……"

That is misleading. Here, Medical Diagnosis is Associated with Inductive logic. A good Diagnostician uses Natural Deduction in a search for True Meaning and does not use Inductive logic. Truth must be Detected. The term "physicians tend to go for the hypothesis that best explains the patient's symptoms" has the Appearance of Being Reasonable and even the Appearance of Necessity. But Abduction is Inductive and Not Justifiable. It is not possible to count how many physicians use Inductive reasoning and thus guess the conclusion while we cannot tell the difference between those who use Induction and those who use Natural Deduction to Calculate Meaning from data. The foundations of Medicine are the Natural Sciences. When Natural Deduction within the Process of Pure Science leads to one conclusion, the conclusion is Justifiable and if a conclusion has not been Verified then further Testing occurs. Pure Science is a matter of Qualitative Analysis in terms of Reactions.

Inductive logic is Invasive and Defences have not been in place to prevent its spread. Abduction is the predominant mode of reasoning for diagnosis when the Quantitative Method is used for inference. The Quantitative Method takes a dimensional approach and calculates statistical probabilities by entering coded input into mathematical equations. Quantitative Algorithms lead to Errors of Judgment.

When doctors use Quantitative Algorithms to reach a diagnosis, it is not they who make the diagnosis. The Statisticians do that for they designed the statistical equations. Statisticians are specialists in Artificial philosophy. They work on any Meaning that comes their way. That which is Unjustifiable should not work on that which is Justifiable. The Action should be the other way round. That which is Justifiable should work on that which is Unjustifiable. An Unquantifiable number of doctors use Common-sense and their experience when interpreting the numbers that they are gifted through mathematical equations.

For those who have acquired the habit of relying on Assumptions, Inductive "reasoning" to the "best" explanation is common. Douven notes that:

> "Sometimes our reliance on abductive reasoning is quite obvious and explicit. But in some daily practices, it may be so routine and automatic that it easily goes unnoticed."

According to Douven:

> "Some philosophers of science have argued that abduction is a cornerstone of scientific methodology...... .. McMullin (1992) even goes so far as to call abduction the inference that makes science."

Induction is the Method of science and Abduction is part of that Method. Inferences in science are dependent on Premises, involve generalisation from visible, static characteristics of things of which we have had experience, and often involves Selection of one Meaning for Assertion as a conclusion. Imagination is Active along the way. Inferences in science are inherently Unstable. But Inductive science is pseudo-science and is not Pure Science, nor does it bear any resemblance to the type of Action that takes place within the Process of Pure Science other than by taking the name science and dropping the difference in the Point of Reference from view. Abduction is Certainly Not the inference that makes Pure Science.

Even if deductive rules are applied to a range of possible Inductive inferences in order to arrive at a single conclusion, the rules cannot change the Quality of Uncertainty carried into the Structure by Premises. Assertion of a Single conclusion gives the Appearance of Certainty where there is none. Prefacing an Assertion by entering possibilities into the Structure of deduction provides the Appearance of Justification. The "proofs" in Inductive "reasoning" bear some resemblance to Proof by Reasoning. Those who use science mimic many details of the Actions of those who engage in the Process of Pure Science. The Universal exception in the detail is the direction of the search. Quite simply the purpose is different. Expert witnesses who use Induction and rely on Abduction for their Opinion are not different in type to lay witnesses. Yet they are treated in a different way on the basis of their Qualifications. Qualifications do not Justify an Opinion. Opinions are Justified by Reasoning in which details and the relations between them are Tested for Truth.

PART THREE of Chapter three

DEDUCTION AS MASTER AND SLAVE

Slavery was customary during Ancient times. Something that is customary is habitual and has a high frequency of occurrence. A Master governs a slave and determines the slave's every Action. In general, that which is governing determines the Meaning and Action of all the parts within its sphere of influence. Even opposition cannot override its influence. When Induction determines the Meaning on which deduction Acts, then the Action of deduction is not free to determine the conclusions. You are unlikely to see the difference in Meaning between Inductive deduction and Natural Deduction if you rely on what you are told about the Meaning of the word deduction. The common definition of deduction in our dictionaries is misleading:

deduction: 1 a. deducting b. amount deducted. 2 a. inferring of particular instances from a general law or principle. b. conclusion deduced. [Latin related to deduce]

That is deduction governed by Induction. That is not Natural Deduction. The Action of Inductive deduction is described as:

deduce: [Latin duco duct - lead] infer logically

The word logic is also misleading. The word logic covers with one word different types of Cognitive Action as does the word reasoning. Specialists in philosophy and those who have been drawn under their spell drop differences in Meaning by failing to Mark its type. Thus True Meaning is hidden. I will show you the definition of logic in the Penguin Dictionary of Psychology.

logic: That normative branch of philosophy that deals with the criteria of validity in thought, the canons of correct predication and the principles of reasoning and demonstration. Logic concerns only the reasoning process, not the end result. Incorrect conclusions can be reached through logical means if the original assumptions are faulty.

That definition of logic is restricted to Formal logic. Natural Logic does not rely on assumptions. The definition of logic refers us to the definition of reasoning we saw before. It is the Cognitive Processes that are of concern in philosophy. It is the working of the human mind that is the Focus of Attention. Yet that is not how it Appears to be. It is a Fact that incorrect conclusions can be reached through logical means if the original assumptions are faulty. Particular conclusions follow necessarily from that Fact. If you can determine the assumptions that are feeding the Cognitive Processes in the mind of an audience, then you can lead them to False conclusions. That is exactly what Premises do. There is another side to that coin. If you ensure that assumptions are Tested for Truth and are not accepted without Justification then you are not feeding the Cognitive Processes in the mind of an audience with ideas that will lead them to False conclusions.

The Judgment of what is correct in terms of effects and outcome depends on the Point of Reference for the Judgment. And that which has the Appearance of being Trivial can be Highly Significant. Now, if we distinguish between effects and outcome then we have identified one Factor that helps to make Sense of self-destructive Actions. Effects are immediate and if those have the Quality of Reward, then the Action becomes systematic. The outcome is delayed in time and is surrounded by the Uncertainty that Formal logic creates. The Qualities of the outcome do not catch our Attention and Habits are hard to shift once they become established. But even drawing the outcome to Attention at an early stage cannot alone counteract the pull of immediate effects. Beliefs are Active determinants of Actions.

We Do Not Believe that the End Will Come and False Prophesies back up that Belief. Some False Prophesies have been Quantified and made Testable – the World will end on 22nd December in the year xxxx, or the Dystopian State of the World will become a reality in 1984. Quantification gives the Illusion of Testability and the Illusion of Certainty that the proposition is False once the time has passed. And one Bias Acts against a different Bias. The Bias of immediate pleasant effects overcomes the Bias of a possible delayed, unpleasant outcome. Outcomes are not made Testable by assigning a number to a happening. But then we remember Halley's Comet and the Illusion of Prediction by Numbers returns to haunt us. However that was not a Falsification. So, perhaps Verification is easier to achieve than Falsification. And the prediction of the return of Halley's Comet was not precise and was not a prediction of a Change of State. That was a good prediction of a systematic regularity in the Action of an Object in Motion governed by External Forces. It was based on statements that were Testable Propositions. Halley's Comet was something that continued to be moving in a cyclical pattern as it previously had been and its position recurred. But good predictions are not precise Quantitative descriptions and vice versa.

The Quantitative Approach Acts against the Perception that there are differences in type of State and against the Concept of Changes in State. What if the Dystopian State of the World did become a reality in the vicinity of 1984? It was after all a Change of State that was predicted. Perhaps that has become a Reality and we just did not notice when that was, for Changes of State happen gradually. Note a Fact to carry with you. The Quantitative Method is a tool that fits some types of happening, but it does not fit all types of happening and it does not fit the task of making Qualitative Distinctions. Crucially, the Quantitative Method cannot produce Causal Knowledge. Only Causal Knowledge enables prediction of the happening of Changes in State and their detection. And if we have Causal Knowledge the Factors that lead to destructive habits and the Changes in State related to them can be controlled. Beliefs can be changed. That Fact is known. In the Context of Formal Induction, Beliefs are changed through Manipulation of the Mechanism of Emotional Association to serve particular purposes.

It is clear to see that Errors in Judgment are not an issue for Specialists in philosophy and that where those Specialists lead many have followed, including those who openly specialise in the study of human beings. The number in each field of study who do and do not rely on Formal logic cannot be counted while the differences in Type of Cognitive Processing remain hidden. Also from the Penguin Dictionary of Psychology we have:

deduction: Reasoning that begins with a specific set of assumptions and attempts to draw conclusions or derive theorems from them. In general, it is a logical operation which proceeds from the general to the particular. Deductive inference is, in and of itself, an abstract process which requires no verification other than logical consistency. The proof of the pudding is in the appropriateness and demonstrability of the theorems and conclusions which are deduced. Contrast with induction……

That pudding is not to my taste. The definition with which we are told to contrast deduction is:

induction: 1. A process of reasoning in which general principles are inferred from specific cases. A logical operation which proceeds from the individual to the general; what is assumed true of elements from a class is assumed true of the whole class. The experimental method is basically inductive in nature in that conclusions about populations are drawn from observations of individuals and small samples. Contrast here with DEDUCTION…………

It is clear to see that Specialists in psychology have picked up and accepted False Meaning gifted by Specialists in philosophy. They seem to have lost sight of the Fact that philosophers are human beings. Those definitions are not challenged as they are set before us. That which is assumed true of a whole class is assumed true of all parts of the class in a Quantitative Structure. Common-Sense tells us that the Action of Assumption leads to problems in Social Dynamics – the Evidence for that is scattered throughout history. That which is True in terms of Meaning of a whole class is different to that which is True of each part in terms of its Meaning. And the "experimental method" is only "inductive in nature" when it is Induction and then it is Not Actually Experimental. The Experimental Method within the Process of Pure Science is different in type of Meaning and Action.

Natural Deduction is not defined anywhere, so though it can be seen in Operation in many places it cannot be recognised. We cannot pin that name on it. It is peer agreed that there is one type of deduction and any visible differences are viewed on a Single Dimension. Natural Deduction does not determine Meaning in Specialist arrangements and that Fact is never stated. Truth is Not an Issue in Specialist arrangements and that Fact is never stated. A justificatory procedure does Not Justify what happens and that Fact is never stated. There is peer agreement to ignore Facts and hide them from view. Natural Deduction is a particular type of Action on Meaning and although its existence is ignored, it is impossible to entirely eliminate that Action from Social Systems. Since that type of Action is not specified elsewhere I will do that. The Action and the output of Natural Deduction can be represented by the term Common Sense. Are you puzzled that Specialists have dropped the Action of Common Sense from the word reasoning? Or perhaps you did not notice it is not there. If nobody has noticed its absence then that is puzzling. Puzzles are grist to the mill of Pure Science. So, what is the Meaning of the term that is not included in Formal reasoning? The Meaning as given in Collins Dictionary is:

common sense: ordinary good sense or sound practical judgment

It could Reasonably be concluded that Specialists do not use ordinary good sense or sound practical judgments when they use Inductive logic, which they call science. That conclusion will of course be taken as a proposition and Tested thoroughly. In essence that conclusion follows from their agreement not to use Natural Deduction on Sense data. It has been agreed to begin deduction with a specific set of assumptions and it is a Fact that incorrect conclusions can be reached through deduction if the original assumptions are faulty. That Fact is known and clearly stated. So why is the potential for Error built into the Formal Structure of "reasoning", rendering Reasoning impotent? That is not accidental. Chance is not a Factor in Operation in Social Dynamics. I will add at this point that what is True of a class of people is not necessarily True of all the individuals of which the class is composed, but the Ethos of the class is a Bias that affects the Quantitative distribution of different Qualities within the group.

Common Sense inferences are related to Reality. Common Sense Judgments are practical, in the sense that they Function well. The Penguin Dictionary of Psychology defines common sense thus:

common sense: 1. In the Aristotelian sense, the capacity to comprehend the qualities of an object through use of the other senses. 2. In Thomas Reid's revival of this basic idea the focus was on the ability to apprehend qualities common to all the senses (time, space, numerosity) 3. Beliefs, opinions, practical understanding of things shared by the 'common man'. 4. Colloquially, good reasoned judgment.

Note that time, space and numerosity are referred to as Qualities. The word numerosity is defined in that dictionary.

numerosity: Quantity, 'manyness'. Young children (and some nonhuman species) can detect fairly subtle distinctions in numerosity in two displays of objects without any indication that they have the abstract notion of number.

The Concept of Quantities is different to the Construct of Quantities. The effect of lowering the value of Common-Sense is the raising of the value of Formal logic to a higher level. That rise in value is of course Unjustifiable. Peers would necessarily agree with that statement for Justification is an Insoluble problem for Induction and there is peer agreement on that Fact. It is possible that the abstract Construct of Number that is in use Interferes with our perception of the Qualities of time, space and manyness. Certainly Formal logics Interfere with the Operation of Common-Sense. If we do not understand what Reasoning is, then we cannot acquire skill in that Method of Judgment. Good Reasoned Judgment leads to Actions that are based on good foundations. Judgments of Meaning and the Actions that stem from them are likely to be in Error when Reasoning has not been involved. The Bias of agents in Specialisms moves inferences away from Good Reasoned Judgment. Yet the direction of movement is reversed as we look in the mirror they hold up before us. It is Strange indeed that there should have been Action against Common-Sense but the Evidence for such Interference is overwhelming. It is a Fact that where there is a Bias there is a Purpose that is served.

It is a Fact that the word deduction covers two different types of Action and that only one type is shown. It is Denied that there is an alternative to Formal logic. Different types of Action have different Meaning. Meaning in Structures is governed by the Point of Reference and particular Meaning can be seen when the Point of

Reference is visible. I need to show in detail the difference between deduction in Formal logic and Deduction in its Natural State. I have proposed that within Formal logics, deduction is enslaved by Induction. In other words, Induction governs the Action of deduction by determining the Meaning on which the rules of deduction can Act and thus determining the conclusion. The rules may or may not be the same in the different Contexts, but the same Action has different Meaning when it serves a different purpose. There are no specified Rules for Induction but if there actually was No Point of Reference then Induction would have No Particular Meaning and Action using Induction would be disorganised. Formal logics are Artificial, Inductive and orderly. Thus there is necessarily a Focus of Attention for Operations by Induction. Clearly there are Rules for Induction.

We observed that Induction has no rules that peers have agreed or will show. Natural Deduction has Rules. Those rules have been Formalised as part of Induction. The False Appearance of Deduction gives the False Appearance of Rules to Formal logics. Aristotle's gifts are important parts in Formal logics. Smith (2019) states that:

> "A deduction is speech (logos) in which, certain things having been supposed, something different from those supposed results of necessity because of their being so. (Prior Analytics I.2, 24b18–20)"

Premises are Not Sense Data. Induction begins with Premises which are things that are supposed to be true. A supposition is something that is assumed or it is something that is taken as a hypothesis. Whether you assume or hypothesise makes a different to the Action that follows. Hypotheses are statements for testing and assumptions are not. Premises are not hypotheses. Valid deductions are those in which the conclusions necessarily follow from the Premises and are not necessarily correct. The common definition of the word valid from the Pocket Oxford Dictionary is:

<u>valid</u>: [….<L. Validus, strong < valere. To be strong] 1. Having legal force; binding under law 2. Well-grounded on principles or evidence; as an argument; sound. 3. effective, cogent etc. 4. Logic correctly derived or inferred according to the rules of logic

The visible rules of Formal logic are rules taken from Natural Deduction or bear some resemblance to those. So, why does Inductive logic lead to Errors in Judgment? We have seen the answer to that question. Humans select the input to which the rules are applied and there are different types of input. Some types of input lead to Error in output. Peers agree to assume particular things, and agree that their Selected Assumptions should be Acted upon by the Structure of deduction. Thus as a Whole the Action of Inductive deduction is Not Natural and has different Meaning to the Action of Natural Deduction as a Whole.

A hypothesis is a potential answer to a question. When the input to Natural Deduction is a hypothesis, it is a statement for Testing. Each possible answer is input for the Structure of Natural Deduction. The Structure of Natural Deduction does Not Test each answer for its Truth. Truth can only Emerge Relatively. The Rules of Deduction show where each answer leads in terms of a conclusion. In Pure Science that is done for all potential answers. Seeing the conclusions enables their Testing by Comparative Analysis until the correct answer is found. In Formal logic the potential answers are Restricted. Thus, the Potential for the Discovery of Truth is Restricted.

Meaning becomes clear when it is Represented in different forms. I will do that. Take two Facts into account. A difference makes a difference. And Meaning is Relative. Different propositions lead to different conclusions. The different conclusions can be compared to each other and in relation to other things. Testing Meaning for Truth is a Comparative Process. When peers agree on Premises, they are ruling out other possible statements and the rules are not applied to those, thus they arrive at only the conclusion or conclusions to which the Premises lead. Testing is ruled out at the beginning of Induction and at the end, and does not take place in the middle where the rules are.

Rules are not Tests of Meaning, and are products of Action elsewhere. The type of Action that produced the rule makes a difference to the type of rule. The rules of Natural Deduction are products of Nature and Function well in service of the purpose of Nature. But man takes the products of Nature and uses them to serve purposes of his own. The purpose determines the Meaning of the Action.

It is a Fact that not all Inductive deductions that are valid are True. We cannot tell the difference between True and False conclusions thus we have no idea how many valid deductions are Not True. Natural Deduction has a particular Structure. A Structure has parts in relations. The parts of the Structure and the Relations between the parts are variable and are specified for that is necessary in order to activate the Structure. Once the parts have been put into the Structure and a relation specified, the rules are applied and the conclusion follows necessarily. The necessity of the conclusion gives it the Appearance of Truth. The rules give the necessity and thus they give the Appearance of Truth to the Meaning Inducted.

It is a Fact that Structures are powerful determinants of the conclusions that people draw. Smith (first published in 2000) states that:

> "Aristotle's definition of deduction is not a precise match for a modern definition of validity. Some of the differences may have important consequences:
>
> 1. Aristotle explicitly says that what results of necessity must be different from what is supposed. This would rule out arguments in which the conclusion is identical to one of the premises. Modern notions of validity regard such arguments as valid, though trivially so.
>
> 2. The plural "certain things having been supposed" was taken by some ancient commentators to rule out by definition arguments with only one premise, and Aristotle himself says in some places that nothing new follows from just one premise.
>
> 3. The force of the qualification "because of their being so" has sometimes been seen as ruling out arguments in which the conclusion is not 'relevant' to the premises, e.g., arguments in which the premises are inconsistent, arguments with conclusions that would follow from any premises whatsoever, or arguments with superfluous premises."

Aristotle provided many ideas and there was vagueness in his writings. We saw what happened when Bridgman failed to pin Meaning down. Aristotle is used as an Authority for interpretations of his ideas. Interpretations have particular effects. The modern definition of validity does not rule out arguments in which the conclusion is identical to one of the Premises. And it is a Fact that such arguments are not always treated as being Trivial, though they do have that characteristic in relation to Truth, and it is a Fact that they have a high frequency of occurrence. That which is actually Trivial in terms of Meaning is given weight in various ways. In due course I will present Evidence for that statement. Inductive arguments are composed of many deductions from things that have been supposed. An Enthymeme is a valid or acceptable argument in which the conclusion has already been set by Premises and is arrived at through precisely regimented steps that can take Meaning nowhere else. An acceptable argument is a plausible argument, or is it just True that a plausible argument is one that will be accepted? The delivery of False Meaning is aided by the development of skill in Rhetoric. Rhetoric gives additional Force to the movement of Meaning from one agent to another. It is a Fact that the Enthymeme delivers False Meaning. Meaning that is Not Tested for Truth and is not Verified is put into relations that suit particular purposes that are different to the purpose of finding and presenting True Meaning. But it is claimed or suggested that the Meaning is Verified and thus is True. That is what Premises are; they are claims to truth. Inductive deductions are presented as proof of the truth of Premises, and those are Not Proof of their Truth. An Inductive argument presents an interminable procession of truths that are no such thing. Lakatos saw and expressed that Fact.

To perceive False Meaning and accept that as being True Meaning is a disadvantage to your Operations in the Real World. It is not a disadvantage to perceive False Meaning when you can tell the difference between True and False. The perception that Meaning is False is an advantage. I allow the possibility that Specialists in philosophy do not understand that. They have dropped the difference between True and False from view. Unless Meaning is Calculated through Natural Deduction, we are susceptible to Illusions. There are legions of Illusions of Meaning and they are all Active in Social

Dynamics. We have all been overcome by their Power. It is time for Revival.

Natural Deduction is rule based. The Point of Reference that gives the rules of Natural Deduction particular Meaning was dropped somewhere along the line of philosophers and the Action was formalised thus dropping its Quality of Flexibility. Rules must be related to something in order to have particular Meaning and become Active, and rules serve particular purposes. If the question of whether the Premises are True or False is Ignored, the Structure of the deductions guarantees that the audience will accept the conclusions built from them. I will rephrase and repeat that Meaning. The Structure represented by Inductive deduction does not guarantee anything in relation to Truth and only guarantees that the conclusion will be accepted by the audience. Human beings can be drawn to agree to anything irrespective of its Truth using Inductive deduction.

I aim to find True Meaning in relation to doings and happenings and that is what the Process of Pure Science was designed to do. As I understand it, Pure Science is a Natural Process and was not designed by human beings. That which is Natural can be misused or well used - the type of use is determined by the user to suit their purpose. I will pick up and include the particular part of the Natural Action that Aristotle found and described, for that is part of the Process of Pure Science. The Syllogism is a small and important tool of Natural Deduction and has a particular Structure in which parts are related to other parts. The word syllogism represents a Simple Unit of Natural Deduction when that word is used in the Context of Pure Science. It is possible to understand the Action in Complex Systems through the construction of Simple Models using units of Natural Deduction. The Syllogism is flexible tool and can be used for the Calculation of Meaning. Thus we need to understand what the tool is and how it can be used and misused. I begin with the common definition of syllogism from the Pocket Oxford Dictionary.

<u>syllogism</u>: n. reasoning in which a conclusion is drawn from two given or assumed propositions [Greek logos reason]

That definition is too vague to be useful here. There are detailed specifications in the Specialist literature that are useful and can be

Tested. They show the Action of the syllogism. The Syllogism is a small Structure through which Meaning is moved by rule from one thing to another. The accuracy of the Action of the rule depends on the accuracy of the relation between parts within a particular Syllogism. In a Single Syllogism, the movement of Meaning occurs in a specified sequence through a particular type of relation, which is also specified and is a constant for the unit in its basic form. I have proposed that the tool is misused in the context of Induction and that proposition is Testable. The general Structure of the syllogism is given Formal expression in the literature as:

major premise:- All As are B's.
minor premise: - a is an A.
conclusion : - Hence, a is a B.

Premises set the syllogism for Action as part of Induction where Truth is not an issue. The Latin root of the word Premise is praemissa which means set in front. According to the Oxford Pocket Dictionary, in logic (i.e. in Formal logic) premise is spelt premiss and is a previous statement from which another is inferred. A premise/premiss set in front of an argument enables Meaning in the argument to be built in a particular way. One inference leads to another inference and so on. A Premise is an Assumption in conjunction with an Assertion of its truth. Here you see the syllogism described as using two Premises from which the conclusion is derived. A different statement follows of necessity from Premises when they are Acted upon by the Structure of the syllogism. Aristotle showed that Fact. The two Assertions of truth in Inductive syllogisms are worked in conjunction with each other, and something different results of necessity "from what is supposed". The Whole is different to the parts.

Assertion is an Active Mechanism for the Installation of False Meaning and is systematically applied when Induction is in operation. Assertion is the common thread of Action that holds Inductive arguments together. Assertion is a necessity when Truth and Justification are not there to bind Meaning Naturally. Assertion is not part of the Action when the Process of Pure Science is engaged and is not part of the Mechanism of Natural Deduction. But the Syllogism is a unit of Natural Deduction so the description we have been given must be changed in order to reflect the difference in type of Action. That is easily done. As a unit of Natural Deduction within the Process of Pure Science, the Structure of the Syllogism is:

major statement :- All As are B's.
minor statement : - a is an A.
conclusion : - Hence, a is a B.

I need to add a condition. Within the Process of Pure Science, the statements of a Syllogism are never Premises. Otherwise the Structure of the syllogism is common to Formal logic and Natural Logic. The Syllogism is an arrangement of two statements and one conclusion, all of which have two parts. The symbol 'B' stands for a part that is present in the major statement and in the conclusion – that part is named the major term. The term that is present in the minor statement and in the conclusion is named the minor term; the small symbol 'a' stands for that. The middle term, 'A', is present in the major and minor statement and is not present in the conclusion. The middle term forms the channel through which Meaning flows between the different parts. One particular relation is threaded through the arrangement. There is always one type of relation per unit but different types of relation can be used in different units and I note that one type of relation can take different forms. Each unit is a Single Syllogism. Meaning in the parts of the Single Case that each unit is, leads to a Single Conclusion that has different Meaning to each of the parts. The parts become a Whole and the Meaning of the Whole is different to the Meaning of the parts. That is a Simple Natural Principle of Meaning.

Meaning is Relative and Qualitative. The conclusion to the Syllogism is a new statement to which the Quality of the original statements is transferred. If both statements are True the conclusion is True. If the Truth of one or both of the statements is Uncertain, then the Truth of the conclusion generated is Uncertain. It is impossible for Meaning in the two statements in a Syllogism to be True and for the Meaning in the conclusion to be False under the condition that the relation is also True. That is the necessity of the rule. I note that the conclusion is restricted and does not go beyond the Meaning that emerges from the parts through the Action upon them, but its Meaning can be taken further when the conclusion is entered into a new arrangement and related to a different part. Parts become a Whole when they are related and Wholes are a part that can be related to other parts. Relations are Active on the Meaning in the parts. That is Simple to understand.

A True statement is one in which the parts and the relation between them are True. Facts are Truths, though some statements that bear the same name are Not Truths. Popper suggested there is no such thing as a Pure Fact and that is misleading. We can either stop using the word fact for Meaning that is False or propositional, or we can use the adjective Pure as a Mark of statements that are actually True. The necessity is to mark differences in Meaning; how we do that requires agreement and consistency in application. I will just use the word Fact to Mark that something is True, with the small detail of a capital letter as a mark. I am not inclined to agree that the word can be used where statements are Not True. The Syllogism can lead to Certainty in the conclusion when the statements are Facts. Two Facts in their Intrinsic Relations lead to a True conclusion, which is a different Fact. The Syllogism is also useful to generate new propositions. When either or both statements are propositions then the conclusion is a different proposition. Propositions are statements waiting to be Tested for Truth. Syllogisms are part of Testing Meaning for Truth and Justification when used in a particular way. Statements can be entered into the Structure of the unit in various combinations. Their Meaning can be viewed from different angles in different relations and in relation to different things using more than one Syllogism. You will see the importance of that type of Action during the course of this narrative. Attention to detail is vital when Calculating New Meaning using Syllogisms Relatively.

I need to mention another type of statement which can be input to Natural Deduction - an Artificial Rule. Artificial Rules are different to Premises, Facts and Propositions and are Not Natural Principles. I would categorise Natural Principles as Laws of Nature that govern Actions in the Real World. Thus, I pick up a difference between Artificial Laws and Natural Laws. Artificial Laws are Rules derived from Artificial Principles - the word artificial represents man-made Constructs. Artificial Rules are man-made prescriptions for Action.

The Quality of the material determines the Quality of the product that man makes. I would pick up a difference between an Artificial Rule and a Premise (all Premises are Artificial, so that adjective is redundant when the word Premise is used). The difference is that the Truth of an Artificial Rule is Not Asserted. That difference is subtle and Significant. Artificial Rules are Pure Assertions. They do not need to be given the Appearance of Justification and they are enforced whether they are accepted or are not accepted by the audience. Truth in terms of fit may be open for Testing in Single Cases when there is an Artificial Rule, but often it is not open. However, the word rule is Formally defined to cover different types of thing and Premises are a type of Meaning they cover. The truth of Premises is not open for Testing but they must have the Appearance of Justification in order to take effect. The effect of Premises is dependent on their being accepted by the audience.

So, Meaning can be attached to things and detached from them to pass on. The output from one Syllogism can act as a statement for the next. Chains of Reasoning are built from these basic units. That which is Complex is Simplified and moves on. Meaning can be transferred from Single Cases to General Principles, from General Principles to Single Cases, from one Single Case to another and between General Principles. All of that Action can be accomplished in Natural Language. Within the Process of Pure Science all the Action is directed towards the Discovery of True Meaning. Thus all Actions are Tested for Truth and Justification. Aristotle showed that mimicking the Action of Searching for True Meaning can deliver False Meaning and serve particular desires. That is what is done within Inductive logic. In order for a type of Action to be mimicked, it has to be displayed. Aristotle did that.

A particular example of a Syllogism is commonly used as an example of deductive inference. You will quickly grasp that the example they use is not deduction as part of Induction and is actually an example of Natural Deduction. That is a Trick and it is one that is repeated systematically in support of Inductive Systems. This example makes Sense and the idea that Sense is made is transferred to Formal logic where Sense is not made. This familiar demonstration of the Syllogism does not begin with Premises and does begin with Facts. The specified relations are True. The conclusion is True:

All men (middle term) are mortal (major term).
Socrates (minor term) was a man (middle term).
Socrates (minor term) was mortal (major term).

This Syllogism shows the Operation of Natural Deduction and also shows the Fact that Natural Deduction can begin with Sense Data. There can be No Doubt about that. The Proof is short. Mortal has particular Meaning given through particular experience and abstracted. It has been Observed without exception across the Centuries that all men die. Dying is a True part of the definition of man and is only part of that definition, for other living things die and that is readily Observable. The word Mortal as a noun is restricted to the death of human beings. There are recorded Observations of the life and death of a particular man named Socrates. The detail in descriptions of the man, his doings and the happenings at the time varies and may not be True but taken as a Whole they are witness to the existence of a man with that name in Ancient Greece. The two statements are Facts and the conclusion is True. In this case Meaning does not move far. The principle is demonstrated.

I can prove a different point using the word Mortal. The definition as a noun in the Pocket Oxford Dictionary is:

mortal: n. human being

The Latin root is given and is mors mort = death. As an adjective, mortal is given the following Meaning. There are many parts to this definition and the movement of Meaning through the parts is clear:

mortal: adj. 1. subject to death. 2. causing death; fatal 3. (of combat) fought to the death. 4. associated with death (mortal agony) 5. (of an enemy) implacable. 6. (of pain, fear, an affront etc.) intense, very serious. 7. colloq. long and tedious (*for two mortal hours*) 8. conceivable, imaginable (*every mortal thing; of no mortal use*).

All living things are subject to death. Meaning is precise and clear in part 1, and through parts 2 to 4. Then the definition continues, with part 5, "(of an enemy) implacable", which is distantly related to the other parts, for (when we look) implacable is defined as: "unable to be appeased". So, this part covers attempts to appease that cannot succeed and thus it is the effort that must die. Then Part 6 spreads Meaning from death, which is one type of thing, into degrees of injury, which is a different type of thing through the arrangement: "(of pain, fear, an affront etc.) intense, very serious". So, Meaning is at least diluted in that part and is moving towards Quantification. The definition of injury is:

injury: n. 1. physical harm or damage. 2. offence to feelings etc. 3. esp. *Law* wrongful action or treatment.

In terms of Meaning, injury can lead to death or not, but death is a particular State of Being that is different to the State of Being Alive and Injured. The difference in State is between Life and Death. Different types of State have different Qualities. Injury can relate the two different States. Recall Part 7 of the definition of the adjective mortal: "colloq. long and tedious (for two mortal hours)". Part 8 is also colloq.: "Conceivable, imaginable (every mortal thing; of no mortal use)". As you can see Meaning is allowed to spread in definitions and as it spreads, the original particular Meaning fades from sight. Meaning is moved away from death towards Representing a Quality of life and feelings. Life and death become part of one dimension. One dimension is necessary for Quantification. It is possible that there is actually more than one dimension in the Cosmos and it is Certain that there is more than one State. Either way there are Qualitative Differences that the Quantitative Construct cannot deal with. It is a Fact that while we are alive, we are not dead.

The Quality of Life can be said to resemble the State of Death, but that is Not a Testable Proposition - we cannot Verify that. And Life does not have the same type of Meaning as Death. I note that once again happenings with different Qualities are lumped together so that they can be divided into numbers and entered into mathematical equations for the manipulation of the Perception of Meaning by audiences. That leads to Errors of Judgment. The Ancients were clever fellows and they knew many Facts but they did not have Knowledge. It Appears that they were searching for a Method of producing Knowledge and there is Evidence that they were doing that. What I cannot know is whether they found that Method and hid it from the multitude or failed in their search. They did discover parts of that Method, but they may not have understood the Whole of that.

We can never predict the precise moment at which a change in State will occur for there isn't one. The Quantitative Method cannot deal accurately with changes between or within States. Specialist philosophers have directed Attention towards the search for precision that is not there to be found in Nature. There must be flexibility to accommodate Natural movements. We should instead be searching for Meaning and fitting the Representation to that. The structure of numbers does not have the flexibility that is necessary. Formal logics fit Meaning to the Representation. It is wrong to do that.

FORMAL LOGIC IS AUTOMATED

Intelligence is Natural and is Intrinsically Related to Natural Deduction, which is Reasoning. Those who use Formal logic do not claim that type of Action is Natural Intelligence. The term Artificial Intelligence covers the automation of Formal logic in machines. Or perhaps I should say that Formal logic is automated. When Formal logic is installed in human beings, humans Act like machines. Formal logics are Inflexible. There are Specialists in Artificial Intelligence. Specialists in Artificial Intelligence are Not Specialists in Natural Intelligence.

Portoraro (2014) states:

> "Reasoning is the ability to make inferences, and automated reasoning is concerned with the building of computing systems that automate this process."

Here we have a statement that presents the Illusion of Reasoning by Computer. Natural Deduction cannot be automated. Formal logic can be automated. Formal logic is Not Reasoning. Those are Facts. Portoraro tells us that:

> "Although the overall goal is to mechanize different forms of reasoning, the term has largely been identified with valid

deductive reasoning as practiced in mathematics and formal logic."

Mechanisation is Related to Formal logic – they Appear to be one type of thing. Computing systems operate on data with Formal logic. Computer software is written by human beings using Formal logic. Computers are just Formal mechanical systems and do not change the type of thing that Formal logic is. Mechanisation puts Power into the hands of those agents who specialise in Formal logic. Force is applied to ensure those mechanizations are used by us all. Most of us do not understand what is happening when we use them. All we understand is the necessity of following the procedure rules. Any deviation from the procedure rules is not tolerated by Formal Systems.

Computers can serve various different purposes but the discovery of True Meaning and Causal Knowledge is not a purpose they can serve. That is Impossible for Computers to do and impossible for us to do using computers. It is a Fact that Reasoning Interferes with the Operation of Automated Systems. The Capacity to Reason is enhanced by the use of Natural Language and the attack on that type of Representation is an Attack on the Capacity to Reason. The use of Natural Language has long been under sustained attack. After Centuries of Undermining, the skill of Reasoning is now underdeveloped to a degree that threatens our existence as individuals and as a species. Where it is not underdeveloped, it is suppressed. And we do not actually know the extent of the underdevelopment or the amount of suppression. I suspect there is more suppression than there is lack of skill.

A VALUABLE NATURAL RESOURCE LAYS VIRTUALLY UNTAPPED.

Reasoning Interferes with Automatic Actions and Automatic Actions Interfere with Reasoning. But once Reasoning has taken place and True Meaning has been discovered, then the perception of that Meaning can safely become Automatic. The Qualitative Order of Precedence is in terms of Necessity.

PART FOUR of Chapter three

THEORY AS METHOD

Theory is an Active part of the Process of Pure Science and of Inductive science. So what is that part and how does that part Function in each Context? Howard (2017) comments:

> "If theory choice is empirically determinate, especially if theoretical concepts are explicitly constructed from empirical primitives, as in Carnap's program in the Aufbau (Carnap 1928), then it is hard to see how theory gives us a story about anything other than experience."

Formal logic does not give us a story at all. Telling stories requires the expressive capacity that Formal logic does not have. Natural Language is well suited to delivering Meaning through stories. Howard quotes Einstein (1946) as stating that Mach's basic Error was:

> "He did not place in the correct light the essentially constructive and speculative nature of all thinking and more especially of scientific thinking; in consequence, he condemned theory precisely at those points where its constructive-speculative character comes to light unmistakably, such as in the kinetic theory of atoms."

Einstein saw that theory is related to experience and goes beyond experience in some way. Theory adds particular Meaning to Sensations through speculating with relations between parts. That is how theory gives us a story about something other than experience. Relations between things and happenings draw out particular Meaning from them and different relations give rise to different Meaning as a Whole. The parts of a theory, their positions in the confines of the arrangement and the relations between them are variables. Selecting and putting parts in relations is constructive and the Structure is speculative when changes are introduced and differences in Meaning perceived. A change makes a difference. The difference it makes depends on what is changed.

Speculation is necessary when True Meaning is not known. The word theory signifies that the story about experience is not necessarily True. By definition theory goes beyond experience and tells us about something else. So, what else does it tell us about? Well, what do we want a theory to tell us about? What do we seek when we speculate? When we are Acting Naturally, we are looking for Meaning with the Quality of Truth in relation to experience. If a theory is expressed in Natural Language then it takes the form of a story that we can all understand.

Theories are Puzzles of Meaning. Puzzles are solved when we have all the necessary parts in the right positions and in the right relations with other parts. Then everything fits together accurately. An Intrinsic Relation is a Natural Relation. Before a puzzle can be solved, the parts that will fit must be found and put in the right places. We have Simple Models of puzzles. In order to solve those, it is necessary to have in mind an image of the completed puzzle and when we are searching for Meaning we do not have that. So, in order to fit the parts of a puzzle together do we need to know the conclusion they will make? How else can we solve the puzzle? Well Einstein told us that Speculation is necessary to solve puzzles. We cannot answer a question until we know the answer to that question, and unless we express an answer it cannot be Tested for fit. When we Speculate, we generate many possible answers to a question and Test their fit to the question against a Standard of Accuracy. The Standard is what actually happens in the Real World. We detect what actually happens with our Five Senses. I propose there is a Sixth Sense through which we relate sensations to discover their Meaning. All the Senses are open to Interference. The Action of Pure Science can detect and deal with Interference.

Within the Boundaries of Pure Science, we accept that we do not know the answer; all we have is the pieces of the puzzle and we may not have all of those. That makes it difficult to solve the puzzle. But acceptance that we do not know the answer also makes it potentially possible to solve the puzzle. Since we do not know what the Whole image actually looks like, different possibilities must be imagined. Within the Boundaries of Pure Science, Attention shifts backwards and forwards from the details of the arrangement as it is constructed to the Whole images that are envisaged as solutions and with that movement the Perception of Meaning changes. Theories give direction to searches. Anything that gives direction to Action is a Point of Reference. Natural Points of Reference are not fixed and are flexible. Within the Boundaries of Pure Science, a possible answer to a question is not a purpose to be achieved. Recall that Aristotle is quoted as explicitly stating that what results of necessity must be different from what is supposed. Smith informed us that statement is Ignored by agreement among modern philosophers. Modern followers of Aristotle diverted from the path he indicated by agreeing that the conclusion can be identical to one of the Premises. They agreed not to Speculate. That agreement is a Fault.

So, as part of a search for Knowledge, the word theory represents an expression of Meaning that is essentially constructive and speculative. We have Facts when we Know Meaning that is True for Certain. Once the Truth of a Theory is Verified, then it is no longer Theory and becomes a Model of Meaning that is True. At that point we stop using the word theory. The theory has become Fact. Then the Model becomes part of a new theory. Onwards and upwards we go as we Speculate and Test for True Meaning. Propositions are small theories. There is no Quantitative point at which a proposition becomes a theory and no Qualitative point either. Theories and propositions are one type of thing in the same State, and that is the State of Uncertainty. Propositions are Speculative.

Repetition is a Powerful Factor in embedding Meaning and I use that Factor. In order to repeat Meaning it is not necessary to repeat the form in which it is Represented and is more effective to change the form. Inductive science cannot Verify its theories for those are presented as solutions, and to Act in that way Circumvents Testing. Including Facts or Verifying a small part of a theory can strengthen the acceptance of False Meaning in the Whole. Designers of arrangements know that Fact. When the truth of a theory is assumed and believed, then that which is said to be a theory is not that and is actually a Premise. Premises are constructed and are not speculative. Whole theories can be set as Premises and often are. Then we have Dogma.

Speculation is necessary to produce Knowledge. So let us do that. Is Uncertainty accepted? That would defy what I propose is a Natural Principle of Meaning: - There is a need for Certainty in relation to

Meaning. If that Principle is True then it is impossible that Uncertainty is accepted. Uncertainty Naturally leads to a Search for Certainty. If Uncertainty is accepted, then it would Appear that my proposition is Falsified. But Falsification is not a Simple criterion. Nor is the answer to the question of whether Uncertainty is accepted. We all accept that there are things that we do not know, but that is different. We can be content with not knowing things and driven by Curiosity. We have seen some Specialist Unsettlers of Meaning in Action. It is said to be Certain that we can never be Certain. All we can have is estimates of the Quantity of Uncertainty on a Quantitative Scale of Probability. But, if we can be Certain of nothing, it is impossible to be Certain that we can only be Uncertain.

It is a Fact that we can be content with not knowing things and Certain that we do not know them. Uncertainty is an Emotional State and not knowing things is a Cognitive State. The relation between Emotion and Cognition is Complex and so is the relation between Uncertainty and Knowledge. An important question I have not seen Specialists in Formal logic ask is: Where does Uncertainty lead in Social Dynamics. I presume they know the answer for explorers of psychological territory through the Centuries have found many Facts and those have been taken into other Contexts. I presume they also know that the answers to particular questions can be found through Analysis of Single Cases. That Fact is strongly Denied by Specialists in Artificial Induction. So, I will take the Fact that True Meaning can be found through Analysis of Single Cases as a proposition. That proposition is Testable. I will take several Cases into Analysis over the course of my narrative including a particular happening during the development of cosmology as a science between 1932 and 1948. The question of where Uncertainty leads can be answered.

As the study of cosmology was being specialised, there was a sharp division of opinion on the Method of constructing theories in science. There were theories about how that should be done. However, those theories were not offered as propositions. There was the Certainty of Belief on both sides. Premises had removed the Uncertainty that should have led to Testing. Differences in the Premises led to a battle to Assert a particular definition of science and the Premise that the word science covers one type of Action had taken hold. There was Confusion. The factions were attempting to Mark a

Difference between them and the Method used in argument was the same on both sides. That Method was Induction. Induction cannot produce Certainty that is Justifiable but it does not necessarily produce Uncertainty. Gale (2015) describes the battle and I will come to the details in due course. For now, I note his description of the Division of Opinion:

> "On one side were those scientists who had their roots mostly in the experimental side of natural science. To them, there was one and only one legitimate method for science. Theory construction, they believed, involved two closely-linked steps. First, one began from the empirical observations, that is, from measurements, manipulations, experiments, whose results were evident to the human senses; this is classic empiricist epistemology. Observational results would then suggest possible hypotheses to examine via further empirical testing. When enough data concerning the hypothesis had been gathered, logical generalization could be carried out, thereby producing a theory; this is classic inductivist logic."

Marks draw Attention. A Trivial Difference can be given Significance that it does not have by making a Mark. A close look at the other side of the Division will show the Trivial Quality of the difference that is drawn to our Attention. Gale continues:

> "Opposing these inductive-empiricist scientists were those whose roots were mostly in the theoretical side of natural science, most especially mathematical physics. To them, there was another, more logically sound, method to construct theories. First, hypotheses could be generated in any fashion, although most believed that imagining hypotheses which were based upon very general, very reasonable concepts—that the Universe's physical processes had simple mathematical descriptions, for example—was the best place to begin; this is classic rationalist epistemology. Once the hypothesis had been generated, strict analytical reasoning could be used to make predictions about observations; this is classic deductivist logic. Scientists who held this view came to be called hypothetico-deductivists; their views

> about both hypothesis generation and deductive predictions were each strongly opposed by the inductive-empiricists."

Both sides of the division used Induction. The difference drawn to Attention is the starting point for Induction. The term "strict analytical reasoning" covers the deductive part of Induction. Imagining that the Universe's physical processes have simple mathematical descriptions is not a Concept and is Not Reasonable. People describe things in different ways. Descriptions are not there for the taking. Whatever form descriptions take, their fit to the Meaning of that which is described must be Verified. The Universe's physical processes have Meaning. Induction cannot Verify Meaning. Mathematical descriptions do not fit Meaning. Can Induction produce predictions that can be Verified? Clearly it can do that. Some predictions are Testable and while there is No Bias towards Truth, some predictions are likely to be accurate if chance is in Operation. But Verifying a prediction through Testing its Truth cannot verify the Method by which the prediction was made. The Premise that observation of whether that which is predicted actually occurs will cover the lack of Attention to Truth elsewhere is a False Premise. And the Meaning of any prediction must be Calculated in Qualitative terms.

Now recall that Popper won the argument with members of the Vienna Circle by suggesting that predictions can be Falsified and can never be proved for Certain. Those peers reached agreement that it is possible under certain conditions to prove that a proposition is wrong but it is never possible to prove that a proposition is right. That is a complicated manoeuvre with particular Meaning and it is difficult to grasp the type of Action involved. The parts of the Action relate to each other while the Meaning of the Action as a Whole is different and hidden from view. I note that it is not only hypotheses that are generated by "strict analytical reasoning" in the Context of Induction, Premises, rules and tests are also generated by Formal logic. None of the words that are parts of the term "strict analytical reasoning" actually fit the Method of Induction; the description is misleading.

A large part of the Action on Meaning in Specialist philosophy is through its Representation. It might be supposed that there is no point in Representing No Meaning at all. That Appearance is False.

Representations empty of Meaning stand ready waiting for Meaning to be attached, and if they have none of their own then they can be attached to any Meaning at all. Thus, there is a point in doing that. It is ultimately Meaning that is Acted upon. A conclusion is at the end of an Induction and a conclusion always has particular Meaning. We recall that the conclusions from Induction are not always wrong; we just cannot distinguish those that are right from those that are wrong. The Fact that some of those conclusions will fit subsequent happenings maintains the Illusion of Justification for that Method.

So, the difference represented in the words inductivist and deductivist as described by Gale is the difference in the starting point that we have already examined. Classic inductivist logic starts with sensory experience. Classic deductivist logic begins with the generation of hypotheses in any unspecified "fashion" that presumably is not sensory experience. Yet, it is impossible to conceive of a description of physical processes that is based on nothing at all. In order to generate a description, there must be something to describe. Mathematical descriptions are Constructs and are not Concepts. While mathematical descriptions will fit some things to some degree, they do not fit Meaning or human Actions and they do not precisely fit as descriptions of any Natural thing. Natural Processes in Social Dynamics cannot be described in Quantitative Terms. That was known to the Ancients. Mathematics has enabled the production of Unnatural things. Those are now proving to be dangerous to living things and extremely difficult to dispose of.

Part of Einstein's view on theory was that the "conceptual system … is univocally coordinated with the world of experience" and is "reducible to a few basic laws from which the whole system can be developed logically". Einstein's roots were in the theoretical side of "natural science", and mostly in mathematical physics. It is understandable if he took the view that the Universe's physical processes had simple mathematical descriptions. Mathematics has long been dominant in physics. But Einstein took other views as well. Only the Process of Pure Science could produce a Conceptual System with Meaning that is univocally coordinated with the world of experience. Mathematical descriptions can be part of the description of the Universe's physical processes, but the Whole description will be different in type.

We saw the view that when there is a conflict between theory and evidence, the fit can be restored in a multiplicity of different ways. Popper included that view in his narrative. Howard (2005, 2017) cites Edouard Le Roy as pointing out in 1901 that the law of free fall could not be refuted by experiment because it functioned as a definition of "free fall" and Henri Poincaré (1901) as saying "much the same about the principles of mechanics more generally". We are told that in acting against that which was called the "conventionalism" of Poincaré and Le Roy, Einstein in the 1920's used the argument that Duhem used:

> "Yes, experiment cannot refute, say, the law of free fall by itself, but only because it is part of a larger theoretical whole that has empirical content only as a whole, and various other elements of that whole could as well be said to be, alone, immune to refutation."

So, Specialists in philosophy have suggested that theoretical Wholes are immune to refutation. But Einstein took a different view. Einstein drew us towards Relativity. Now, a Whole Theory leads to definitions and those definitions fit the Theory. Definitions then become Assertions of Truth that support the Theory and enable its development. Thus experiment can only fail to refute the theory while the theory remains unrefuted, and it remains unrefuted while definitions continue to be accepted as truths not subject to Experiment. Einstein and others saw that definitions should be subject to Experiment. Here we also have the idea that the Whole theory has empirical content that is different to the parts of the theory. The Meaning in that idea was perhaps not fully developed.

I will express some Meaning I picked up from the parts I have quoted so far. Within Pure Science, which is Natural, the benefits of theory come from the Fact that the Whole is different to the parts from which it is composed and thus the propositions that emerge from the Whole are different to those generated by the original Sense Data. New propositions generated by the Theoretical Whole are available for Testing by Experiment. Thus new Sense Data is viewed from a different angle. But in terms of Meaning, the Theoretical Whole is not larger and is smaller. Or rather, it is more compact if we define size Qualitatively. When the Meaning of Wholes is different to the Meaning in the parts from which the Whole is composed there is Speculation that can pay off. Meaning in the theory is developed, and in changed form it moves further on to different things. In short, theory can lead to novel predictions that are Tested by Experiment and the results of Experiment can change Meaning in the Theoretical Whole. That is essentially the Meaning of the Unit of the Syllogism written large. All things in the Cosmos are composed of molecules that are composed of atoms that can be split into parts. But is it Natural to split them? If and when Nature does that, the Action, the effects and the outcome are Natural.

So, Meaning is Tested in a particular way by those who engage in the Process of Pure Science. I propose that there is a condition where the fit between theory as a Whole and Sense Data cannot be restored. I will again show my Reasoning. When a particular proposition is constructed from a Theory and the prediction to which it leads fails Testing within the Process of Pure Science, then the relation built on Meaning from that part to another part also fails. That opens a gap in the Theory. The effect of the gap on the Theory as a Whole depends on the position of the part within the Theory. According to MOM, there is a Critical Position that determines Meaning throughout the Structure that the theory is. Falsification of a Theory as a Whole is possible under the condition that the Point of Reference is Tested and Fails Testing. When the part in that Critical Position fails then the Whole Theory fails, with no possibility of redemption as a Whole. The theory collapses when its foundations are removed. I presume that someone along the line from the Ancients understood that is a Fact because Points of Reference are well hidden. Premises are removed from Testing by the Assertion that they are True. The Illusion that they have been tested is created by the showing of proofs that are a Web of Complications composed of many Premises without the Quality of Truth. The Web of Complications has the False Appearance of Complexities.

When True Meaning is the Focus of Attention, theory is an Active part of a search for Knowledge. Outside the Boundaries of Pure Science, theory serves other purposes and Functions in different ways. Only the purpose of Pure Science is Justified and Justifiable. But I grant you that I am defining Justification in a particular way that fits that Process. And I am defining the Process of Pure Science in a way that fits the Concept of Justification. Definitions must be Tested for Truth and for the way they Function in relation to different things before being accepted and set.

According to MOM, Meaning and Action are Intrinsically Related. Types of thing are shown through their Qualities and those are disclosed when one type of thing is placed in conjunction with a different type of thing. The value of things is Judged by a Standard. The Standard that governs Pure Science is Functioning in the World that is created by the Forces of Nature. The Standard of Operationism is Functioning in the Artificial World which is created by the Forces of the Will of particular human beings. Different Standards have different particular Meaning and all Meaning can be Tested for its effects and the outcome as a Whole. Good Standards are products of the Action of Pure Science. Pure Science is a Force of Nature. Human beings are endowed with the Capacity for Reason. Reasoning is Action that is Innate. Thus Reasoning should be Automatic. It follows that something is interfering with Reasoning.

CONCEPTS IN RELATION TO CONSTRUCTS

Operationism restricted the Meaning of Concepts to measurement operations and measurement operations were dominated by Quantitative operations. That practice continues in Specialist psychology and in other Contexts under the name Empiricism. According to MOM, that is a Fault. Observations, Sensations and the results of Experiment can be expressed in Qualitative Terms. Their Meaning cannot be Quantified but even if it made Sense to attempt to do that in Social Studies, it makes no Sense to do that until we know the type of Meaning to be represented. It is a Fact that Sense must be made before and after a code is applied. A Code is Purely Representative. The purpose that guides Meaning and Action should be specified and if it is hidden then it must be Tracked. Once the purpose is specified, it is necessary to Test the Fit of the Representation to the Meaning found. If empirical content has Meaning then the content is Qualitative and can only be accurately represented in Qualitative terms. Natural Language has the Expressive Capacity to deal with Meaning that the Code of Numbers lacks. These Facts are Denied. Specialists in philosophy would have us Believe that Quantification is Natural and can be Universal. The difference between a Concept and a Construct is not the form of representation, nor is it a difference in the number of Meanings represented. The difference is a difference in type of Meaning. Where there is a difference in the type of Meaning, the Action that relates to it is different in type. Differences in Meaning must be Calculated. Action is visible and Meaning is not. Meaning can be Calculated from Actions.

The common definition of the word concept covers all types of abstract idea. The definition of concept from The Penguin Dictionary of Psychology is lengthy and refers to problems of "legendary difficulty" in both philosophy and cognitive psychology. In part that definition is:

concept: 1. A complex of objects all of which share some attribute(s) or property(ies). 2. The internal, psychological representation of shared attributes. Strictly speaking, the term should only be used in sense 2 since it is the mental representation that is the concept and it is the mental representation that is ultimately responsible for behaviour with regard to the outside world.........

The suggestion here is that concepts are Generalisations, which are produced by abstracting common characteristics. But that involves dropping differences from the arrangement. I propose that view is in Error. The definition recognises the Intrinsic Relation between the Structure of Cognitions and Actions on the World. But this definition does not recognise that the particular Structure of Cognitions gives rise to Meaning or that it is Meaning that leads to Action on the Outside World. It does not recognise that Human Action on the Outside World leads to the State that it is in. Nor does it recognise that Meaning leads to Action that also relates to the State of the Internal World of each human being. Instead of the word Meaning we see the term "mental representation". The entry in that dictionary for the word construct is:

construct: 1. The least confusing way to use this term is to treat it as a rough synonym for concept, at least in so far as both are basically logical or intellectual creations. Essentially one infers a construct whenever one can establish a relationship between several objects or events. In common use is the notion of a hypothetical construct, where the process is not presently observable or objectively measurable but is assumed to exist because it (hypothetically) gives rise to measurable phenomena.....2. in testing, a quality or trait........

Empiricism relies on resemblances expressed in Quantitative terms as correlations. Correlations are produced by mathematical equations and are used in other mathematical equations. Statistical formulae establish relationships between objects and events and resemblances are at the core of those Operations. In this definition of construct you see Authority to substitute Constructs for Concepts. This definition is a very confusing way to use the word construct, and probably the most confusing way. The difference between concept and construct is dropped and lost. Here we see Operationism enshrined and Active in Psychology. As we recall, Bridgman did not recognise the interpretation of his ideas by psychologists. In 1935 Stevens had enthusiastically picked up Bridgman's idea and corresponded with him. Stevens, who was a Specialist in psychology, treated concepts in a particular way and we have an example that clarifies the Meaning of that type of treatment. Ribes-Iñesta (2003) states that:

"Stevens adhered to a conception of truth by agreement."

That conception was not a Concept of Truth and was a Construct of Truth that effectively removed Truth from agreements and did not fit any aspect of any previous Concept of Truth. The Concept of Truth is an Ancient one made visible in some of the stories they told. The Construct of Truth is also an Ancient one and is shown in the same way. Narratives are a Natural Action for the expression of Meaning. Truth was constructed by some of the Ancients as a matter of convenience and many who came into existence later have followed their lead. A definition of the Meaning of truth was designed and Asserted. That definition is a Fault that leads to Errors of Judgment. We can see where that definition leads in terms of Action. Ribes-Iñesta quotes from Hardcastle (1995) one of Stevens' views on science:

> "Individuals who disagree are ipso facto excluded from the community without regard to the truth of their claims or the character of their method, which for that matter are also a matter of agreement."

I had become aware of that position during long years of experience in the same profession, but I had not perceived the Source until I began to Track Meaning for my books. It was Observation, Sensory Experience and Experiment that sent me on a Journey of Discovery. I had seen efforts to exclude ideas and the people who would present them. Individuals who disagree are ipso facto excluded from the community of Specialists in psychology without regard to the Truth of their claims or the character of their Method, but only under the condition that those in Authority are allowed to get away with doing that. Some psychologists do challenge the Authority of those who are in Power in that Realm and govern the Specialism. And, even though those who challenge the Authorities may prevent their own exclusion, they cannot enable their inclusion for that is dependent on agreeing with the peers in Authority. Those whom the community try to exclude and find they cannot exclude when the absence of Justification for so doing is shown, will find they cannot fully reach their potential within the profession unless they radically change their ways. Doing that of course limits their potential to produce new Facts and Knowledge. Curiouser and Curiouser said I.

It is a Fact that exclusions happen in Contexts other than psychology without regard to the truth of the claims or the character of the method. Peer agreement is a Powerful Operational Mechanism and part of a generalised philosophical position. It is a position that I find strange, and Evidence will Verify that this position is Not Natural. Truth is Not the Point of Reference for peer agreement. It is clear that Error is introduced to a Justice System when general acceptance in the scientific community is accepted as a Factor in deciding admissibility of evidence. But the idea of general acceptance does have the Appearance of Justification. I propose that the Error lies in the Construct. The production of Good Constructs requires the development of a Concept. Operationism puts the cart before the horse. Stevens (1936) wrote:

> "The 'true' value of a physical constant is true because physicists agree that it is true, and, if someone convinces physicists that the value is not true, it will thereafter be false. Of course each individual may think that he has his own private standards of truth. . . .The only difference is that the scientist's standards.......conform to those of his associates."

Actually, I do not think that is a difference between scientists and the rest of humanity. But some people are nonconformist. And different ways of looking at arrangements are necessary if the Discovery of New Meaning is to happen. Einstein and Bridgman struggled to be heard by those who did not want to hear different Opinions. Stevens used Bridgman as an Authority for his own interpretation of an idea and it is Stevens' version that goes by the name Operationism. In psychology, Stevens was instrumental in creating the monster that Bridgman did not recognise. But it was not an original creation. Stevens took Meaning from other Sources. As quoted by Hardcastle (1995, from Bridgman's correspondence in 1936), Bridgman commented to A.F. Bentley that:

> "[Stevens] has talked with me at length about a couple of his papers before publication and professes to be most enthusiastic for "operational ideas" .... but I simply cannot make him see that his "public science" and "other one" stuff are just plain twisted. I have also discussed with him his "basic act of discrimination without making much impression. . . (Bridgman to Bentley, 4 May 1936, Percy Bridgman Papers, Harvard University Archives, 4234.10)"

It is a Fact that some people do not use the Capacity to Reason and do not listen to the Reasoning of other agents. That Fact can be dealt with once Pure Science is engaged. But Pure Science is not engaged in Social Dynamics and Facts are not dealt with in a way that is Justifiable. Peer agreement cannot Justify Meaning. But Justified Meaning can be peer agreed and that is a different type of agreement. I picked up from a paper by Chang (2009), something Bridgman said that is particularly important and should be noted. Bridgman arrived at the following expression of Meaning:

> "When it came to constructs, "of which physics is full," Bridgman "......said..... "that one concept could correspond to many different operations,".....and.... "suggested that such multiplicity of operational meaning was "what we mean by the reality of things not given directly by experience."....... "....the measurement method can be said to be valid if it coheres with the other aspects of the concept's meaning."

Bridgman used the word Meaning and while there is still Vagueness in his Representation, in that arrangement Bridgman begins to pick up the difference in type between Constructs and Concepts. Please note that though the idea of differences in Quantity is there, the word cohere shows a different idea and that idea should be picked up. In 1959 Bridgman stated that an operational analysis was only:

> "... a particular case of an analysis in terms of activities - doings or happenings, instead of analysis in terms of objects or static abstractions, or in terms of things or static elements."

That is a Significant Statement. That does not tell us the difference between Concepts and Constructs, but it does tell us where to look. We need to use words to distinguish differences in Meaning. I would not use the word concept to cover Generalisations. Those who use Induction drop Meaning in order to produce Generalisations. The products of Induction are Not Concepts. Pure Science is Qualitative. Concepts can be produced by Natural Deduction through direct and indirect sensory experience of the Real World. The Meaning of that experience is gauged. And the Mechanism of Experiment is part of Calculating the Meaning of those readings. Natural Deduction can take and align Meaning from different Sources and Concepts can only be produced in that way. And once produced, Concepts do that in reverse – Concepts align Meaning from different Sources.

The Process of Pure Science enables access to the True Meaning of things and thus to Causal Knowledge through Qualitative Analysis of doings and happenings. Pure Science is a Qualitative Approach, with a few Quantities thrown in. Pseudo-science is a Quantitative

Approach with a few Qualities thrown in for the sake of taking their Meaning. It seems that Bridgman, as did Einstein, moved towards a view that is Holistic. But Bridgman does not in the quotations we have seen make it clear that analysis in terms of doings or happenings involves abstractions of a particular type. He rules out static abstractions and that is fair enough as far as it goes. Einstein went further and saw that theory is abstract and necessary, and that abstractions are part of a Dynamic System for the production of Knowledge. Gestalt is a familiar word for an approach that resembles the Holistic position. But that was a particular approach developed by psychologists and the Gestalt approach does not have the same Meaning as the Holistic approach - the Point of Reference is different. Gestalt psychology is Not Pure Science. Operationism did not move in a Holistic direction towards the discovery of True Meaning. Operationism is Constructive and is Not Speculative in terms of searching for Meaning.

I have yet to show you details of my Model of Meaning and since those are extensive, they will be presented in a separate narrative. But at this point I will note that MOM involves Precise Investigative Excavation of Meaning for which I use the acronym PIE. MOM's PIE involves analysis in terms of activities - doings or happenings. That is the route to the Discovery of Facts, Concepts and Knowledge. Constructs are secondary to those things. Or are they? Actually, PIE led to the development of MOM. And this narrative is a product of those. So please note that Action and Meaning are Intrinsically Related and that any attempt to rank order them in terms of precedence would be Unjustifiable. There could not be MOM without PIE or vice versa. Perhaps there can be no Concepts without Constructs and vice versa. An Intrinsic Relation is a matter of Necessity. Bridgman (1927) noted that:

> "... all sorts of phenomena cannot at the same time be treated simply." (Source Chang 2009)

A phenomenon is a doing or a happening in a particular Context. All phenomena have parts which are in particular relations and also have particular Meaning of their own as a Whole. Meaning is Not Concrete and Static and is Active and Mobile. All sorts of Meaning cannot at the same time be treated simply. But all sorts of Meaning can be treated simply at different times in different spaces. Meaning moves from that which is Complex to that which is Simple and that which is Simple moves on to become part of something that is Complex. That is Natural Action. Round and Round we go in Natural Dynamic Systems. But we do not stay in one place indefinitely within a Natural System. A Natural System is perpetually in Motion.

THE RELATION BETWEEN QUALITATIVE AND QUANTITATIVE ANALYSIS

Pure Science is a matter of Qualitative Analysis of which Quantitative measures can be a part. The order of precedence is clear, but in Pure Science precedence is not defined as rank order. The Concept of Precedence involves Necessity. Chemistry involves engagement in Pure Science. In my youth I picked up the Meaning of Pure Science in that Context. The International Union of Pure and Applied Chemistry (1995) expressed the relative status of Qualitative and Quantitative analysis as follows:

> "The general expression Qualitative Analysis [...] refers to analyses in which substances are identified or classified on the basis of their chemical or physical properties, such as chemical reactivity, solubility, molecular weight, melting point, radiative properties (emission, absorption), mass spectra, nuclear half-life, etc. Quantitative Analysis refers to analyses in which the amount or concentration of an analyte may be determined (estimated) and expressed as a numerical value in appropriate units. Qualitative Analysis may take place without Quantitative Analysis, but Quantitative Analysis requires the identification (qualification) of the analyte for which numerical estimates are given."

And there you have a clear statement of Fact. It might be presumed that the identification of Qualitative Differences takes precedence over Quantitative Measures and it does within the Process of Pure Science. It is a Fact that Qualitative Analysis is necessary before numerical estimates can be of use. I repeat that this is not about a type of precedence expressed as rank order, and is about a type of precedence that is a matter of necessity, which is a different type of order. And that different type of order is a Natural type of order. That is a Significant Distinction between Pure Science and Inductive science, so I urge you to remember that. Chemistry shows Precedence in terms of Necessity. Different types of substance must be identified before any Quantity of the substance can be estimated. Identity is Verified during Experiments. Chemistry is Pure Science and deals in terms of Reactions from which differences in type of Meaning are picked up. Differences in Meaning are visible in Reactions. Reactions are Analysable Properties. Chemical Analysis is "analysis in terms of activities - doings or happenings". Chemistry deals with the topic of Reactions and when it deals with measurement it does so in many different ways. And it is made clear that numerical values are estimates and are not precise. Physics deals with the topic of measurement in one way using mathematics. That is the wrong way to deal with Meaning and it is also wrong to describe things in only one way. Precedence in terms of Necessity is a Complex Concept and has yet to be developed.

Within the Process of Pure Science, the word Analysis represents the type of doing and happening in which Attention is focussed on Experiments in which things are placed in conjunction and on the Meaning of the Reactions. Meaning must be Calculated and we have no Concept of Calculating Meaning and only a Construct of calculation by numbers in the absence of Meaning. The Quantitative Method is analysis in terms of objects or static abstractions, or in terms of things or static elements. If you recall the activity of Painting by Numbers or the activity of Colouring a Picture that has previously been drawn, then you will grasp the difference in Quality of the product and in the type of Action that produces it.

The Inductive overlay of Formal logic in physics obscures the Perception of Meaning. Bridgman had an Intuition that a difference was there but he had not reached Insight. At least he was trying to do that and he did move us some way in that direction. His Speculations can bear fruit. Stevens was more than content to bask in Ignorance and he took others with him into the position named enlightenment where the sun shines brightly on those who lack Knowledge and do not want to discover True Meaning. It is impossible to measure things that are not previously demonstrated to exist. But it is possible to predict their existence through Analysis of doings and happenings within the Boundaries of Pure Science.

Different types of thing Function in different ways. Chang (2009) quotes Bridgman as writing:

> "Mathematics does not recognize that as the physical range increases, the fundamental concepts become hazy, and eventually cease entirely to have physical meaning, and therefore must be replaced by other concepts which are operationally quite different. For instance, the equations of motion make no distinction between the motion of a star into our galaxy from external space, and the motion of an electron about the nucleus, although physically the meaning in terms of operations of the quantities in the equations is entirely different in the two cases. The structure of our mathematics is such that we are almost forced, whether we want to or not, to talk about the inside of an electron, although physically we cannot assign any meaning to such statements. (Bridgman 1927, 63)"

It is a Fact that different things in different relations have different Meaning as Wholes. Bridgman points out that mathematics forces us into statements that do not make Sense. Here he makes a distinction between Mathematical Representation and Meaning, but he does not clarify what that difference is. He was attempting to do that through Speculation. We see misuse of the word concept, so it is perhaps not surprising that Stevens latched onto that. But Stevens did not pick up the idea that mathematics can force us into making statements with Meaning that is Not Justifiable and does not make Sense. Somebody should have done so for that Meaning was there for the taking. Stevens did pick up from Specialists in philosophy the idea that Justification is not necessary. Justification was not an Issue for Stevens. Truth was whatever Stevens wanted it to be under the condition that he could persuade enough of his peers to agree with him.

Heat and its effect of melting are commonly cited as an example of the failure of Quantitative measurement to fit happenings. That example shows us the puzzle that Bridgman was attempting to solve.

But he failed to pick up a difference that Einstein tried to pick up. Concepts do not become hazy - "basic laws are more and more simplified under the press of experience". I propose that Concepts can resolve the Appearance of Chaos into Order, and that is a Testable Proposition. Whether that is attributable to the rule of the individual mind or to the press of experience also needs to be Tested. The Action of conceptualising becomes Speculative in the face of gaps in Meaning. Perhaps Gaps in Meaning could be described as hazy but it is not fitting to do so – they just have no finite limits. But there may be limits that are not finite and are not infinite either. What is Certain is that Gaps in Meaning are Unquantifiable. Within the Boundaries of Pure Science, Constructs can be viewed as Simple Models that enable parts of Meaning in Concepts to be Tested. Constructs should not replace Concepts. Concepts have the Power to relate one Construct to another when they are in place and Active as Wholes.

Einstein criticised Mach for condemning "theory precisely at those points where its constructive-speculative character comes to light unmistakably". I would agree with Einstein and say that those who interpreted Bridgman's narratives made the same Miss-Take as Mach.

When iron melts it changes State. Changes of State do not occur sharply. But there are measurement operations that could apply to that type of change. Those measurement operations are Qualitative. It should now be clear that Qualitative Analysis involves putting things in conjunction with other things and Observing the Reactions, then expressing their Meaning in Qualitative terms. Mathematical descriptions do not fit Changes of State and the Qualities of things within States. Qualitative Analysis is the Action at the core of Speculation for Meaning.

THE EXPRESSIVE CAPACITY OF NATURAL LANGUAGE
IS A NECESSITY
FOR THE REPRESENTATION OF MEANING.

Chapter 4: IN SEARCH OF REASONING - Putting two and two together by way of Narrative

PART ONE of Chapter 4

VAGUENESS IN ACTION

I begin here with the common definition of vague:

vague: adj. 1. uncertain or ill-defined. 2. (of a person or mind) imprecise; inexact in thought, expression or understanding.

The word vague is used for Appearances and for Meaning. The difference between Meaning and its Representation is Significant and the distinction should be Marked. Things that are ill-defined in Appearance can have well-defined Meaning. And things that are ill-defined in Meaning can be given a well-defined Appearance which creates the Illusion of Precision (that is the Action in Operationism).

The fight by Specialists in philosophy against Metaphysics in general and Qualitative Analysis in particular was backed up in many ways. Vagueness was the topic for a paper published by Russell in 1923. Russell stated:

> "I propose to prove that all language is vague, and that therefore my language is vague, but I do not wish this conclusion to be one that you could derive without the help of the symbolism. I shall be as little vague as I know how to be if I am to employ the English language. You all know that I invented a special language with a view to avoiding vagueness, but unfortunately it is unsuited for public occasions. I shall, therefore, though regretfully, address you in English, and whatever vagueness is to be found in my words must be attributed to our ancestors for not having been predominantly interested in logic."

It is a Fact that some Ancients were predominantly interested in logic. And I will presume there has been a misunderstanding of Natural Language. Perhaps at the start of the Twentieth Century the prior Action to Destabilise that type of Representation was not understood. Russell took a particular approach to solving a problem that presumably he saw, based on an assumption that he knew its nature and the cause.

I will now borrow a sentence from a Dystopian novel. Those narratives are categorised as works of Fiction of course. But that is a classification of the Representation and it is a Fact that Meaning is a different matter with particular Qualities of its own. Orwell (1949) wrote in his narrative Nineteen Eighty Four:

> "It's a beautiful thing, the destruction of words."

I would not use the word satire to cover this part of his expression of Meaning. This sentence represents a happening coupled with an expression of irony. The irony lies in the use of the word beautiful to describe the Action. Irony indicates the presence of a double Meaning. Words are only parts of the target of those who specialise in Artificial Systems. In Orwell's narrative these words are put into the mouth of a Representative of those who gained benefit from the destruction of Natural Language. The destruction was a beautiful thing to them. There is overwhelming Evidence that those who designed and use Formal logics have attacked Natural Language. That is a Fact. There is no room for doubt about that. The damage sustained interferes with the use of a Powerful tool in Social Dynamics. The interference with Meaning in words and with their use is not a beautiful thing for most people or for the human species as a Whole. We need to understand the happening that Orwell represented and in order to do that we need to look at it from a different angle.

It is a Fact that Natural Language is not vague. Despite the attacks, in skilful hands Natural Language is still a Precision Instrument for the Calculation and Delivery of Meaning. The numerous narratives of novelists of high calibre are Evidence that Russell's Assertion that the English Language is vague is False. Regretful Russell may have been, but despite that he wrote many lengthy narratives using the English Language. Words convey Meaning from one place to another and he must have known that. Whatever vagueness is to be found in his words must be attributed to him. And the vagueness in his narratives carried particular Meaning to audiences. Enthymemes are Not Vague when you Focus Attention on Meaning. And that is what human beings do.

There is overwhelming Evidence that the English Language can deliver True or False Meaning precisely. And it can be Verified that the Representation of Meaning can be vague while the Meaning it delivers is precise. Natural Language is flexible and can be used in different ways. When arrangements of words show Meaning in vague terms then the vagueness is present by design. Vagueness is not a property of the tool. The users of tools determine the Qualities of the products they make. The Structure of the tool determines what you can do with it, not what is done. Russell stated:

> "In an accurate language, meaning would be a one-one relation; no word would have two meanings, and no two words would have the same meaning......... The fact that meaning is a one-many relation is the precise statement of the fact that all language is more or less vague."

Here, a Premise is stated and the Premise is False. It is a Fact, for which I will provide Evidence, that a Structure based on one-one relations does not fit the Structure of Meaning. Meaning has a Dynamic Structure composed of one-many relations. A One-Many Relation is a necessity for the expression of Meaning. Language is "more or less vague" when the conditions are set by human agents for that happening. The Evidence shows that those conditions have been set by Specialists in philosophy. The conditions can be changed and the effects can be overcome. Words do have more than one Meaning and that is not limiting. That actually increases their Potential to deliver Meaning within the Structure of Natural Language. Particular parts of the Meaning of words are drawn from them through their arrangement with other parts. It is perfectly possible for more than one Meaning to be held in a word and for only one Meaning to be in Action at a time. Intention is not an Issue in this narrative. I presume that Russell did not understand relevant Facts or how to relate them. He did not have Knowledge.

Vagueness is produced when the part of Meaning taken into Action is not specified through the Context. While the lack of specification can lead to a State of Uncertainty in the audience it can also lead to a State of Certainty in the audience when assumptions are Active. People supply their own Points of Reference when there is a gap in that position. Vagueness is designed to serve particular purposes. Russell states a Belief:

> "…that most of the problems of epistemology, in so far as they are genuine, are really problems of physics and physiology; moreover, I believe that physiology is only a complicated branch of physics. The habit of treating knowledge as something mysterious and wonderful seems to me unfortunate."

That is vague. He delivered what he promised. Russell does not specify his particular Meaning and yet the arrangement is heavy with suggestion. This is Action against Metaphysics. Physics is heavy with the Bias of Mathematics. Inductive physics is Complicated and Artificial, and physiology is not. Physiology is Complex and Natural, and that is a different State of Order. Russell's thinking was aligned with the Quantitative Method and was in tune with the thinking of other Specialist philosophers. They use Formal "logics". My thinking is not in tune with that. And I do not like to guess what people mean or have to decipher their coding system when they will not give up the code in which Meaning is Represented. But it is not necessary to guess. Using the right tools, Meaning can be drawn into view. Knowledge gives us the Power to Act effectively on the World. Knowledge is precious and like other precious things is sought and is protected once it is possessed. Knowledge is wonderful and I will gradually remove the mystery that has surrounded it. The only unfortunate thing about Knowledge is the limits humans have placed on its production and use. Specialists in philosophy do not produce Facts or Knowledge. Kelly (2008) wrote:

> "Conceptual analysis is attractive to philosophers for a number of reasons. One is that it makes sense of a good deal of philosophical practice—what George Bealer (1998) calls the standard justificatory procedure. Philosophers are always constructing thought experiments and eliciting intuitions. If this practice makes sense, then there has to be an understanding of what philosophy is that would vindicate its utility. Conceptual

> analysis is supposed to provide just what's needed here. Intuitions can be said to be of value to philosophy precisely because they help us to get clearer about our concepts, especially concepts of intrinsic philosophical interest (Justice, Knowledge, etc.)."

Sometimes True Meaning is thinly veiled. Here we see clearly that the Insoluble Problem of Justification for Inductive Methods is ignored by philosophers. The term "standard justificatory procedure" cannot be a procedure that provides Justification. It is Reasonable to ask why that term is used but I will not do so, for Specialists in philosophy do not use Reasoning and could not give a Justifiable answer. Being unable to supply the Real thing, Specialist philosophers give the Appearance of Justification to their Actions. Their "standard justificatory procedure" is composed of assumptions and assertions. The philosophical literature is full of complicated debates about concepts with particular names. If there is ever a burst of clarity it is rare and short-lived. There is not even agreement about the Meaning of the word concept or the nature of intuitions. The terms "thought experiments" and "eliciting intuitions" are used vaguely and are generalised without constraints. Though Specialist philosophers are always "constructing thought experiments and eliciting intuitions" those activities have not produced Certainty and have produced Uncertainty. Their Actions are not Experiments and are pseudo-experiments. Their Intuitions help us to become Confused about Concepts. That is a Fact. You can Verify that statement by taking samples of their narratives.

Observation reveals much Confusion of Meaning and its Representation in Specialist philosophy. If the value of their intuitions is to be judged by the effect of helping philosophers to get clearer about their concepts, then their intuitions have proved to have no value at all. Let me qualify that last statement. I see no value there for I have a particular purpose. My purpose is to enable the development of clear Concepts of Justice, Knowledge, etc. That purpose is not served by Specialists in philosophy. There would be value in their intuitions if their purpose was to interfere with such development. To do that would be Irrational. And Evidence leaves no room for doubt that their arrangements Function in that way. It is a Fact that the Premise that Formal logic is Rational is False. They set the Premise and there will be a standard justificatory procedure to deal with any challenges to that Premise. But doings and happenings can be Tested for their Meaning and for Justification. And Verification is possible under certain conditions. If the Actions of Specialist philosophers have any particular value then the type of value and who benefits can be specified. The weakness of their standard justificatory procedure is the Absence of Reasoning and the lack of that type of Action leads to the Absence of Justification. That is a Simple Fact.

Experiments in Reasoning can provide Certainty of True Meaning. Since Specialist philosophers can only vaguely express Meaning in Natural Language, I will help them out in Conceptual Analysis by Testing how their Actions Function through Tracking True Meaning in some of their arrangements. It should be clear by this stage in my narrative that the philosophical practice of Specialists does not make Sense. It is standard philosophical practice to Unsettle Meaning and to gift to audiences the Illusion of Uncertainty. As it is currently constructed, philosophy is a form of Art. But Artificial philosophy is different to Pure Art. Pure Art is Justifiable. I would include Justification as part of the definition of Pure Art. Only Art that is Justifiable is Pure Art and that type of Art is part of Natural Philosophy. Justification is a Qualitative Standard. Justification is a Concept that is Intrinsically Related to Functioning in the Natural World against which doings and happenings are judged. That which is Justifiable enables Stable Functioning within a Natural Dynamic System and of the Natural System as a Whole. Artificial philosophy is the Art of Persuasion without Justification. That type of Art can lead to Malfunctioning.

In order to understand Action in the Real World it is necessary but not sufficient to experience the Real World and to observe the experiences of others. The word empirical has been defined as the Method of Induction, so if that definition is allowed to stand Pure Science is not an empirical approach and is rooted in the Real World and Sensory experience. Pure Science is Not Inductive and is Not deductive. Pure Science is Natural Deduction used in a search for True Meaning in relation to past, present and future Sensory Experience. All the parts of Pure Science are old and Natural. In order to use a tool, it is necessary to know what type of thing it is and what can be done with it. Within the Boundaries of Pure Science, the value of Intuitions lies in their supplying Meaning for Testing.

NATURAL LANGUAGE IS NOT CONFUSING

It is a Fact that Inductive philosophy is a form of Art. Russell was an Artist. Other Artists disagree with his view of Natural Language.

> "Words can be like X-rays if you use them properly -- they'll go through anything. You read and you're pierced." (Aldous Huxley 1932 Brave New World)

The category Fiction covers different types of Art. It is a Fact that Specialist philosophy is a type of Art that is Not Pure Art. I regard it as a Fact that the Dystopian Novels are Pure Art, as are the works of Aristophanes to which I refer and the work of many other writers across the Centuries in the general category of Fiction. Evidence from Testing Meaning for Justification provides Verification of that statement, but you need definitions in order to see the Meaning of that term. I define Pure Art as a type of Art that aims to improve the condition of the audience and can enable the Perception of True Meaning. The Perception of True Meaning can enable Functioning in the Natural World. But whether it does so or not, depends on some Factors Operating in Conjunction. A condition is specified by Huxley: - Words must be used "properly" to have the specified effect. But what is the proper use of words? Natural Language can deliver Meaning precisely if you use that tool with skill. Russell did not develop the skill of delivering Meaning precisely. Or did he? And, what was the aim of his type of Art?

I propose that the Structure of Natural Language fits the Calculation and Communication of Meaning and was designed to do that. Human beings did not design Natural Language. But Natural thing grow and develop, adapting to changes in the environment. Words are the units of Natural Language and I have demonstrated that human beings have interfered with the definitions of those units. In doing so, they interfered with Meaning and its movement. But Specialists in philosophy recognise the Power of words to penetrate the minds of audiences and they use words improperly. I define improper use as using words in a way that disrupts the Calculation and Communication of True Meaning by audiences. Improper use Installs False Meaning in minds and can do so precisely. And, however they are used and whether they are written or spoken, words enter the human mind through the Senses and then have the Power to affect doings and happenings, and thus they have the Power to affect the State of Being of other things. Round and Round we go in Dynamic Systems. Words in relations carry Meaning into Structures. Operational Analysis is part of Pure Science. Action with Natural Language is part of the doings and happenings that are subject to Operational Analysis in Pure Science. Operational Analysis within the Boundaries of Pure Science involves Testing Meaning in Single Cases using a particular Qualitative Standard. And Action with Natural Language is Human Action on Meaning. We all pay Attention to Human Action.

Natural Language has been thrown into Confusion. Confusion is not a Natural State of Being, but it is a Natural State in the short term. When there is Confusion there is Disorder. According to MOM, Disorder is part of Transitioning from one Type of Order to a Different Type of Order. Disorder is an Unstable State. Confusion can be created and can be enduring but when it endures, it is maintained Artificially to serve a particular purpose. Energy must be applied if Disorder is to be maintained. It is a Principle of MOM that Disorder Naturally tends to move towards Order. Order does not tend towards permanence in Dynamic Systems but that does not contradict the Principle. For there to be different Order, the Order that exists must become Disordered. The Action of Disordering enables development in Natural Systems. And Natural Language is a Natural System. There has been a persistent misunderstanding of the Meaning and Action of the Process of Nature and its products. The Structure of Natural Language is Aligned with the Structure of Meaning. Natural Language is flexible and accommodates movement as Meaning develops. Natural Language is Not Naturally Confusing and has been made so by human beings for particular purposes. The purpose of some human beings is not the purpose of producing Knowledge and their Actions conflict with its production. Knowledge is produced through Calculating Meaning in Qualitative

Terms guided by the Search for Truth and the Standard of Justification.

So, Action with Natural Language is part of the doings and happenings that are subject to Operational Analysis in Pure Science. And the only way to discover the True Meaning of any object, event or happening is to Observe the way that object, event or happening Functions in relation to other things. True Meaning is Calculated from Observing the Reactions of things in conjunctions. That is what happens in Chemistry. Pure Science is Analysis in terms of Doings and Happenings. So, we need to Observe what is done with Natural Language and what happens in relation to that. Each Single doing and happening is a Whole composed of parts each with their own particular Meaning. Testing for True Meaning does not require repeated Observations of the same doings and happenings or of identical happenings. The Simple Fact is that no doing or happening is identical in its details. An effort to make the conditions for happenings identical in experiments is Active against the possibility of finding True Meaning. It is a Fact that there are always differences in arrangements. Details are not precisely repeated while Meaning is. Taking different views of particular doings and happening is necessary for Testing Meaning.

Those who designed the Art named science built into their Structures requirements that are known to be impossible to fulfil. And while Attention is occupied in attempting to do that which it is impossible to do, it is not turned towards doing that which is possible. The Action of Experiment is designed to discover the Meaning of differences. And that happening is not enabled by attempting to eliminate them. It is through dealing with differences that Meaning is recognised. The Action of Experiment involves varying particular details within an arrangement while not varying other parts. The conditions are controlled and the parts are Observed as they React within the Context of the Experimental arrangement. Experiments in Reasoning are conducted in that way. The Structure of Natural Language enables us to conduct Experiments in Reasoning. It is Natural to represent Meaning in different ways and Natural Language enables us to do that. The Expressive Capacity of Natural Language is the Source of its Power.

It Appears that there are no grounds on which to mount a challenge to the Misuse of Natural Language in Social Dynamics. Specialists in philosophy have Asserted that any word can represent any Meaning that agents want the word to represent and have Asserted that Truth is Not an Issue. That is a Licence for Misuse. Of course, there has to be some agreement on the Meaning represented and there is. The agreement is in terms of the direction in which Meaning is moved. Without any agreement on Meaning there would be total Confusion as we Act in the World and engage with other agents through that medium. Total Confusion is Chaos. Recall the "conception of truth by agreement". That is not a Concept and is a particular Construct that does not represent any aspect of the Concept of Truth. All Constructs can be Tested for their fit to the Concept it is claimed they Represent. But oddly enough, Specialists in philosophy and their acolytes do not claim that they search for or tell the Truth. By Formal definition the 'true' Meaning of words in dictionaries and in use is that which a community of agents agree is true. The standard justificatory procedure for that involves showing there is otherwise no such thing as Truth. That is backed up by ensuring that Truth is not told and can be overridden by agreements when it is told. Crucial questions arise concerning the Method through which agents are drawn into agreements. Answers to those questions can be found.

Interference with the use of Natural Language Functions in particular ways and serves particular purposes, and purpose is a Point of Reference that guides Human Action so we need to discover the type of purpose that is served. The Points of Reference that are Active can be discovered by Observing different Actions relative to each other. All Actions relate to other Actions and have Meaning. Popper did not accept that a single happening can be sufficient for the Perception of True Meaning – according to him the happening must be replicated. And even replication does not lead to Truth. He accepted only that a single happening can be sufficient for the Perception that Meaning is False. Popper Unsettled the prospect of Verification under all conditions and accepted the prospect of Falsification under some conditions. Other philosophers Unsettled the idea there can ever be Falsification. Thus, two Intrinsically Related parts of the Whole that is Pure Science were surgically removed in Formal Operations in conjunction with the Action of blunting the most effective tool we have for the Calculation of Meaning. That is a pincer movement and that is an effective Strategy in Warfare. Wars are waged in order to seize Power. This War against Reasoning has been waged across the Centuries. I propose that this is the War as a Whole that emerged from many different wars in Ancient times and gives rise to other wars in the times that came after.

Once Power has been seized, Attention turns to maintaining the position. Exclusion of individuals who engage in Comparative Analysis from a scientific community, removes the possibility that the Opinions of their peers will be Tested for True Meaning. Alternative propositions are a necessary part of Comparative Analysis and those are Restricted. Criticism is an expression of disagreement. Comparative Analysis is necessarily an Appraisal involving criticism using a Standard. Standards are Points of Reference and have been restricted to Quantitative Constructs. That is a Fault leading to Error in Social Dynamics. Comparative Analysis engaged in a search for True Meaning enables Verification and Falsification through Experiments. Within the Process of Pure Science, disagreement is a Stimulus for Testing Meaning for Truth. The Focus of Attention of Inductive scientists has been the topic of Human Actions and Reactions. So how much do they actually know about Normal Psychological Processes? That can be Tested.

The Action that has thrown Natural Language into Confusion is part of the Action to restrict the Operation of Pure Science. I will summarise some Facts: Every happening is unique and the differences in the details of Appearance make it so. The Meaning of things that are different in Appearance can be the same. Meaning is the part of doings and happenings that repeats systematically. Meaning is Abstracted from Single Cases and moves on. Pure Science is a Natural Process and is persistent despite its restriction.

PHILOSOPHY, SCIENCE, PHYSICS AND METAPHYSICS

An unquantifiable number of agents have been involved over the Centuries in trying to create Order in Social Dynamics where there is Disorder and only the Illusion of Order. While harmony has eluded us, there is a wealth of material on which to draw. Much of that material was classified as Metaphysics and Cast into the Wilderness or Consigned to the Flames by those who did not want to be Unified. While some agents have been involved in trying to create Order where there is Disorder, many more have been creating Disorder where they perceive Order. And as you have Observed that which is Not Unification was defined as Unification. We need to look very carefully at the definitions we have been given. I am confident that Evolution of the Natural World will take place with or without human beings and I am content with the happening of Evolution one way or another. If I can do nothing to change things, I have to be content for otherwise I will be in an Unsettled State and that type of State interferes with Functioning in relation to other things. Evolution is a Natural Phenomenon. But it might be possible to avoid Extinction. In order for human beings to evolve in a Natural way alongside other Natural things on which the existence of our species depends, they have to change their doings. People will not change how they Act if they are not motivated to change. I can and do aim to change Motivation. I cannot play you music to fit the parts of this narrative while you are reading and I am not presenting visual images in this narrative. Thus, I cannot tug at your heartstrings using those tools. If Reasoning in Natural Language is not sufficient to draw forth positive Emotions in relation to Pure Science then I will fail to motivate you to call for engagement with that Process.

Universally, changes in Structures lead to changes in doings and happenings. Changes in Cognitive Structure lead to changes in Action. I propose that we need Concepts if we are to enable Human Evolution and the continuing existence of the Human Species as part of the Natural World. That is an Ancient idea to which we should return. The Significance of Concepts was recognised by Specialists in philosophy. Margolis & Laurence (2012) note that "the nature of concepts - the kind of things concepts are - and the constraints that govern a theory of concepts have been the subject of much debate" by

Specialists in philosophy. They describe the empirical/inductive view thus:

> "Traditionally, empiricists have argued that all concepts derive from sensations. Concepts were understood to be formed from copies of sensory representations and assembled in accordance with a set of general-purpose learning rules, e.g., Hume's principles of association ……... On this view, the content of any concept must be analyzable in terms of its perceptual basis. Any purported concept that fails this test embodies a confusion."

Empiricism is Inductive. Here it is suggested that all concepts derive from Induction. The statement that Induction relies on the rules of learning does not get us far towards understanding those rules or the idea of assembling Sense Data according to them. The Mechanism of Association is at the core of Induction. Hume recognised that Fact but he does not seem to have understood that humans were endowed with the Capacity for Reason which has the Potential to regulate that Mechanism. Hume is said to have specified some principles that govern Learning by Association. Learning by Association is Active in primitive life forms. It is a Fact that Learning by Association is a Natural type of Learning that does not involve the Capacity to Reason. Association is not merely the coming together of things. The Action of Association has a Powerful effect - there is the potential for a Reaction when things come together. So, the conditions for Activating the Mechanism of Association are Simple. And the type of effect is determined by which parts are positioned together. The designer of the arrangement determines which parts are positioned together.

Margolis & Laurence then quote from Hume. This shows where the Specialist philosophical position that gained prominence leads:

> "Thus David Hume ends his Enquiry with the famous remark: When we run over libraries, persuaded of these principles, what havoc must we make? If we take in our hand any volume; of divinity or school metaphysics, for instance; let us ask, does it contain any abstract reasoning concerning quantity or number? No. Does it contain any experimental reasoning concerning

matters of fact and existence? No. Commit it then to the flames: For it can contain nothing but sophistry and illusion."

That Action and Meaning aligns with that of Logical Positivism. These remarks attributed to Hume are fallacious and divisive. They are inflammatory. That is a Fact. That remark is Incitement to Burn things. He may or may not have been speaking metaphorically but others took that Meaning literally and into Reality. Kristallnacht springs to mind as a practical example of the Meaning gifted. Hume knew these remarks are Not Justifiable. He Asserted his Opinion on what should be destroyed and what should be preserved. This famous remark contains the suggestion that any volume that contains quantity or number is not sophistry and illusion. The term "experimental reasoning" allows that some arrangements that do not include quantity or number are also not sophistry and illusion. That allowance makes room for the narratives of Specialists in philosophy. The suggestion is Testable under the condition that Meaning in the parts is specified. The term Experimental Reasoning has particular Meaning within the process of Pure Science and Hume is not referring to that type of thing. Specialists in philosophy use Induction for the arrangements they represent by the term "experimental reasoning". Induction relies on Association for its effects. And Induction by number is the Quantitative Method that Hume Asserts is not sophistry and illusion. The overarching Premise that Formal Induction is not sophistry and illusion is Testable. I propose that Premise is False. And the Premise that the Mechanism of Association is the core of Induction is Testable. I propose that Premise is True. Sophistry and Illusion are characterised by the Absence of Justification for Meaning and Action. It is a Fact that the Meaning in Illusions does not fit what actually exists or happens in the Real World. The Sophists use the Art of Deception to persuade us that the Illusions they create do fit what actually exists or happens in the Real World.

Dictionaries are far more interesting than they Appear to be. It is time to view some more definitions in conjunction:

philosophy: n. 1. use of reason and argument in seeking truth and knowledge of reality, esp. knowledge of the causes and nature of things and of the principles governing existence. 2.a. a particular system or set of rules reached by this. b. personal rule of life. [Greek: related to PHILO-, Sophia wisdom]

metaphysics: n. 1. branch of philosophy dealing with the nature of existence, truth and knowledge 2. Colloq. abstract talk; mere theory [Greek. As having followed physics in Aristotle's works]

metaphysical: adj. 1. of metaphysics 2. Coloq. excessively abstract or theoretical 3. (of esp. 17th-c. English poetry) subtle and complex in imagery

There is only one type of Meaning in the definition of philosophy - part 2 is said to be produced by the Action described in part 1. The same Meaning defines metaphysics. The reference to poetry as an example of subtle and complex imagery that is metaphysical is interesting, and certainly poets have attempted to deal with the nature of existence, truth and knowledge through their Art. But Art does not provide Evidence and so the word theory is fitting, but the word "mere" is a downgrade of Pure Art that is Not Justifiable. Theories have Powerful effects. I would drop the word mere for all types of theory. In these definitions, theory is part of the definition of metaphysics and another part is philosophy dealing with the nature of existence, truth and knowledge. Here we have a definition of a branch of philosophy as metaphysics and no other branch of philosophy is specified.

I am puzzled by the empiricists ban on a branch of philosophy dealing with the nature of existence, truth and knowledge. I presume others will be puzzled by that. That is particularly puzzling since the definition is applied to the Whole of philosophy. It Appears that Specialists in Formal logics are banishing philosophy. Yet they have retained that name. And in retaining the name they retain the Association with that particular Meaning. Thus, whatever took the place of philosophy is now defined as "dealing with the nature of existence, truth and knowledge". But I note the word wisdom is at the root of the word philosophy. The Source of the different Meaning that became attached to philosophy is indicated. The definition of wisdom will be shown and Tested for Justification in due course.

Recall the Greek root of the definition of metaphysics – "having followed physics in Aristotle's works". Aristotle is acknowledged as being the leading father of Formal logics. The Formal philosophical position followed him and thus the word metaphysics is linked by the root to all Formal logics. The fathers of Specialist philosophy had a mathematical Bias. Physics is excessively mathematical. Mathematics is excessively abstract and theoretical. It is a Fact that "abstract talk; mere theory" underlies the Quantitative Ideology. Probability Theory is the basis of Statistics. It is puzzling then that the word metaphysics is not applied to the Quantitative Method. But the fit to one particular part of the definition of metaphysics is there and plain to see. The Quantitative Construct is excessively abstract and theoretical. And that Construct does not fit the other part of the definition of metaphysics.

The Ideology of science flourished on the ground that had been cleared of all Abstract reasoning that did not concern quantity or number, or "experimental reasoning concerning matters of fact and existence". That which was planted was peer approved Inductive reasoning defined as deductive and experimental reasoning. The Evidence Verifies that Inductive reasoning cannot produce matters of Fact in relation to existence and has been Active against Experiments in Reasoning that produce Facts and for which Natural Deduction is necessary Let us look at what grew on that cleared ground. The definition of science from the Concise Oxford Dictionary is:

science: [Latin: scio know] 1. branch of knowledge involving systematized observation and experiment 2 a. knowledge so gained, or on a specific subject b. pursuit or principles of this. 3. skilful technique.

That is a definition of Inductive science as being knowledge and leading to knowledge. There is no definition of Pure Science and no recognition that there is such a thing. Inductive science does not lead to Knowledge. That is a Fact. Nothing can be Verified using Inductive science, so the definition of science is an Error of Representation given that the word represents only Inductive logic. The Proof of that is threaded through my narrative. And the Fact that Inductive science cannot lead to Knowledge is peer agreed. But there is no actual lie, there is only Deception. In that definition, science is said to be a branch of knowledge which carries the suggestion it is part of a tree of knowledge. I would cut that branch off for it is interfering with the growth of the tree of Knowledge. It is Natural to prune back dead wood and trim shoots that are growing in the wrong direction. Doing that promotes healthy growth of the Whole. But we can use the parts that are cut off in Comparative Analysis. Nothing is wasted in Natural Systems. Natural Systems are Dynamic and self-supporting.

The use of subtle and complex imagery of the type poets' use is included in the common definition of metaphysics and that definition Associates narrative art with dealing with the nature of existence, truth and knowledge. The Action of dealing is different to the Action of discovering, and is related to that. Something must be discovered before dealing can take place. And subtle and complex imagery, of which poetry is an example, is not classified as Fact and is classified as Fiction. Fiction is defined as Not True and Not Knowledge. There is a contradiction. At the risk of complicating matters further for my audience, I will point out that the Quantitative Method involves subtle and complex imagery in the form of numbers and equations. We do not perceive that to be imagery but it is True that it is. Numbers and mathematical formulae do not exist in the Real World and cannot be Sensed. That type of Imagery does not come easily to many of us; thus, we do not see it as being of the same type. Formal logics are Inductive and cannot discover the nature of existence, truth or knowledge. Formal logics deal with that which is discovered in a particular way which is different to the way Pure Art deals with it. That Meaning is True and peer agreed. And the coming into being of Artificial logic has changed the Structure of the Natural World and affected the existence of all things within it. That is a statement of Fact amidst all the Confusion.

Pure Science has been muddled with science in order to achieve Dominance for a particular group of theoreticians in Social Dynamics. I take my hat off to these clever fellows for they have succeeded where others have failed. However, I will not bow down before them. It remains to be seen if it can be Verified that their approach to Social Dynamics has brought misery to the doors of the majority of the population. That is a viable proposition and Testable.

I propose it is time for a change to Natural Philosophy. But the change would not bring misery in its wake. If it did so, there would be no Qualitative change. There has been more than enough suffering in the World and suffering will not end quickly. But none should be added as we move towards its reduction. Pure Science focusses Attention on Action that is Justifiable.

Having seen the words sophistry and illusion used by empiricists we can turn to Collins dictionary, or any common dictionary, where those words are defined for us in this way:

sophistry: 1. unsound or misleading but subtle argument or reasoning. 2. the methods of the sophists.

illusion: 1. a false idea or conception. 2. an unreal or misleading appearance or image. 3. a false perception or interpretation of what one sees. 4. a hallucination.

The Calculations presented in my narratives aim to detect unsound or misleading Representations no matter how subtle their disguise might be. All forms of Representation are an Appearance or Image of Meaning. An Illusion is an Appearance or Image of False Meaning. There are different types of Illusion. There are different Determinants of different types of Illusion. Some Illusions are created by the doings of human agents and some are not. But all types of Illusion have a particular Quality of Meaning in common – they all show False Meaning. Illusions Deceive our Senses. Empiricists claim that "abstract reasoning concerning quantity or number" together with Inductive "experimental reasoning concerning matters of fact and existence" are not sophistry and illusion. Empirical means the same as Induction - Action that is empirical is inductive. So, it is claimed that Induction is not Sophistry and does not lead to Illusion. That is a False Claim. The Evidence that their claim is Not True is detailed and overwhelming.

We have no accurate definition of Artificial philosophy - the Meaning of Natural Philosophy was taken and attached to that. Those agents who have been Active in constructing definitions have been choosy about what they will and will not define. Choices are governed by purpose. Selections Function in particular ways. This narrative aims to Mark the Meaning and Action of Pure Science. The definition associated with the word philosophy is not sufficient to define the Action and Meaning of Pure Science but will serve well as a general definition with some small adjustments to the parts. One of the adjustments I will make is to remove the Association at the root with wisdom. The Justification for doing that will become clear later in this narrative. So, at a general level I define Pure Science as:

Pure Science: use of Natural Deduction in seeking Truth and Knowledge of Reality, esp. Knowledge of the Causes and Nature of things and of the Principles governing existence, through a Flexible System of Observation and Experiment guided by questions and propositions related to them. Note: Pure Science is opposed to science for there is no common Point of Reference.

According to MOM, Natural Philosophy, Natural Psychology and Pure Science are Intrinsically Related. Pure Science does deal with the Nature of Existence, Truth and Knowledge, and I propose that the subtle and complex imagery of Pure Art is part and not the Whole of that. Simple Meaning can emerge from subtle and complex imagery expressed in Natural Language. Natural Language is a beautiful thing and its destruction is not. The same Meaning can be expressed in different forms within Natural Language. There is a One-Many relation between Meaning and its Representation. Truth is the Point of Reference that governs the Process of Pure Science and is the Mark of that type of Action. Art is part of Pure Science when the purpose of the Artist is to Represent True Meaning that has been Verified elsewhere or to Speculate in relation to Truth; then the adjective Pure can be applied to Mark its type. Pure Art does not Test the Meaning it represents for Truth and Justification. Art is the Action of Representation and a Mechanism of Communication. Anything that is Justifiable is aligned to Truth. There is potential danger in subtle and complex imagery that is set free to Act without the benefit of Justification. Justification involves Calculation of the True Meaning of things in relation to the Real World. Only Pure Science does that. Some things should not be set free to Act on other things directly or through other minds. The Ancients knew that Fact.

I can move towards the Emergence of the Meaning of Pure Science by showing you the type of Action that it is not while engaging in the type of Action that I propose it is. So Observe some Action that is not Pure Science and is said to be philosophy conducted by scientists. This is an interesting example because the protagonists were physicists and it might be thought that they knew what Pure Science is and how to Reason. Yet neither of those Intrinsically Related doings is happening in this example. This Action is Not Constrained by the Capacity to Reason. This is a splendid example of a type of Action that has a high frequency of occurrence in Politics. Politics is of course the exercise of Power. The happening described next was a fight to gain Power in a community.

A FIGHT FOR BELIEFS

It is time for a Chorus from Aristophanes. You are again invited to join in:

<u>Chorus</u>: "Words are their weapons: watch out, as the armour-clad syllables hurtle, helmeted, crested and plumed, from the lips of the Poet Most High! Wait for the clash and the din as the metaphors mingle and jumble, the sparks as the particles fly!"

I will demonstrate Action that fits Meaning in that description. According to Gale (2015), "Sometimes, philosophy drives science". I would not argue with that. I would remove the word "sometimes" from his sentence, specify the type of philosophy as Formal and specify the type of science as Inductive. Under the condition that those adjustments are made, you have a Fact. He provides an example of Formal philosophy driving Inductive science into a field of enquiry. The example is a short tale of happenings during the development of Cosmology between 1932 and 1948. Gale informs us that:

> "… explicitly philosophical considerations directed the evolution of a modern science during a crucial period of its development."

The philosophical considerations were made visible. There was a sharp division of Opinion in relation to Method as Cosmology developed. But Attention was focussed in the wrong direction. The Premise that Gale seems to accept (and I do not accept) is that inductive empiricism and hypothetico-deductivism are two opposing Methods. I included Gale's description of the differences between them in Chapter Three. These differences are Not Differences in Type of Action and Type of Meaning. Between 1932 and 1948, a Specialist Context to contain studies of the Cosmos was being given form. The main protagonists attempting to shape a particular approach to Cosmology were Milne for hypothetico-deductivism and Dingle for inductive empiricism. They were both highly respected in their field. Others, also highly respected as specialists, joined in the fray. Eddington seems to have opposed both Milne and Dingle, but there is difficulty in distinguishing which side people were on - when differences in type are dropped there is Confusion. According to Gale, battle commenced following an address to the Philosophical Association by Milne. Gale includes Milne's description of the two supposedly opposing methods, inductive empiricism vs. hypothetico-deductivism:

> "Strictly speaking, physics has no philosophy. It has method...... Now the methods of theoretical physics seem to be reducible to two species, the method of starting with concepts and the method of starting with things observed. ...When a subject is developed from concepts the concepts play the part of the terms occurring in the axioms of geometry.... The concepts are undefined save as being governed by propositions of which they are subjects."

I note that Naturally speaking there is no philosophy without Method and vice versa for Meaning is Intrinsically Related to Action. I also note that as it has been constructed Physics is Not a Natural Science. The word concept is misused when it covers Mathematical Constructs. Milne used the word concept where the word construct should be. And he did recognise that Constructs replace Concepts and are Active as definitions, without recognising that Error is introduced to Judgments by doing that. The Quantitative Constructs of Physics have, strictly speaking, No Meaning. Physicists have been heavily influenced by Specialists in Artificial philosophy. The study of those things in the Natural World to which physicists pay Attention took a Quantitative route. But it is the type of argument that took place that is of interest here. How did these men of science react to the expression of different Beliefs amongst them? Well, they did what humans commonly do under such circumstances; they fought for their Beliefs. Eddington challenged Milne's arguments. As Gale tells the tale:

> "For Eddington and his colleagues, the equivalence of gravitation and spacetime geometry was a genuine reality, a feature of the physical world, just as real as suns and moons and stars."

That is a simple statement and Meaning is clear there. Eddington and his colleagues believed that mathematical descriptions precisely fit the physical world. If so, then the Natural World was designed by a mathematician. But Evidence shows that cannot be True. What is True is that the Artificial World was designed by mathematicians and imposed on the Natural World. Natural Laws are Not Mathematical Rules. That is a Testable Proposition, as is the alternative proposition that they are – that alternative has been set as a Premise by those who use the Formal logic of mathematics. I will provide Evidence that their Premise is False. It is a Fact that there are relations between physical aspects of the world. It is a Fact that Significant Variables are ignored. Mathematics is an approximation that leads to judgments that are in Error. But the Appearance is that mathematics is precise and Nature is imprecise. That Appearance is Deceptive and the Illusion took hold of human beings. There was the brief Appearance of recognition by some of those who joined the argument that a different type of description might be required to precisely fit the physics of the Natural World. But the issue that occupied these men was where to start the search for the Mathematical Descriptions, not whether to do that or what part such descriptions might play. Gale's description continues as follows:

> "In a series of lectures given at Harvard during late '32 and published early in the new year as The Expanding Universe, Eddington denied the efficacy of both operationalism and

> hypothetico-deductivism, and not only defended explanatory realism, but strengthened his ontological position heroically: the theoretical entities of relativistic cosmology were not just plausible, they were so necessary to understanding the universe, that cosmological knowledge was essentially impossible without them (Eddington 1932, p. 19)."

As far as I can see, the term explanatory realism has the same Meaning as Operationism – the construct is the concept. The theoretical entities were mathematical constructs. Theory had gained the Certainty of Belief. The descriptions of the positions adopted in the argument are more confusing than enlightening. A total absence of Action to provide Justification has that effect. Justification is a unifying force and Meaning fragments without that. The Concept of Justification awaits development but it is a Fact that where there is No Justification, only the Force of Will is in Operation and Beliefs rule the direction in which that Force is applied. The argument that ensued is Evidence for that statement.

Before observing Dingle's challenge let us look at the Context in which Gale puts the argument. In Gale's opinion:

> "The participants are almost universally scientists, and not philosophers. Yet this does not much affect the level of philosophical thinking going on; these scientists knew their philosophy well, and they wielded philosophy's weapons and defenses with great skill."

If that is so, then we will see the best that philosophy's weapons and defences can do as we observe those with great skill in philosophy in Action. But remember that differences have been dropped routinely and must be picked up if we are to track Meaning so I will clarify the type of philosophical skill. These scientists had great skill in Inductive philosophy which is Artificial. Natural Philosophy is not the same as that and its tools are at the core of Pure Science. So, we will also see whether or not these scientists wielded the tools of Pure Science as they debated, given that you accept my definition of that Process.

> "Milne's metaphysical views were based in positivism, most especially in operationalism: only those objects whose properties could be directly revealed by some observational procedure, or operation, were to be counted among the real."………………..."Milne held that 'what you saw was what you got'; since "curved space" of various geometries couldn't be observed, space was just as it looked - Euclidean."

Clearly Stevens was not alone in his misinterpretation of Bridgman's ideas. Constructs are mistaken for Concepts in Operationism. That is known to be wrong but that which is right is not known, thus there is room for argument. It seems they had not picked up the notion of Unified science. Gale quotes Dingle's (1937) summing up of the difference between his own position and that of Milne as a question:

> "Whether the foundation of science shall be observation or invention?"

That is the wrong question. That question led them astray. Differences in type of abstract thought are the Significant Variable, not the starting point which is Trivial, and those differences are produced by different types of Action on the input to the Senses. And, the question of where the "abstract description" comes from when you say you start with the abstraction hangs in the air and remains unanswered. Abstraction disconnects Meaning from particular things, and in order to disconnect there must have been some connection in the first place. These physicists had picked up the idea of Relativity. Or rather, some of them Appeared to have done so. Gale tells us that:

> "… because of its birth within the context of Einstein's theory of relativity, the new cosmology became quickly and broadly known as Relativistic Cosmology".

There is No Evidence that they understood the Concept of Relativity. Gale also tells us that:

> ".. the so-recently won consensus about relativistic cosmology and its philosophy had dissolved into confusion and controversy over Milne and his methods".

The Formal philosophical position systematically has that effect. Those who use Induction are Combative. According to MOM, Observation and Invention are both parts of Pure Science and Intrinsically Related; sometimes one Appears to come first and sometimes the other Appears to come first. And it is recognised within MOM that Appearances are Deceptive. Observation and Invention have no strict order of precedence – there are only necessities and those fluctuate. A difference in Meaning becomes Trivial where there is an Intrinsic Relation between parts, for then Meaning is drawn into alignment. Pure Science is engaged in a search for Knowledge and the common Point of Reference for the searches draws Observation and Invention into a close relation that unifies those Actions. We have Knowledge when we find True Meaning and we find that in relation to particular happenings.

Milne held that 'what you saw was what you got'. But that which is abstract is not what you see. This form of Operationism is not empirical, since empiricism begins with Observation. But, Operationism makes concrete that which is abstract. That which was an invention and is abstract, is given concrete form and then that is what see and what you get. Empiricism begins with Observation, moves into the abstract and then returns to Action on the ground. Empiricism moves from what you see is what you get to you get what you see, after having determined in-between what it is that you will see and get. That which is given the Appearance of being Real in either form of Operationism, is not actually Real in the sense of what is Observable and Natural. It is difficult to understand the contortions. But it is easy to understand that you are viewing contortions. Contortions are acrobatics within and by a form that is static as a Whole.

Gale suggests that "talk about 'foundations' is always philosophical talk". That may Appear to be a Trivial comment but it is a Curious one. It is a Fact that Appearances are Deceptive and it is a Fact that Suggestion is a Powerful Mechanism in the Art of Persuasion. The Significance of a statement is not visible until the way that statement Acts is Calculated. Specialists are given Authority in the Context to which they devote their Attention. There are agents who specialise in philosophy. The views of Specialist philosophers take precedence over the views of other agents when talk is described as philosophical talk. If we accept the Assertion that "talk about 'foundations' is always philosophical talk" then philosophers become the Authority for the foundations of all areas of enquiry. That is a Fault. We Should Not Accept Assertions of Meaning.

Without Justification and with no possibility of there ever being Justification for the claim, Specialist philosophers have been accepted as an Authority in discussions of the theoretical foundations of every discipline. Those in Authority determine what happens in a Context. Philosophers have determined the type of foundation in every Context in Social Studies by the gift of Probability Theory and that Theory necessitates the use of the Quantitative Method of Statistics. That is Clever! That is Unscrupulous.

Now, recall Einstein's words:

"THE PHYSICIST CANNOT SIMPLY SURRENDER TO THE PHILOSOPHER THE CRITICAL CONTEMPLATION OF THE THEORETICAL FOUNDATIONS".

Then take Einstein's Meaning to heart and mind. That should have been done long ago and was not. That Meaning was Represented and was Ignored. None of us in any field of enquiry should surrender to philosophers. Yet we have done that which we should not do. The Action of Befriending is mistaken for the Action of Friends. Since those who use Formal logic claim to be philosophers and to have Authority when talk becomes philosophical, we will judge that claim on its merits through Observation of the way in which it Functions in relation to other things. Specialist philosophy as it has been constructed simply adds a particular Bias to talk about foundations. It is a Fact that the Bias of Artificial philosophy is not in the direction of Truth. How did the argument progress in the absence of the particular Bias towards Truth? Well, they did not talk in numbers, or use any other code. They necessarily used Natural Language thus accessing its Expressive Capacity. But the Fault here is not the tool. The Action is not hidden from view and Meaning is easily Calculated. The skill in Formal philosophy led to Confrontation and Disharmony as you will see. According to Dingle (1937):

> "... the first step in the study of Nature should be sense observation, no general principles being admitted which are not derived by induction therefrom.".

But, Milne (1937) had suggested that:

> "... it is, in fact, possible to derive the laws of dynamics rationally...without recourse to experience" (Milne 1937, p. 329; Source Gale).

Dingle had strong Beliefs. He thought he knew the difference between right and wrong and he could not Verify the Truth of what he knew was right. It is of course part of the Formal Construct of philosophy that there is no such thing as Verification. If Dingle knew his philosophy, then he will have realised that there is no such thing as Truth and no such thing as Proof. That alone is a Curiosity for a contradiction is present. Dingle was Certain that his position was the only true position. He had skill in philosophical talk. He had observed the practice of Unsettling in Formal philosophy and practised what he had seen modelled there. Instead of attempting to Verify and thus Justify his own position, he aimed to Falsify the opposing position. But how do you do that when Truth cannot be found and is Not an Issue? Well, the protagonists in the argument were not thoroughly versed in the teachings of the fathers of Formal logic. There is a Method in the Context of Formal logic that Dingle did not use and could have used. I will show that Method later in this narrative.

Dingle was an Authority in the area of physics in which he specialised. Beliefs can be given to others and when they are given by someone in Authority they must be received and accepted by the audience. The philosophical position is in short: When an Authority tells you that something is true then it is true by definition. That is a simple statement of the Operation of Premising. The problem for Dingle was that there was a gap in the new science where, as he saw it, he and his Belief should have been. He needed to establish the Premise and transfer his Authority from the old to the new science. He told the audience that:

> "Eddington, Milne and Dirac have got it exactly backwards. They first assume an hypothesis, then, and only then, go about collecting data. .... to [the Aristotelians'] modern representatives it seems as though a fancy is no sooner in the head than it is on paper and sent for publication" (Dingle 1937, p. 785 Source: Gale)."

Publication is one criterion by which prestige is measured in the professions and by which position is allocated there. Gale comments that:

> "cosmology is a data-poor science: observations are hard-won and rare, and they frequently must be run through elaborate theoretical manipulations and corrections in order to make sense at all."

Those who can have their fancy accepted for publication are likely to publish more frequently than those who have to first find and then work laboriously through data. The hare would win the race while the tortoise lagged behind. Dominance in the new hierarchy was dependent on the outcome of a Battle of Wills. The hare must be shackled if the tortoise is to win. As I recall the tale of The Hare and the Tortoise, in Aesop's Fable the hare was overconfident and fell asleep while basking in the sunshine. The tortoise won. It Appears that Dingle did not realise that there are particular strategies whereby the hare could be sent to sleep, along with the audience watching the race. The strategies Dingle used were arousing. Nonetheless they were part of Inductive philosophical skill. In Gale's opinion:

> "Dingle is simply wrong; it never occurred to his opponents that hypotheses would not be followed immediately by attempts at deductive prediction of observational consequences."

But it also may not have occurred to them or to Gale that deductive prediction is part of Induction. Truth is not an Issue when one has skill in philosophical talk but the perception of True Meaning is necessary for accurate prediction. Observation is necessary for the perception of True Meaning. Dingle was not "simply wrong" but he could not Represent the Meaning he perhaps felt was there. And feelings of doubt will not be accepted by men of science. If Truth had been an Issue on both sides and Pure Science had been engaged, then the course and the tone of the debate would have been different. That did not happen.

Dingle's assumptions and assertions were challenged and the argument according to Gale went as follows:

> "lively debate is far too gentlemanly a description for what now occurred. After holding his ire somewhat in check for - as he saw it - already far too long, Dingle finally erupted."

Neither Dingle nor Milne engaged in Natural Deduction during their argument. There were No Calculations of Meaning. Bear in mind that these scientists were exercising skill in philosophy as Specialists define that. It is time to chant again the Chorus from 405 BC:

Chorus: "Words are their weapons: watch out, as the armour-clad syllables hurtle, helmeted, crested and plumed, from the lips of the Poet Most High! Wait for the clash and the din as the metaphors mingle and jumble, the sparks as the particles fly!"

Gale tells us that this volley of fire in mid-1937 continued thus:

> "Dingle, his stew having finally boiled over, wrote privately to the editor of Nature, first castigating the rampant cosmological 'mysticism' passing itself off for science, and then offering to produce an article taking the sword to the mystics themselves. His offer was immediately accepted. The result was Dingle's notorious "Modern Aristotelianism", a polemical diatribe chiefly against Milne, but aimed as well at Eddington and Dirac on account of their "betrayal" of the scientific method of Newton

> and his fellow members of the Royal Society (Dingle 1937). ...The article is remarkable both for its style and for its content. Dingle's style in the article is vituperative. Thus, emotionally-loaded terms such as "paralysis of reason," "intoxication of the fancy," "'Universe' mania", and the like frequently appear, these to be topped only by references to "delusions," "traitors," and, of course, "treachery," each associated with one or more of the guilty parties (Dingle 1937, p. 786)."

Aristophanes described such conduct in the following way Twenty-Five Centuries earlier. I invite you to join in a different chorus:

Chorus - Ah how impressive the rage that burns in the heart of the Thunderer! Vainly the fangs of his rival are bared in a gesture of hate! Note how superbly he raves, with what fine independence his eyeballs in divers directions gyrate."

While I will not repeat that particular Chorus in this narrative it can be applied as a fitting description of many interactions in Social Dynamics. There are many struggles to gain Power in the absence of Justification. Aristophanes supplied a useful tool that can be used to defuse a situation if used with care. However, it does also have the potential to heap coals on the fire. People do not like True Meaning being displayed in a manner that catches Attention. In Fact, they do not like True Meaning to be displayed at all and have Acted against that for Centuries. The followers of the Ancient Sophists developed great skill in the Art of Personal Attack on any particular human who shows the True Meaning of the Actions of those who have a need to hide that Meaning. I could apply the word Satire to that chorus for Satire is acceptable by peer agreement. But I note that extremes are realised on occasions. I use Aristophanes' Representation here in order to demonstrate that during the course of Twenty-five Centuries we have learnt little about dealing with differences of view in a civilised manner and nothing about the Calculation of Meaning. There has been no perceptible change. Under Natural conditions change would be expected. Over the course of my narrative I will show you the Causal Factors that determine the Enduring Sameness and I will demonstrate the solution to the puzzle of its maintenance.

CAUSATION IS A MATTER OF PROCESS.

And I note that I take a particular view of the Meaning of the word civilised and that I would define it in a particular way as a Concept. The Concept of Civilisation awaits development. The Construct of Civilisation is in terms of material objects and technology. Operationism has Infected the Perception of Meaning. Time has not changed the Actions of human beings for there is a Warp. In a State of Frustration, Dingle resorted to Assertion, Ridicule and Rhetoric. He is quoted as writing:

> "Nor are we dealing with a mere skin disease which time itself will heal. Such ailments are familiar enough; every age has its delusions and every cause its traitors. But the danger here is radical. Our leaders themselves are bemused, so that treachery can pass unnoticed and even think itself fidelity. It is the noblest minds that are o'erthrown…the very council of the elect can violate its charter and think it is doing science service."

Whatever happened to the numbers, mathematical equations and experimental reasoning that philosophers and physicists are so fond of? Could that tirade have been adequately represented in the terms of Formal logic? Well no it could not for, as Specialists in philosophy know, it is a Fact that Artificial logics lack the Expressive Capacity of Natural Language that Meaning requires. Dingle was expressing his Emotions. Emotion and Meaning are Intrinsically Related. Dingle Acted in the old tradition of the Sophists that was dominant before the new tradition of the Sophists was developed by Plato and Aristotle. Both forms of Sophistry are still Active in the pursuit of Power in Social Dynamics. The Discovery of Causal Knowledge in the Context of Human Action and Reaction is delayed by the failure to understand the differences in type of Emotion and their Intrinsic Relation to Meaning and Actions. But that failure to understand has been engineered and is Not Universal. Much has been Verified during the study of Emotions. Different types of Emotion lead to different types of Action and vice versa. Facts are known and their knowing is Denied. It is a Fact that the Expressive Capacity of Natural Language is necessary to deliver Reasoning and it is a Fact that it was not used in that way during this argument.

Gale adds that:

> "Along with Milne, Dingle indicts Eddington, and, by implication, Dirac" as "guilty of inventing scientific hypotheses by free mental imaginings rather than by strict immersion in observations and observational data."

Unlike Einstein, Dingle did not Appear to see that "free imaginings" are useful in conjunction with immersion in sense data. Nobody picked up that theme. It is Reasonable to conclude from his Reaction that he felt threatened by the acceptance of "free imaginings" within a community of which he was a part. Dingle aimed to take away any advantage their approach gave them for advancement by equalising the ground for engagement and setting the terms. Perhaps he was not conscious of that aim but if it succeeded, that is how his argument would Function. However, strict immersion in observations and observational data is no protection against free imaginings. That is clear from his Reactions.

I note that agents who Specialise in philosophy seem to barely dip into Observational data and Appear to make no empirical observations of their own. Philosophy is full of apparently free imaginings that issue forth from these Specialists. They sit aloft in Aristophanes Clouds - they occupy the space between earth and the heavens. The problem is that Specialists in philosophy think they are so far aloft that they have become Beings in the Heavens. And the Appearance that they merely watch as others play is Deceptive.

Now recall Gale's Opinion that "these scientists knew their philosophy well, and they wielded philosophy's weapons and defences with great skill." And note that the type of thing we are observing in the interplay here described, is the abandonment by Dingle of all Reason and the giving of free rein to Emotion. Do Formal philosophy's weapons and defences draw on the Human Capacity for Emotion? The Evidence Verifies that they do. Gale's narrative points Meaning as a Whole clearly in that direction making that conclusion necessary without putting it precisely into words. That is the Structure of an Enthymeme. I will put the inevitable conclusion into words. Those with great skill in Formal philosophy do not draw on the Capacity for Reason when they disagree; they draw on the Capacity for Emotion. They were pointed in that direction by the Ancients. And, disagreement has a high frequency of occurrence. People fight and fighting takes different forms. Some forms do not have the Appearance of fighting.

The particular argument in relation to cosmology involved many other agents. Some were relatively calm and some were not. There was no consensus on what constitutes science and yet each individual knew precisely what it is not; they knew what they did not like. Gale describes a response that was published in a fifteen-page special supplement of Nature (No. 3528). Gale states that:

> "Some of the contributors were quite obviously livid with rage and other volatile emotions. Others, such as Milne himself, who had come in for particularly scathing criticism in Dingle's article, were patient and careful in rebuttal."

Yet Gale comments that while Milne suggested that he was:

> "….. amazed at the outcry that his theory and its philosophy caused, Milne here is being a bit disingenuous. In many places in his letters he not only recognizes the outcry, he delights in it, and seeks to provoke it even more (Milne 1932–37, 12 May 35; 28 Jul 36)."

Set aside the Issue of Intention, and focus on Functioning. Milne continued in the absence of Reasoning to challenge Dingle and that produced further responses in the same vein. Neither side could Verify their position or Falsify the other position. It is a Fact that Unsettling is a philosophical tool. The word "provoke" represents the deliberate raising of the emotional temperature of the opponent. I set aside Intention, so I would not use the word deliberate. But Physicists should know that there is an Intrinsic Relation between heat and disorder. Chemists Certainly Know that Fact. It is a Fact that while Emotions run high the Capacity to Reason is Disabled – thoughts are disordered. I must ask some questions at this point. Is provocation a philosophical skill? When your attempts to change the Beliefs of others to your own Beliefs fails, is reacting in frustration a philosophical skill? Are Assertion, Ridicule and Rhetoric philosophical skills? That is what is modelled in the Action described and the Premise is that philosophical skills are on display. While that is a Premise you must not dismiss it on those grounds, for some Premises are actually True. Premises are Testable and their Qualities emerge when the Potential for Testing is Realised. Whether or not it is agreed that the descriptions I have included from Gale's article are descriptions of the exercising of Formal Inductive philosophical skill, it cannot be denied that the skills described on all sides of the divide had the effect of producing frustration and led to Confrontation.

EMOTIONS WERE INDUCED.

Confrontation as described in this particular example is commonplace in Social Dynamics. Just to avoid any hint of doubt that the use of confrontation is systematic, let us observe more of the showing of philosophical skill. In 1937 many advocates of one position or another joined the philosophical fracas, and all the Action was of the same type. Gale reports that:

> "Harold Jeffreys, F.R.S., noted geologist and astronomer, and author of a well-regarded philosophy of science book Scientific Inference (1957), led off with a nice ad hominem: "Without using induction, Milne and Eddington could not order their lives for a day, and what they are really asserting is that they are entitled to use special axioms in physics, for which no need has been shown."

The book came later. Jeffreys was recognised by Specialists in philosophy as having skill in philosophical talk of the type in which they engage. Gale adds:

> "The problem, as Jeffreys sees it, originates in the perpetrators' "belief that there is some special virtue in mathematics." L.N.G. Filon, F.R.S., vice-chancellor of the University of London agrees on this point, noting that "some men of science appear to think that they can solve the whole problem of Nature by some all-inclusive mathematical intuition." R.A. Sampson, the Astronomer Royal, focusses upon the rationalistic aspects of the 'modern Aristotelians', to wit, for their "framing a theory independent of experience, such as is denounced in Dr. Dingle's article", which produces work not unlike that "of a poet or other humanist, who gives us at most a number of illustrative cases."

Filon located the problem as being the Premise that there can be an all-inclusive mathematical intuition to enable understanding of Nature. Sampson dragged the problem back to metaphysics by picking up a resemblance between those who use the Quantitative Method independently of experience and the metaphysics of those who use Natural Language. Those particular comments are Justifiable, depending on the part of definitions in use. But Mathematical intuition does not give us at most a number of illustrative cases, it gives us formulae to apply to a large number of static abstractions drawn from static elements in a large number of cases and deals with them in a particular way that is unlike that of a poet or other humanist. And it is a Fact that poets and other humanists draw on experience. Poets and other humanists frame theories dependent on experience. And theories are not necessarily True but they are necessary in order to access Truths.

Nobody so far has been able to Verify that the Method of Pure Science is different to Inductive science. Recall that Einstein was kept in the philosophers' fold despite the differences in view. There was no attempt to frustrate him and provoke confrontation. That is a shame for it is likely he would have been more than their equal if challenged. The difference in view is important. I propose that there is no special virtue in the Quantitative Method. Yet, one particular all-inclusive mathematical intuition is now governing Social Dynamics in our World and is set to govern all Human Actions beyond that if the men and women of Inductive science can get to other planets. And they cannot do that unless they ride on the products of Pure Science. Probability Theory is one product of Inductive science and is unlike the work of a poet or other humanist in that there is no humanity there. That statement is Testable and will be Tested within this narrative. I use the word humanity in the sense of morality. And as I have demonstrated here, Single Cases are particular Contexts that are illustrative of Meaning through doings and happenings.

I will not Test Probability Theory for Justification in this particular narrative, but I have Tested that and I will show you the Tests in a different narrative. Here I will demonstrate there is a common lack of humanity where Formal logic is used. The work of Aristotle and those who follow him does not have a Moral Dimension and the work of poets and humanists does have that Dimension. I note the Premise that a number of Single Cases is not sufficient to lead us towards solving the "whole problem of Nature". And I note that Nature is Not Actually the Problem and that understanding it is. I agree that if Single Cases were at most illustrative then they would not lead to solutions but it is a Fact that they are not at most that. I propose that the Calculation of Meaning requires Single Cases and does not require large numbers of them in order to gain Causal Knowledge. That proposition is Tested in this narrative and its Truth is Verified. I do not claim that my narratives are valid for they are not that. I suspect that there is No Licence to tell the Truth, the Whole Truth and nothing but the Truth outside Court. And as we have seen, there is Verification in a number of Single Cases that Truth is not invariably told or disclosed by testing there. It is not possible to claim that Meaning is True in any Context while Truth is defined as that which peers agree that it is. I do not make that Claim to Truth.

The Illusion that there is No Method of Testing Meaning for Justification and Truth not only allows huge amounts of Error to enter the conduct of Social Interactions, it also allows firing of the Emotions and Error is Intrinsically Related to the Firing of Emotions. The Illusion that Meaning cannot be Tested and Verified is Active upon us all. Hume accepted that there can be no Justification for Induction and it is commonly known that both empiricism and hypothetico-deductivism are of that type. Since those who knew their philosophy well will have known that they could not solve the problem of Justification, it is unlikely that these scientists would have searched for a quality of Meaning that they knew their Opinions did not have. Or at least they should have known that, but perhaps all they knew was that they could not show the Justification that they believed was there. In the absence of Justification, there is necessarily resort to Force of a different type. It could be said that Hume's statement of an Insoluble Problem for Induction is part of philosophical Action against Pure Science, for if there is one thing that philosophers do know it is that Meaning once taken is generalised. Hume Certainly knew the Power of the Mechanism of Association. That which is an Insoluble Problem only for Induction becomes a Universal Insoluble Problem when nothing but Induction is put in the field of view and deduction is in Association with and Controlled by that. That which is wrong and is open to criticism cannot be replaced while the Illusion that no alternative exists remains Active. That Fact is known. The number of people who know these Facts is Unquantifiable and we cannot identify who they are.

There are no Testable Propositions in view in the philosophical argument between "scientists" as described by Gale. Yet, the received view in Formal philosophy is that even if the protagonists had formulated propositions in testable form and tested them, there could still be No Certainty. The Adversarial Process is accepted for testing arguments and is aimed at Unsettling them. Are we endlessly doomed, sometimes in a state of high Emotion, sometimes calmly and serenely, to hurling insults back and forth when we engage in argument? Do we actually Believe that is wielding philosophy's weapons and defences with great skill? Agents Act on the basis of their Beliefs. My Observations indicate that people do Believe that the hurling of insults back and forth is both a skill and necessary during engagements with other agents. I have shown you the Causal Process that leads to that Belief. So, while I find that to be a Strange Belief, I am no longer puzzled by its presence. It is puzzling though that these eminent people who had experience in the Sciences of the Natural World did not pick up the Virtue of Experiment and apply that Action in the exchanges between them. Do you now see that Formal philosophy draws all parts into an Unstable State irrespective of their experience, and that it necessarily leads to Instability in Social

Dynamics? I am sure you will require more Evidence before you accept that as being the True Meaning of their Artificial Structures.

Milne's work led to a proposition called the Cosmological Principle. Later, Bondi entered the debate and, working with others, suggested a correction to that Principle. The corrected version was called the Perfect Cosmological Principle. Gale tells us that:

> "In 1948, a young mathematician, Herman Bondi, in concert with two close friends Thomas Gold and Fred Hoyle, proposed a radical new cosmological theory, the Steady State theory................ Bondi came to his notion of the steady state primarily from his commitment to the philosophical components of Milne's work, most especially the methodology of rationalism plus hypothetico-deductivism; additionally, Bondi coupled to these Milnean notions some ideas taken directly from the philosophy of Karl Popper."

Bondi took from Popper the notion of Falsification and this example demonstrates what that notion enables.

> "...... according to Bondi, deductivism can be a scientific approach in cosmology only if its postulates (or axioms) are candidates for disproof."

According to Gale, Bondi tied "a cosmological theory to Popper's falsificationist principle as a guarantee of scientific acceptability". Gale tells us Bondi made that Action explicit in 1990. He quotes Bondi as stating:

> ".. 'I think the person from whom we had most help on the philosophical side was Popper. His analysis of science encouraged one to be imaginative, and encouraged one to go for something that was very rigid and therefore empirically disprovable.' (Bondi 1990, p. 194) 'For the correct argument has always been that the steady state model was the one that could be disproved most easily by observation, therefore, it should take precedence over other less disprovable ones until it has been disproved.' (Bondi and Kilmister 1959, p. 55–6)"

I would say that there is a contradiction in terms here. Something that is very rigid is not something that is imaginative. And to use Popper's falsification principle as a guarantee of scientific acceptability is Misuse of a Principle of Pure Science. So, is that what Falsification was kept in place to do? It really does depend on what you mean by falsification and the Meaning is determined by the Context of use. Within the Process of Pure Science, Falsification does Not Function in the way described by Bondi. That is how the criterion Functions in Inductive science. Meaning is different when the Point of Reference is different. Giving weight to one proposition relative to another simply because it has the Potential for being Tested at some point in the future is Not Justifiable and is Non-Sense. The Steady State theory to which Bondi contributed was Falsified in 1965 and dropped from the range of possibilities. That is not a problem. But, until then, a False Theory was given weight over others that were possible. That is a problem. The False theory became fashionable. I repeat the words of a Judge with whom I agree on this point: "Any tendency to dogmatise should be met with an answering challenge". I do not agree because he is a Judge, I agree because there is Evidence that Verifies the conclusion he showed.

The argument about Observation and free invention necessarily lingers on and will continue to do so while Reasoning is Disengaged. Operational Constructs are technological developments. Within the Process of Pure Science Operational Constructs enable Conceptual developments. Pure Science and Technology work hand in hand when they are governed by the same Point of Reference. Technology is the Action of Modelling Meaning from Concepts. Technology serves a different purpose when taken outside the Context of Pure Science.

Gale states:

> "Kinematics, in Einstein's Special Theory, is especially concerned with observers trying to make measurements of the behavior of objects moving in systems relative to themselves. According to Milne's understanding, these measurements could only be made by the observers signalling among themselves

> using light or other electromagnetic radiation, using clocks to time the signals. In the end, all measurement - even space and velocity - reduced to timing the signals as they went from one observer to another. This idea of measurement represented a revolutionary simplification of previous ideas, and held philosophical interest for that reason alone. There were practical consequences as well. As Bondi later remarked, "Milne's idea led straightaway to the radar speed gun!"

This idea of timing the sending of signals from one observer to another was not revolutionary in the Context of Social Dynamics and was already Simple there. Poetry and all forms of Pure Art use that Natural Principle in delivering Meaning. Those who use the Art of Deception have also known that Fact across the Centuries. Signals can be delayed or speeded up and differences in the timing of stimuli and responses produce different effects and outcomes. The Ancients knew that Fact. And it is False to suggest that "in the end, all measurement - even space and velocity - (is) reduced to timing the signals as they (go) from one observer to another". The measurement of Meaning does not reduce to that. And, Technological developments have particular Meaning and Meaning can be Calculated. A failure to Calculate the Meaning of Technology in relation to other things is a Fault with huge Potential for Error, and a failure to respond to the Meaning Calculated Compounds the Error. There is Failure to Respond to Meaning. The Potential Error is not yet fully Realised and the point at which that will happen cannot be represented in numbers.

Training in Formal logic Interferes with Reasoning and that Interference is Active when those who have acquired skill in Formal logic do try to deal with the nature of existence, truth and knowledge. Thus, there is Confusion, and even when Common-Sense is shown their peers can Act against it with the tools provided by the fathers of Formal logic. The Ancients designed tools that are Powerful Disruptors of Meaning and Powerful Installers of Meaning.

PART TWO of Chapter four

ANALOGY AND REASONING

The word analogy is familiar to us all. We all use analogies. Or at least we think we do. We have Observed that different types of Meaning and Action are covered by single words. So what does that word mean? The definition of analogy in the Pocket Oxford Dictionary is:

analogy: [Greek analogia proportion] 1. correspondence; partial similarity 2. arguing or reasoning from parallel cases

There is a gap between lines that run parallel. The common definition of the word parallel is:

parallel: 1a. (of lines or planes) continuously side by side and equidistant……2. (of circumstances etc.) precisely similar, analogous, or corresponding 3a. (of processes) occurring or performed simultaneously……….

Now part 3 a. would be part of the definition of Analogy within the Process of Pure Science and things that occur or are performed simultaneously have effects that relate to each other. That which is parallel becomes united. But that part is here being associated with parts that are not within the Process of Pure Science. Within Induction that which is parallel remains separate indefinitely. Meaning is being transferred without Justification. The definition of analogy given in Collins Dictionary has more parts than that in the Pocket Oxford Dictionary:

analogy: 1. similarity in some respects; partial resemblance. 2. a comparing of something point by point with something similar. 3. Biol. similarity in function but not in origin and structure. 4. Linguis. the formation of new words based on the pattern of older (often unrelated) ones……. 5. Logic. the inference that because one idea, thing, etc. resembles another in certain respects it does so in other respects.

Part 5 describes only the Action of analogy in Formal logic. Formal logic is Induction and we can recall the difficulty of defining Induction. Induction deals Meaning through resemblances. A resemblance is a similarity. Now we see that in most parts of common definitions analogy is defined as dealing Meaning through resemblances. When it does, analogical reasoning is Inductive reasoning. The Truth of inferences reached by Induction is Uncertain. The Truth of Inductive analogies is Uncertain. And Inductive analogies have the Insoluble Problem of Justification. At part 3 we have a reference to Biology and the word similarity is there too. Biology is categorised as a Natural Science and here it is associated with the Action of Induction for that is the only type of Action described in other parts. Where Pure Science is engaged the Action is Not Inductive and Attention is focussed on differences and their Meaning while commonalities and their Meaning are taken into account. Analogies are not based on similarities in the Context of Pure Science. Attention in that Context is focussed on Functioning. Origin and Structure are related to Functioning. And I note we can clearly see that there is a systematic regularity of the Association of that which is Unjustifiable with that which is Justifiable in definitions.

Inductive analogies are described in some detail in the philosophical literature. An article with the title "Analogy and Analogical Reasoning" by Bartha (2013) begins thus:

> "An analogy is a comparison between two objects, or systems of objects that highlights respects in which they are thought to be similar."

A highlight gives a Contrast. When similarities are drawn to Attention, differences fade into the background. He continues:

> "Analogical reasoning is any type of thinking that relies upon an analogy. An analogical argument is an explicit representation of a form of analogical reasoning that cites accepted similarities between two systems to support the conclusion that some further similarity exists. In general (but not always), such arguments belong in the category of inductive reasoning, since their

> conclusions do not follow with certainty but are only supported with varying degrees of strength. Here, 'inductive reasoning' is used in a broad sense that includes all inferential processes that 'expand knowledge in the face of uncertainty' ……., including abductive inference."

Actually, Abduction contracts Meaning in the face of Uncertainty and does not expand knowledge. That is a Fact. It is peer agreed that Induction does not lead to Knowledge: – Verification is not possible; "conclusions do not follow with certainty but are only supported with varying degrees of strength". And yet it is peer agreed to use the word knowledge in the Context of Induction. There is agreement among peers in Specialist philosophy to misuse many words. Bartha indicates that analogy does not always belong in the category of Inductive reasoning so there is a suggestion that the conclusions of a different type of analogy follow with Certainty. But the idea that there is a different type of analogy is dropped in Specialist philosophy.

Reasoning based on partial similarities is a tool in the Art of Persuasion. Aristotle was a prolific developer of tools for use in persuading audiences to take particular points of view. Aristotle was one of the founders of Specialist philosophy and thus of Inductive science. Bartha tells us that:

> "The argument from example (paradeigma) is described in the Rhetoric and the Prior Analytics: Enthymemes based upon example are those which proceed from one or more similar cases, arrive at a general proposition, and then argue deductively to a particular inference. (Rhetoric 1402b15)"

That looks like a description of the Method of Induction in general and Abduction in particular. In this case the Appearance fits the Action precisely and Meaning is the same. Inductive analogies are parts in Enthymemes. Bartha continues:

> "Aristotle, in common with just about everyone else who has written about analogical reasoning, organizes his analysis of the argument form around overall similarity…… horizontal

> relationships drive the reasoning: the greater the overall similarity between the two domains, the stronger the analogical argument. Hume makes the same point, though stated negatively, in his Dialogues Concerning Natural Religion: Wherever you depart, in the least, from the similarity of the cases, you diminish proportionably the evidence; and may at last bring it to a very weak analogy, which is confessedly liable to error and uncertainty….." Most theories of analogy agree with Aristotle and Hume on this general point. Disagreement relates to the appropriate way of measuring overall similarity."

Just about everyone else who has written about analogical reasoning is aligned with Artificial philosophy, or is brought into alignment with that. The Greek root of the word analogy is proportion. The weakness of the similarity is said to affect the strength of the evidence. Actually, in Induction the similarity is the evidence. That is a Fact. The strength of the similarity cannot be measured quantitatively, so mathematical proportions cannot be obtained for the strength of similarity which is the same as saying that mathematical proportions cannot be obtained for the strength of evidence. Yet that is routinely done by Inductive scientists. The underlying Premises that support the Action in science are False. Inductive science is an Illusion of Science. And in an Inductive argument the number of parts identified as being similar has a Qualitative effect on the Judgment of similarity as a Whole. Recall the word numerosity human beings and other active organisms can make quite subtle distinctions of quantities without using the Construct of Numbers. The Construct of Quantity by Number replaced the Concept of Numerosity. That is Operationism which is known to be wrong. That Construct does not capture other aspects of Meaning in the Concept or align with their Meaning as a Whole.

Frequencies in terms of numbers dominate Formal Inductive inferences. A conclusion based on identifying points of similarity between things and counting the number of those should not increase the strength of the conclusion to which the argument leads, but that Error systematically occurs. Using mathematical equations does not change that Fact. Those Miss-takes in Judgment are made. And crucially, similarity is not a measure of Meaning. Meaning is

Qualitative and Relative, and can only be Calculated using Reasoning. Significance cannot be measured Quantitatively – one tiny difference in detail can be Highly Significant to Meaning.

Bartha describes the general Structure of Inductive reasoning by analogy using:

> " …… a simple, quasi-formal characterization (versions may be found in most elementary critical thinking texts, e.g., Copi and Cohen 2005). An analogical argument has the following form:
>
> 1. S is similar to T in certain (known) respects.
> 2. S has some further feature Q.
> 3. Therefore, T also has the feature Q, or some feature Q* similar to Q.
>
> (1) and (2) are premises. (3) is the conclusion of the argument. The argument form is inductive; the conclusion is not guaranteed to follow from the premises."

But 1 and 2 do not have the Appearance of Premises. They do not state an assumption and assert its truth. However as you will see in due course the Structure is one that depends on underlying Premises, and it is a Structure on which Premises can be hung. The Structure provides the hinge for drawing conclusions from Asserted truths and Authority for using the conclusions as further Assertions. The Structure looks similar in Appearance to the Structure of a Syllogism and the Action is different. The Inductive Structure is a pseudo-syllogism - it resembles a Syllogism to some degree. And the conclusion of an argument by Inductive analogy has the Potential to become a Premise. This Structure is not the deductive part of Induction. Here we see the Action of generalisation by Abstraction from Sense Data of features in common that leads to input for the deductive part of the Process of Artificial Induction. This describes a unit of Induction from which arguments can be built. The Appearance of Justification for Induction is given by applying part of the Structure of Natural Deduction to Inductions.

Clearly the elementary texts on "critical thinking" do not teach thinking that is critical if this example is illustrative of their content.

This is Sophistry and Illusion. Should we consign those texts to the flames as Hume suggested? If the parts concern quantity or number that would not make a difference in their type. But it would be foolish to destroy them when we have a clear example of what Reasoning Is Not, for we can use that well. We perceive differences in Meaning relatively. In order to see differences in Meaning it is necessary to have things to compare. Perhaps Hume knew that Fact and perhaps he did not. We cannot question him now in a search for his Intention. We will never be sure of that. And we do not need to know Intention in order to Calculate the Meaning of Human Actions.

Natural Deduction deals with Meaning and we know that. The Premise is that Induction does not deal with Meaning. Attention is focussed on the Representation. But, the Meaning of a similarity is the hidden Premise from which Inductive arguments are grown. The hidden Premise is: Things that are similar in Appearance have the same Meaning. That Premise is False. It is differences that are Significant in terms of Reactions, and Meaning Is Disclosed by Reactions. Premises lead to Errors of Judgment and in this case, it is likely that the Error in analogical argument by Induction would be seen if that hidden Premise was seen. Then someone would have Calculated its Quality and discovered that Illusion is built upon it.

The word analogy has vague Meaning in the Context of Induction. Bartha shows that:

> "Despite the confidence with which particular analogical arguments are advanced, nobody has ever formulated an acceptable rule, or set of rules, for valid analogical inferences. There is not even a plausible candidate."

Is that True? Actually, it is Not True that a single acceptable rule that applies to all types of Inductive inference has never been formulated and agreed. There is a rule that is not specified. The rule is: That which is valid is whatever you Assert as being true, given that you have the Power to do that. That is a rule that removes rules for those who use Assertion and the Art of Persuasion. Where there is one rule there is another. The Power to Assert what will be true comes from the development of skill in using the Art of Persuasion. The Art of Persuasion is necessary to draw other human beings into agreements when Justification is Impossible. It is a Formal rule that the Art of Persuasion is used to induce agreement and it is a Formal rule that Power can only be gained using the Art of Persuasion. So, there is a set of rules for Induction and analogical inferences are part of putting those rules into Action. Another accepted rule is that any similarity that is plausible is a candidate to enable the transfer of Meaning between parts. But I grant that, though these are rules that are accepted they are not acceptable, so in that sense Bartha is right. Perhaps that is why they are hidden. Curiouser and Curiouser said I as I felt myself drawn into tumbling down tunnels. There are sides to every tunnel of Meaning but they are not formed as we understand that Structure in its physical form.

But an acceptable set of rules has been formulated and specified for Inductive science and has been accepted by agreement. Numerical Correlations represent similarities in the form of numbers. Statistical Correlations are Quantitative measures of similarity. Inductive analogies are part of Statistical Inferences. The use of similarities is hidden by coding, but Statistical Correlations are expressions of those. Mathematical equations are valid rules for deriving conclusions from correlations. The word valid signifies that there is peer agreement to draw conclusions in that way. The tables of Significance to which scientists refer once they have their numerical correlations are valid and are based on valid equations. Significance is determined merely by peer agreement and the Standard for Judgments of Significance is Quantitative. Tables of Significance give the False Appearance that Quantitative Rules are True and Real. The Quantitative Rules have the False Appearance of Justification and the Illusion is supported by the absence of any plausible candidate with Justification. When there is no other type of candidate to vote for, the type of conclusion is predetermined and only its form is variable. There is an Illusion of Choice when one type of thing is presented in different forms at a point of Selection.

The resemblance to deduction at some point within Induction creates the Illusion of Calculating Meaning. The conclusions that are reached using Statistical Significance are Unstable. The lack of Justification leads to Judgments with that Quality. And at the end of the manipulation of data in the code of numbers and the weighting of the answer by tables of significance, the Meaning of the conclusion is expressed in Natural Language. Few of us can comprehend the Structure that led to the conclusions. Up to the present day there has been no alternative to acceptance of the conclusions that mathematicians help us to reach. The procedure rules are peer agreed and fixed. We should be careful before we accept help from other people. That which has the Appearance of Help may be an Illusion. But we do need help so do not be disheartened. Help can be Truly Given. All we have to do is to ensure that Help is Truly Given. Well I say that is all that needs to be done, but we have to first overcome Resistance to That Doing.

So, Bartha's Premise taken as a proposition has been Falsified. The rules of Formal logics have been accepted by peer agreement even though there are no rules that can be Justified and not even a plausible candidate for a Justifiable Rule. It is a Fact that Meaning does not run parallel in the sense of running continuously equidistant. Meaning is Represented in Qualitative terms and converges to a point that is also Qualitative. There is one Meaning that emerges here. Lawlessness prevails and is hidden by the False Appearance of Rules that Appear to constrain Action where there are No Real Constraints. And that which is hidden has been found by many as they tested for the limits of Meaning through Observation of Reactions in Single Happenings. Once it is discovered that there are no Real Constraints on Actions then those who discover that Meaning will Act in accordance with the Fact that they will not be stopped from doing whatever they want to do. And many in the audience have discovered that.

But you might respond by pointing out that our prisons are full, and it is True that they are full to overflowing. They are full of those who visibly went too far and were caught doing so. Having perceived no stopping point they did not stop and their Disorderly Actions inevitably increased in strength and frequency to a point where Attention was drawn to them. And within that group dealt with by the Courts are an Unquantifiable number of those who did not go too far but otherwise drew Attention and were caught in the net. There are limits to the lawlessness that can be tolerated even by those in Power for the Appearance of Rule must be maintained. The Courts are there to do that. But with care those limits of tolerance are never reached. It is a Fact that the lawlessness of clever fellows is not stopped in its tracks, and if it does draw Attention the consequences are relatively minor for them. Clever fellows are perceived to be of good character for that is how they have agreed to define themselves. Definitions must be Tested for Truth.

The Action of Analogy in Pure Science has been muddled with the use of analogy in Induction. I need to pick up and Mark the difference in type. Bartha (2013) quotes Priestley:

> "According to Joseph Priestley, a pioneer in chemistry and electricity, analogy is our best guide in all philosophical investigations; and all discoveries, which were not made by mere accident, have been made by the help of it….."

The term "philosophical investigations" is restricted to Action of a Formal kind. Specialist philosophy is Inductive. Artificial logic does not aid searches for Meaning and does not lead to discoveries except by accident. If Priestly did mean that Inductive analogy is our best guide in searching for Meaning then he was wrong. But Priestly was a pioneer in Chemistry and electricity and that leads me to the conclusion that he was talking about a different Method. The Differences were Masked by Specialists in philosophy and the similarities used to suit the purpose of those who rely on Induction and make No Discoveries themselves. The difference between Qualitative and Quantitative Analysis is drawn to Attention in Chemistry. Priestley will have seen that and he will have understood the distinction. Those who advocate the use of Induction drop differences in Meaning and seek shelter under a single cover. Then the original type of Meaning is kicked out, or that type of Meaning is enslaved by chaining it to a new leader. That Action has been systematically repeated throughout history. We have failed to understand the Meaning of that type of Action. But the effects are visible and the outcome of that Systematic Error is approaching with increasing pace. There is a need to enable the Calculation of Meaning through Activating the Process of Pure Science.

There is no clear line dividing the Operational and the Conceptual limbs of Pure Science and there is no need to make a distinction between them when they are of the same Body for then they Function as One. But when technology is severed from the body of Pure

Science and taken for use in Artificial Systems without the Constraints of the Central Controller, it's Meaning changes in type.

I will define Analogy in Pure Science. Within the Boundaries of Pure Science, the tool of Analogy is part of Reasoning and the type of Action represented by the word Analogy in that Context can be specified. When the word Analogy is defined as the Action of Modelling Meaning through the creation of different arrangements to represent it, then Priestley would have been right to say that Analogy is our best guide. Meaning can be Tested through the Action of Modelling. Models are Meaning made Operational. Thus, we can see the appeal of Operationism and why it was so eagerly accepted. Operationism has the Appearance of being that which Pure Scientists do. But the Appearance of Operationism is Deceptive. The Appearance is an Illusion of Science. Operationism is not fit for the purpose of discovering Knowledge. Nor is it designed to do so, as Observation of the statements of Stevens makes clear. The Action and Meaning of Analogy in Pure Science is different in type to that in Inductive analogies. There is some partial resemblance and no more than that.

All discoveries that were not made by mere accident have been made with the help of Experiments and Models are part of the Action of Experiments. Analogies within Pure Science are Models of Meaning and are part of the Calculation and Communication of Meaning. In Induction, the model is said to be that which it is supposed to be a model of, and imagination plays a large part in the supposition. The Calculation of Meaning through Experiment is a part that is dropped by Inductive scientists. The Mechanism of Experiment is a necessary part of the Action in the Process of Pure Science and is threaded throughout that Process. That Fact is known and the word experiment is misused to describe Action of a different type outside the Context of Pure Science. Misuse and abuse are enabled by presenting a similar Appearance to that which is good use, thus hiding differences in Meaning. You have Observed that it is valid to hide differences in Meaning. It does not matter if the word analogy is used for different types of Meaning under the condition that the type of Meaning in use is made visible by Marking that difference. That is not done and the failure to do that is a Fault that leads to Errors in the Perception of Meaning and in the Judgments that follow them. Meaning is Relative. In order to do right we have to know what doing wrong is.

WE LEARN THROUGH ANALYSIS OF SINGLE CASES

In order to distinguish differences in type of Meaning it is necessary to see things in Action. I will use particular examples from a paper with the title "Precedent and Analogy in Legal Reasoning" by Lamond (first published 2006). He stated that:

> "An analogical argument in legal reasoning is an argument that a case should be treated in a certain way because that is the way a similar case has been treated."

An "analogical" argument that Single Cases should be treated in a certain way because they are similar is the Method of Argument by Induction. Induction drops differences. In Case Law differences are picked up. In Case Law, Single Cases are the Focus of Attention and Natural Deduction should have the freedom to Act on the details in the Case. I note that Freedom is of course Relative. Now, the freedom that the Structure of Natural Deduction gives in working with Meaning is not present when deduction is in harness as part of Inductive Action. And I note that Natural Deduction is constrained by Rules and that those Rules give the freedom to Act. That Complex Relation between Freedom and Control must be taken into account when the Concept of Freedom is developed in the Context of Relativity. I have shown that the Action of Natural Deduction can be restricted by determining the type of thing it Acts upon. I have shown that only one type of deduction is Marked, and that is Inductive deduction. Meaning cannot be Calculated through the Inductive Method. Meaning can be Calculated using Natural Deduction. The Freedom to Calculate Meaning is one to which there should be a Right. That Right was developed in Nature.

Lamond mentions in his narrative a difference between the Action in the Courts and the Action in Formal logics but does not pick it up as a difference. That potentially Significant difference is made using the Concept of Justification as a Standard. Justification is a Qualitative Variable. I say it is potentially Significant because the Potential for Justification is hard to realise and thus underachieved. The difference is present when Justification is sought by Judges in the Courts. A Presumption that Justification is sought by Judges in the Courts is Reasonable. And it is clear from the work of many people, including Garrett, that the Appearance of Justification can be strong enough to be accepted in Court in place of the Real Thing. In an Unquantifiable number of Single Cases, the search for Justification is not rewarded by its finding. Induction has the False Appearance of Deduction and is Miss-Taken for Reasoning. But Judges are astute and Induction is not easily mistaken for Reasoning. Considerable effort has been expended in producing an Illusion that will be accepted by audiences. Those who put in most of the effort have long departed from Earth.

The Natural Mechanism of Transferring Meaning by the Action of Association is a Simple one known to the Ancients. That Natural Mechanism has been misused in Social Engineering across the generations for particular purposes. Inductive analogies are used to transfer Meaning from one thing to another and the Transfers of Meaning by Artificial Induction are Not Justifiable. Justification is an Insoluble Problem for Induction but giving that which is Unjustifiable the Appearance of Justification is not a problem at all. Hume knew the Power of Association and so do all those who use Formal logics, and they rely on that Mechanism to gain Power for their Method. Lamond's published narrative is part of the effort to associate Induction with Justification and thus to facilitate the flow of the Meaning of Justification from that which is perceived to be Justified in the direction of that which is Not Justified and never can be Justified. That is misuse of the Mechanism. He is merely doing what he has been trained to do. We commonly do what we have been trained to do. Justification is an integral part of a particular type of Standard that is Marked by the word Moral. The Meaning of the word moral has suffered from Interference. That conclusion is derived from Testing Meaning and will be Justified over the course of this narrative. In other words, I will provide Evidence for that statement in due course. That will take the Meaning of the Concept of Moral towards Truth as far as I aim to go in this narrative.

Lamond's argument leads Meaning in a particular direction. The direction is clear and not specified. The Meaning you pick up depends on the angle from which you view the arrangement. He distinguishes between precedent and analogy.

> "Precedent involves an earlier decision being followed in a later case because both cases are the same. Analogy involves an earlier decision being followed in a later case because the later case is similar to the earlier one."

That statement is in Error. Lamond does not state what it means to be the same. It is a Fact that no two Cases are the same in detail, and it is a Fact that Meaning can be the same when details are different. When Meaning is the same and the details are different, that is Analogy within the Boundaries of Pure Science. But here Lamond restricts the word analogy to Induction which involves resemblances. It is wrong to do that.

The philosophical talk of those with skill in Formal philosophy rarely mentions Meaning. Meaning is the Focus of Attention in Court but it is not the Point of Reference for all parts of the Action in Court. Inductive science is brought into Court and that is Formal logic. Formal logic is Not Logical. That which is Not Logical is Acted upon by that which is Logical and vice versa. But while we cannot tell the difference, there is the Appearance that logic is Acted upon by logic and that all logic is logical. There is the False Appearance of One Dimension and the Illusion that there are merely different Quantities of logic has taken hold of audiences. Appearances can be Deceptive. In Fact, Deception has such a high frequency of occurrence in Social Dynamics that it is Reasonable to presume that Appearances are Deceptive until and unless there is Evidence otherwise. It is Not Reasonable to assume that Appearances are Not Deceptive and fail to Test them. Yet that is what we are told to do. And we are told to do that by those who are engaged in Sophistry and the production of Illusion. That Action is not even puzzling. We all understand the True Meaning of that.

Inductive analogies are the Focus of Attention for Analysis at this point in this narrative. Lamond suggests that Inductive analogies are involved in matters other than precedent. As I have indicated, different types of Action are represented by the word analogy and Lamond does not point to that difference, nor do our dictionaries. So, the inevitable conclusion from the part quoted from his arrangement is that the Judges in Court use Inductive analogies. That conclusion is inevitable if we accept a False Premise that analogy is one type of thing. Lamond is attempting to associate the Action in Court with the Action of Induction.

The Appearance of Rules for Induction can be strengthened through the Association of Induction with the Action in the Courts for that is determined by Rules. The Appearance of Justification given by the Rules is strong and surrounds the Justice System. Lamond is drawing the strength of that Appearance into proximity with Inductive logic where the Mechanism of Association will ensure its passing to Induction. Lamond is a Specialist doing what he is trained to do and he is not looking at Meaning as it Emerges Naturally. He wrote:

> "Two questions arise about analogical reasoning. Firstly, by what process does a decision-maker identify the 'common characterisation' between the case at hand and the analogous one? Secondly, what type of justificatory force does the common characterisation provide? On the first question, just as no two cases are identical in every respect, so no two cases are such that some common characterisation of the facts cannot be found. But not every case is thought to provide an analogy, so what limits or directs the selection of analogies? The answer to this question flows into the issue of the justificatory force of analogies. What sort of reason does an analogy provide for deciding the instant case in the same way? It is widely agreed that the existence of an analogy depends ultimately upon the justification for the analogical decision."

If Justification is an Issue in the Courts, and I have presumed that it is, then it is a Fact that analogies in that Context cannot be Inductive, or at least not all of them can be. However, the wide agreement that ultimately there must be Justification for a decision in Court does not Verify the Truth of a statement that Justification is there. Only Testing can do that. It has been Verified using Pure Science that there was No Justification for the Guilty Verdicts in Criminal Trials in 250 Cases. And Garrett's sample was not exhaustive of the happening of that type of Error, nor does it include Cases where Miscarriages of Justice have been Verified without DNA Evidence. Verification is characteristic of the Action of Pure Science. Pure Science is Not Restricted to DNA Evidence.

Dropping words that signal differences in type of analogy is Active against the Perception that a decision using Inductive analogies is not soundly based, is in Error and is likely to be wrong as a Whole. Attaching a word to describe an Action signals particular Meaning. But using a word that is not well defined will not be an accurate signal of Meaning. Specialists with skill in philosophy of a Formal type know that Fact. They signal to each other in code and they Assert they can give accurate signals only by doing that. That Assertion does not Appear to be Testable. They Assert that the Structure of Natural Language, which is a form of representation that we do commonly understand, does not enable them to signal Meaning accurately to us. The rest of us are sitting in the audience and do not have the benefit of knowing the code of Formal logics. Thus, those who use Formal logics are Asserting they cannot signal Meaning accurately to those who are not part of their peer group. The Premise that follows as a conclusion from that arrangement is that the audience will never be able to understand their doings or relate their doings to the happenings in the World of which we are all part. Their Premise is False.

When they do give us Meaning in Natural Language then that Meaning can be Tested for Truth and Justification. Then they can present their Defensive Premise: We cannot tell you what we Mean using Natural Language, so the Meaning you Tested is not the Meaning that should have been Tested. They again present their codes as proof. Thus, the Opposition collapses. Round and Round go the Opposition until they become dizzy and fall to the ground. But actually, codes can be broken and the Meaning that is coded is then disclosed. I propose that we do not need a Mathematician or any other Specialist in Formal logic to do that. In Fact, they refuse to do that. We need Pure Science and a Focus of Attention on Human Action in order to do that. Those who use Induction cannot find Justification and have settled for a Construct of Justification by Agreement. We saw that where the Judges cannot find Truth and Justification, they have settled for a Construct of Dispute Resolution by Agreement in Criminal and Civil Courts. An Appearance of No Difference in type of Opinion is an Active Force taking us towards that position. Curiouser and Curiouser said I. Everything is upside down and/or back to front in Wonderland. But in all Contexts, there are Truth Seekers and Justification can be found.

Justification makes a Significant Difference in type of Meaning and its showing leads to acceptance by an audience of Meaning and the Action that is related to that. Justification is aligned to the discovery of Truth. It is Natural to search for True Meaning. And in order to do that it is necessary to be able to distinguish True from False Meaning. Inductive analogies are Not Justifiable. Justified Analogies are produced when Pure Science is engaged and True Meaning is sought. Within the Boundaries of Pure Science, Analogies are composed of two or more different arrangements which include Meaning that is the same either as a Whole or in part. The Analogy happens only where Meaning is the same and its Representation is different. I propose that Natural Philosophy aids conceptual development and that Artificial philosophy interferes with that. Natural Philosophy is Natural Psychology; there is No Significant Difference in Meaning and Action for those terms. Philosophers Act on human minds. A Significant Difference is a Meaningful Difference. Inductive science does not systematically discover Meaningful Differences. It is not designed to do that. That is a difference between Pure Science and Inductive science. Pure Science is designed to discover Meaningful Differences and to relate them.

Philosophical reasoning is supposed to aid conceptual development so let us see if Artificial philosophy does that. Lamond gives some particular examples to illustrate the use of analogy in legal argument, as he sees that. Now I will pick up for analysis two of Lamond's examples. Remember that this is talk in the Context of philosophy and psychology about Factors in decision making. This is not Judges talking about the Method they use in the Legal Context. The first example is:

> "knives may be analogous to guns if the issue concerns weapons, but knives may also be analogous to teaspoons if the issue concerns cutlery."

Induction deals in resemblances. Use of the word analogous carries the suggestion that something more than similarity is involved. Recall from the common definition of analogy the part Meaning in Biology. In this example the "similarity" is in Function and not in origin or structure. Knives are "similar" in terms of use to guns if the issue concerns weapons and are "similar" in terms of use to teaspoons if the issue concerns cutlery. But actually, all weapons have the same type of Meaning and cutlery also has one type of Meaning. So, the word similar does not fit. An Analogy is different to a similarity. That difference is dropped by Inductive scientists. As a group of things, weapons serve the same purpose and cutlery serves a different purpose. The word but is used to relate the parts in Lamond's statement where the word and would fit better. The word but carries a suggestion of Uncertainty to analogies and Lamond picks up that suggestion later in his paper.

Recall the definition I use for Analogy: - Analogy is defined as the Action of Modelling Meaning through the creation of different arrangements to Represent it. So, is this an example of Analogy as the word is used within the Boundaries of Pure Science? Or is this an Inductive analogy? Weapons are a different Point of Reference to cutlery. There are No Premises here. There are underlying definitions of those words and they are Justifiable. In Action, a knife and a gun are different models of the Meaning of a weapon. The word analogous fits that part. And a knife and a teaspoon are different models of the Meaning of cutlery. The word analogous fits that part. The Focus of Attention for the Action with a tool is different when we fight or eat and we choose our tools to fit the type of Action that suits our purpose. Two Points of Reference are clearly stated. Meaning is clear in the parts.

So, the "similarity" to which Attention is drawn is in terms of the type of Action that is enabled by the part and the Points of Reference are words that represent a particular type of Function that is modelled by the words gun and teaspoon to which the word knife is separately related. The resemblance between the parts is not in their Appearance. Here Meaning is the same and the words resemblance or similarity should not be applied to that happening. And Lamond did not use those words in his statement. Meaning is the same when the Point of Reference is the same. A difference makes a difference. Meaning is different when the Point of Reference is different. This example demonstrates both those Facts. So, this example is Analogy in both parts of the statement and is not Analogy as a Whole statement.

I repeat that in this example the word analogous fits where it is placed. Meaning that a knife has in common with two different things is picked up by different Points of Reference. The commonality is not one that is shared between all three of the parts. In other words, there is not a common Point of Reference for all three parts. Since all things are related one could be found but distance is a Significant Variable in terms of the Power to Influence other things so we have the relevant Points of Reference in view. The relevant Points of Reference are clearly stated; those are the ones that are Active. Since we are clear about the Meaning of the Points of Reference, the models of guns and teaspoons are redundant - knives are particular examples of both weapons and cutlery and we know that.

Is this an Inductive analogy in part or as a Whole? I will recall the general structure of Inductive analogy as described by Bartha and Test the fit of the example to that.

1. S is similar to T in certain (known) respects.
2. S has some further feature Q.
3. Therefore, T also has the feature Q, or some feature Q* similar to Q.

First, I will Mark the parts in our Simple Model of Inductive analogy with the code:

Knives (S) may be analogous to guns (T) if the issue concerns weapons, but knives (S) may also be analogous to teaspoons if the issue concerns cutlery. The Points of Reference are weapons and cutlery and they indicate what the further feature Q is through the models. In this case the symbol Q stands for the Meaning of the parts. The coded parts can be entered into their positions in the Structure.

1. Knives (S) are similar to guns (T) in the certain (known) respect of being a weapon.

2. Knives (S) have the further feature Q that is characteristic of cutlery and that feature gives them similarity to a teaspoon which is an item of cutlery.

The conclusion would be:

3. Therefore, guns (T) also have the feature Q that is characteristic of cutlery.

But the conclusion does not follow from the statements - guns do not have a feature similar to teaspoons in terms of Function, they have different Meaning. Bartha stated that "the argument form is inductive; the conclusion is not guaranteed to follow from the premises." And the conclusion here does not follow. So, this demonstrates that conclusions that are Not True can be drawn by Artificial Induction from statements that are True when the argument form relies on similarities. Similarities in Action and Appearance can Deceive an audience into accepting Meaning that is Not True. In this example it is easy to see that the conclusion is False. The conclusion to which the statements lead would not be accepted if its truth was Asserted. The Inductive Structure has drawn Meaning that is Justifiable from two parts and produced Meaning that is Unjustifiable as a Whole. The Whole is Qualitatively different to the parts. Inductive analogies take Meaning too far. Meaning that is taken too far is Meaning that is taken beyond what is Justifiable.

Now let us look at a second example from Lamond, which was juxtaposed with the first. I note the movement within Lamond's argument from a Simple model with physical objects using familiar definitions that are Justifiable to a Complex abstract arrangement. And bear in mind that the Unjustifiable conclusion was not stated in the original arrangement. This example has a different type of part and the same Structure:

> "Duress may be analogous to provocation if the issue concerns defences, but duress may also be analogous to incitement if the issue concerns complicity."

The difference in the type of part in the two examples is Significant. Is this an Inductive analogy? Here, there is the Appearance of three parts - duress, provocation and incitement. Those are not familiar physical objects as they were in the example that immediately preceded it. The parts are abstract ideas. In essence, Meaning is being moved between parts and here it is difficult to grasp what is happening. I will insert the parts in the Structure we were given for Inductive analogy as I did before.

S = knives = duress
T = guns = provocation
first point of reference = weapons = defences
not coded and representing something with feature Q = teaspoons = incitement
second point of reference = cutlery = complicity

"Duress (S) may be analogous to provocation (T) if the issue concerns defences, but duress (S) may also be analogous to incitement if the issue concerns complicity."

1. Duress (S) is similar to provocation (T) in the certain (known) respect of defence.

2. Duress (S) has the further feature Q that is characteristic of complicity and that feature gives it similarity to incitement which is an example of complicity.

The conclusion would be:

3. Therefore, provocation (T) also has the feature Q that is characteristic of complicity.

And I will specify that the feature Q is Meaning. The Action of building arguments from small units is common to Natural Deduction and Induction and the difference is in the type of unit. The conclusion here does not have the Appearance of being in Error, and much more is going on in this example than the drawing of a conclusion. Now, a similarity is not a relation and does not even resemble one. The similarity is drawing parts together where Meaning will pass by Association. And Meaning can flow backwards as well as forwards in a channel. In Induction, the conclusion can Act as a Premise and support the other terms. This is not an example of Analogy in part or as a Whole as the word is used within the Boundaries of Pure Science. Premises are in Action in this example. There is a Premise that provocation is a model of the Action and Meaning of defence. And a Premise that incitement is a model of complicity. Those Premises are Not Tested for Truth. There is No Justification for the Premises or for the transfers of Meaning from the Premises that are the hidden parts of this Induction. And when we draw the common definitions into view, we see that incite is part of the definition of provoke.

provoke: v. rouse or incite (provoked him to fury) 2. call forth; instigate; cause…… 3. irritate or stimulate. 4. tempt; allure [Latin provoco call forth]

incite: v. urge or stir up n. incitement

If there is no difference in Meaning between the two words there is no need to use both of them, and doing so gives the Appearance of a Difference in Meaning where none exists. Now the term Q was given particular Meaning by the word teaspoons in the first example. And it was Meaning that was given - teaspoons are a practical example of cutlery and are not a particular characteristic of the parts in that group. If I use the word provocation in both parts of the second example, then we realise that we have lost our teaspoons and none of this makes Sense as a unit. So, we could just note that this is an example of Formal logic and that it is peer agreed that an Unquantifiable number of the conclusions reached using Formal logic do not make Sense. But I draw your Attention to the Verified Miscarriages of Justice as showing the particular Meaning of conclusions that do not make Sense. And this example also shows that the conclusions of Induction may or may not make Sense, but when they do it is not the Action of Induction that has made it. This demonstration is Evidence that the reasoning in Induction does Not Make Sense and thus is Not Reasoning, and that the lack of Sense is determined by the gap left where particular Meaning could be modelled and by the use of Premises. I will leave these examples of Inductive analogies at this point.

The products of Induction are inherently Unstable for their Truth is Uncertain. Recall that the use of the word "but" in the examples carries a suggestion of Uncertainty to the word analogy. Lamond observes that:

> "Analogies do not bind: they must be considered along with other reasons in order to reach a result."

Within the Boundaries of Pure Science, Analogies are not reasons. Inductive analogies are used as reasons; they do not bind Meaning for they are based on similarities and are not based on Meaning that is Tested for Truth and Justification or is in the course of being Tested for those Qualities. Where there is No Truth, conclusions can be Unsettled in various ways. But it is a Fact that Inductions are binding under certain conditions and the Fact that the conditions can be changed without Justification does not change that Fact. Peers can agree to remove an agreement and put another in its place. Now, remember that Truth has been taken out of the Equation in Social Dynamics. Having muddled different types of Action together and taken them to be one type of Action, Specialists in philosophy Assert through the terms of agreements that those who enter into agreements that peers have drawn up are bound by those agreements. But agreements are commonly drawn up by small selected groups of peers who use the Method of Induction. The Unjustifiable agreements of small selected samples of a population are imposed on the many. Force is necessarily involved in that. The Calculation of Meaning is necessary to reach agreements that are Justified.

There always is Force involved in Dynamic Systems so the Fact that Force is involved is Trivial. The Fact that agreements are reached by the Actions of a few people is also Trivial, for Attention must be Focussed in order to Calculate Meaning. What is Significant is the type of Force involved and whether or not the agreement is Justified. Binding people to agreements that are Not Justified using Force that is Not Justified is of course Unjustifiable as a Whole. The Whole is different to its parts, but its Qualities are dependent on the

Qualities of the parts. The Fact that agreement has been reached is used to give the Appearance of Justification to the binding of people to the Meaning that is agreed. Round and Round we go in Dynamic Systems. Curiouser and Curiouser said I as I watched things spinning faster and faster. I necessarily felt dizzy. Only Curiosity stops me falling to the ground. And to add to the puzzle, since the Whole is Qualitatively different to the parts it is possible that an agreement reached without Justification may be Justifiable. But if so, the Justification comes from elsewhere and does not lie in the details of the Inductive argument. The conclusions of Natural Deduction follow of necessity and do have the Quality of Truth.

It is a Fact that the Meaning of words and the Meaning of arrangements of them as a Whole, changes as the Point of Reference changes. Meaning is the same when it aligns to the same Point of Reference and different when it does not. Now Points of Reference are a Complex Active System in their own right – they come in parts and Wholes. What distinguishes a Point of Reference from anything else is not a particular characteristic. Any type of part can become a Point of Reference, and which one does is dependent on the conditions that prevail.

Parts are also Wholes and vice versa. A Whole Point of Reference is composed of parts in an arrangement with Meaning that aligns to a Whole Meaning. Meaning is the same when it aligns to the same Point of Reference and unless the common Point of Reference is disclosed then Meaning that is the same has the False Appearance of being Different Meaning. And vice versa: - Unless the Point of Reference is disclosed then Meaning that is different has the False Appearance of being the Same Meaning.

PART THREE of Chapter four

THOUGHT EXPERIMENTS AND THE ART OF PERSUASION

When people conduct Experiments within the Boundaries of Pure Science, they ask questions, formulate propositions, introduce changes, observe what happens, draw Meaning from their Observations and attempt to answer the questions. Then the answer is Tested for Truth using more Experiments. The word experiment is misused by those who already know the answer to everything and have a theory to back it up. Specialists with philosophical skill of the Formal type use the term thought experiment to describe any arrangement that arrives at a pre-determined conclusion or at least one that is aligned to their purpose. Enthymemes are described as thought experiments and Inductive analogies are part of such arguments. The Enthymeme is an Ancient Structure that is a tool of the Art of Persuasion. Specialists in philosophy use units of Natural Language vaguely as they deliver their Meaning precisely. Rapp (2010) tells us that:

> "Although Aristotle employs the term analogy (analogia) and talks about analogical predication, he never talks about analogical reasoning or analogical arguments per se. He does, however, identify two argument forms, the argument from example (paradeigma) and the argument from likeness (homoiotes), both closely related to what we would now recognize as an analogical argument."

It is common practice to use different forms of argument together. It is the type of argument that is the Significant Variable. The difference in the Context of Induction between an example and a likeness is Trivial. If both forms rely on similarity, then they are of the same type. Attention is drawn to similarities in order to persuade an audience to accept a particular conclusion. An Enthymeme is constructed on the basis of resemblances. Rapp tells us:

> "Aristotle calls the enthymeme the "body of persuasion", implying that everything else is only an addition or accident to

> the core of the persuasive process. The reason why the enthymeme, as the rhetorical kind of proof or demonstration, should be regarded as central to the rhetorical process of persuasion is that we are most easily persuaded when we think that something has been demonstrated."

Enthymemes are at the core of Rhetoric. Premises are part of Enthymemes. Premises lead us to think that something has been demonstrated. The "proofs" of Induction are also Enthymemes and are composed of Premises that lead to a particular conclusion that can Act as a Premise elsewhere. Aristotle taught that an Appearance of Justification is sufficient to persuade an audience that there is evidence of fact when there is No Evidence of Fact. The Perception of Meaning leads to Action. Over the Centuries, Active Agents have worked on giving the Appearance of Justification where there is No Justification and thus have drawn audiences into Action in which they otherwise might not have engaged. Truth was not an issue for Aristotle. Aristotle did nothing to change personal Points of Reference in the direction of Truth. I propose that Truth is an important addition to the core of the persuasive process. I also propose that True Meaning can be found and that its finding makes a Significant Difference in terms of Action.

The rhetorical process of persuasion enables acceptance of Meaning by an audience. If those who seek to persuade others do not know the Truth, then they will know that they cannot provide Evidence of Fact and that they seek to persuade others of Meaning in the absence of knowing the Truth. Justification is a sign of Truth. If Justification is absent, then the only way to persuade an audience to accept Meaning is to create an Illusion of Justification. An Illusion of Justification is the core of Inductive persuasion. Where there is Belief in truth, there is the feeling that Truth is there. The feeling that Truth is there must be transferred to the audience. As Aristotle realised, the audience must think that something has been demonstrated if they are to accept gifts of Unjustifiable Meaning.

<u>In short</u>: Aristotle taught that inferences must be given an Appearance of Justification and how to do that. Similarities Appear to provide evidence within the "body of persuasion" and are part of the core in

Designing an Illusion of Truth. Stating, or otherwise showing, a similarity gives the False Appearance of Proof that Meaning is the same. Similarities are an integral part of the Associative Mechanism. The Action of Displaying Similarities is the Action of Positioning Meaning in the Mind of an audience so that transfers occur.

I note from Rapp's review of Aristotle's teachings the following quote:

> "Enthymemes based upon example are those which proceed from one or more similar cases, arrive at a general proposition, and then argue deductively to a particular inference. (Rhetoric 1402b15)"

The part of Induction that has a similar Appearance to Natural Deduction leads to acceptance of Meaning by an audience. The False Appearance of Deduction makes us think that something has been demonstrated when it has not been demonstrated. Rapp continues:

> "It is natural to ask when the degree of likeness between two things is sufficiently great to warrant inferring a further likeness. In other words, when does the argument from likeness succeed? Aristotle does not answer explicitly, but a clue is provided by the way he justifies particular arguments from likeness. As Lloyd (1966) has observed, Aristotle typically justifies such arguments by articulating a causal principle which governs the two phenomena being compared."

When the Method in Action is Induction, to articulate a causal principle is merely to express one in the absence of Justification. The word attribution covers that type of articulation of a causal principle and that Action is notoriously prone to Error. Only Experiments can lead to the discovery of Causal Principles. That Fact is known. Enthymemes are commonly described as thought experiments and have a similar Appearance to Experiments in the Mind's Eye using Natural Deduction. Illusion leads to acceptance that conclusions have the Quality of Truth when they do not have that Quality.

We are informed by Rapp that according to Aristotle an enthymeme is different to Induction in the following way:

> "Aristotle notes two differences between this argument form and induction (69a15ff.): it "does not draw its proof from all the particular cases" (i.e., it is not a "complete" induction), and it requires an additional (deductively valid) syllogism as the final step. The argument from example thus amounts to single-case induction followed by deductive inference."

The difference specified here by Aristotle, according to Rapp, between the enthymeme and Induction is Quantitative not Qualitative. We are told that an enthymeme takes similarities from one or a few Single Cases. But it is not accurate to state that Induction takes its proof (in terms of similarities) from all Single Cases – the impossibility of doing that is recognised. Induction takes similarities from a large number of Single Cases. And both the enthymeme and Induction use deductive rules as the final step in drawing conclusions. And here we again see a line being drawn between Induction and Inductive deduction. The suggestion that there is a difference in type of Action is a common Premise and that Premise is Testable. The Evidence that the Premise is False is overwhelming. There is no difference in type of argument between an enthymeme and induction. The argument type is Inductive. Inferences drawn by Induction are Not Tested for Truth and are Not Justifiable. Applying the Structure of the syllogism within Induction changes the Appearance of the Action as a Whole and produces the Illusion of Calculating Meaning by Reasoning.

Plato and Aristotle gave the Formal Structure to philosophy that has endured. A Philosophy is a Human Cognitive Structure. Particular people in the Ancient World set the Points of Reference and designed Cognitive Structures for humans that are still in place. That is one of the Wonders of the World. But the Wonder is how they could succeed in doing that regardless of the consequences. It is a Wonder that their doings are not recognised as a Fault given what has happened across the Centuries. Once upon a time that Fault was recognised. Aristophanes was Ignored. And what happened thereafter was not a fairy story. We did not live happily ever after.

Existing human beings are not the most Powerful Determinant of happenings in Social Dynamics. Structures are the Powerful Determinants of happenings. And human beings designed the Artificial Structures that are Active in and on the World. Natural Structures have a different type of designer. It would be strange indeed if human beings could override the Forces of Nature indefinitely. But it is True that they can override them in the short term. Nature made allowance for adaptation. And Time is Relative.

Meaning that has been accepted must be Unsettled if it is to be replaced with different Meaning. The term "thought experiment" is used to gift to Induction the Appearance of Testing Meaning for Justification. Brown and Fehige (2011) state that:

> "To be effective, counter thought experiments needn't be very plausible at all. In a court of law, the jury will convict provided guilt is established "beyond a reasonable doubt." A common defence strategy is to provide an alternative account of the evidence that has just enough plausibility to put the prosecution's case into some measure of doubt. That is sufficient to undermine it. A good counter thought experiment need only do that much to be effective."

A Verdict of Not Guilty is Not a Verdict of Innocent. It is not necessary to Falsify a proposition in order to cast Doubt upon it. Attention is not focussed by the Defence on a search for Truth and is focussed on Unsettling arguments for the Prosecution. An argument that has order is set to be disordered and establishing new order for Meaning is not required. And it takes skill to design a good counter thought experiment. The type of skill involved makes a Significant Difference to the Quality of the effect. Evidence from Miscarriages of Justice suggests that the agents for the Prosecution do not Focus Attention on a search for Truth either. Their Attention is focussed on establishing an argument that is Resistant to Unsettling. Presumably they think that is the same Point of Reference, since True Meaning cannot be Unsettled. And the Judges cannot select at the gate of the Court that which is not selected for presentation at the gate. For Truth and Justification to become the Point of Reference in a Court of Law, Truth and Justification would have to become an Issue outside the gates through which experts pass into Court. It is a Fact that the Demarcation of Pure Science has Legal, Political and Moral Significance.

Unsettling enables the Action of one thing by disabling the Action of another thing. Unsettling is agreed to be sufficient to dislodge propositions in many Contexts. Falsification is not required. The word doubt represents the State of Uncertainty and that is the effect of Unsettling Meaning. We have been led to Doubt our own Senses. Specialist philosophers would have us Believe there can be nothing but Doubt when Judgments are made in any Context. That is an Illusion. Specialist philosophers know that there is something beyond doubt. Beyond doubt there lies Certainty. Beyond Disorder there lies Order. But different types of Action lead to different types of Certainty and it is necessary to distinguish differences in type for different types of thing have different Meaning. The Gaps in Meaning left by Unsettling are necessarily filled, and if they are not filled with True Meaning they are filled with Beliefs. The Ancients knew that Fact. They showed how to create gaps in Meaning and how to determine the particular Meaning that filled them. They showed how to do that without giving the Appearance of Doing So; that is an extraordinary feat. The Ancients were clever fellows. They have not been surpassed in cleverness by those who came into being later.

If the Beliefs of the audience are not already in line with those of the designer or cannot be drawn into alignment, then Uncertainty must be produced in the audience before new Certainty can be Installed. Rapp (2010) tells us that:

> "In Topics VIII.1, 155b4–5 Aristotle says: 'we must find the location (topos) from which to attack', where the word 'topos' is obviously used to mean a starting point for attacking the theses of the opponents."

In order to Observe the effects and the outcome of Artificial philosophy in Action it is necessary to Calculate Meaning in Single Cases. When Pure Science is engaged, Single Cases are the starting point for the Discovery of Knowledge. I will now analyse an

Inductive argument that is a particular Persuasion on a particular topic.

A FAMOUS "THOUGHT EXPERIMENT" BY JUDITH THOMSON

Judith Thomson, an American philosopher, published an argument to support the Action of Abortion. Or rather, her article was accepted by a publisher and published. Her article was highly acclaimed. Thomson had learned the Art of Persuasion. Brown and Fehige claim that her argument is a thought experiment and the word Inductive is missing from that term. They introduce her article in this way:

> "Judith Thomson provided one of the most striking and effective thought experiments in the moral realm (see Thomson, 1971). Her example is aimed at a popular anti-abortion argument …"

I will establish the location from which I Test the Meaning in Thomson's article. The particular topic of Abortion is of great importance in Social Dynamics. I am not acting as an advocate for either side. It is the Method of Argument that is of interest here and not the particular topic to which it was applied by Thomson. I will Verify that the type of argument used in Thomson's article on Abortion does not just fall short of Moral Reasoning, it is different in type. However, the type of argument demonstrated by Thomson is the same in type as the argument used by the Anti-Abortion lobby. The same type of Inductive argument can lead to different conclusions. Argument in Social Dynamics is commonly nothing more than a Fight for Beliefs. The failure to Verify Meaning is a Fault that leads to confrontation without enduring resolution. The conclusions from Formal logics are in Error with a frequency that cannot be counted. That statement is agreed by Specialists in Formal logic. Justification is an Insoluble Problem for Formal logics. That statement is agreed by Specialists in Formal logic. But that is what they use. The alternative is to Calculate Meaning. Unless the Process of Pure Science is fully engaged, we do not see Justified Calculations of Meaning.

A Judgment cannot be Moral if Justification is not there. Formal logic is Action that does not fit the Moral Realm. Those two statements are not agreed by Specialists in Formal logic and are Testable. The lack of fit can be Verified. Thomson's article had the Structure of an Enthymeme. Taking her lead from Aristotle her argument was based upon example and moved from one or more similar cases, then argued deductively to a particular inference. She used Inductive analogies relying on partial resemblances for the Appearance of Justification. Thomson understood that we are most easily persuaded when we think that Truth has been demonstrated and she understood that it is not necessary to demonstrate Truth. She also understood that it is not necessary to tell the Truth, which takes Meaning further. The Illusion of Truth can be substituted for the Real thing. A resemblance is all that is required. Those are all Errors that amount to a Fault.

This Single Case allows me to demonstrate the type of Action that is Certainly Not Pure Science. It is necessary to Analyse the Action in detail. I will begin here and continue in the final Chapter. Through gaining Knowledge of the type of Action that falls outside the Boundaries of Pure Science, progress can be made towards identifying the type of Action that is distinctive of Pure Science which will enable movement towards Marking the Boundary where there is a change in type. I invite you to join in a familiar Chorus.

Chorus: "Words are their weapons: watch out, as the armour-clad syllables hurtle, helmeted, crested and plumed, from the lips of the Poet Most High! Wait for the clash and the din as the metaphors mingle and jumble, the sparks as the particles fly!"

I begin the analysis of Thomson's article on Abortion by quoting the summary description of Thomson's Persuasion by Brown and Fehige (2011).

> "Her example is aimed at a popular anti-abortion argument that goes something like this: The foetus is an innocent person with a right to life. Abortion results in the death of a foetus. Therefore, abortion is morally wrong. In her thought experiment we are asked to imagine a famous violinist falling into a coma. The

> society of music lovers determines from medical records that you and you alone can save the violinist's life by being hooked up to him for nine months. The music lovers break into your home while you are asleep and hook the unconscious (and unknowing, hence innocent) violinist to you. You may want to unhook him, but you are then faced with this argument put forward by the music lovers: The violinist is an innocent person with a right to life. Unhooking him will result in his death. Therefore, unhooking him is morally wrong."

Thomson used her imaginings and clearly those were not free. Breaking and Entering is a Criminal Offence. And what she described bears what those who use Induction would refer to as a close resemblance to the Action of Rape, which is also a Criminal Offence. I would say the Meaning she represents is the same as that of Rape. We see the use of evidence to support the particular Action. Here the evidence is in the form of medical records. And we see carefully placed the envisioning of Good Intentions on the part of the perpetrators to mitigate the Judgment of the Action. Good Intentions do not change the type of Action or its effect. Whether evidence does that depends on the type of evidence and that difference in type is made by the type of Action that produces it. The type of Action and its Meaning are Intrinsically Related and dependent on the type of Point of Reference.

Brown and Fehige reflect:

> "However, the argument, even though it has the same structure as the anti-abortion argument, does not seem convincing in this case. You would be very generous to remain attached and in bed for nine months, but you are not morally obliged to do so. The parallel with the abortion case is evident."

Females are not confined to bed for nine months during pregnancy, or if there is such confinement it is not as a consequence of that. And the argument is not convincing in the case of the violinist because of the differences in type of Meaning between the cases. There are similarities and differences between Real Cases of Abortion and the imagined disconnection of a fictional character intimately and externally connected to you by the Actions of other people. Only taking the differences into account enables the Calculation of True Meaning. If we apply the definition given in the Pocket Oxford Dictionary, Thomson has created an analogy for it tells us there that analogy is arguing based on some resemblance. The resemblance is in Appearance and to an unspecified degree. Arguments based on some resemblance do not lead to the Certainty of Truth and do lead to Uncertainty in relation to that. In order to undermine the Anti-Abortion argument, Thomson's alternative account needed enough plausibility to put the opposing case into some measure of doubt. Brown and Fehige continue:

> "Thomson's thought experiment is effective in distinguishing two concepts that had previously been run together: "right to life" and "right to what is needed to sustain life." The foetus and the violinist may each have the former, but it is not evident that either has the latter. The upshot is that even if the foetus has a right to life (which Thomson does not believe but allows for the sake of the argument), it may still be morally permissible to abort. **Those opposed to Thomson's view can either dismiss her thought experiment as useless fiction or provide a different version of the same scenario to challenge the conclusion.** It is a very intriguing feature of thought experiments that they can be "rethought" (Bokulich, 2001)."

I have highlighted the copying of Meaning that we saw gifted by Popper. He had obtained that gift from the Ancients and merely passed it on. I turned to look at Thomson's original paper and you can do the same. First, I will describe the simple steps in the Persuasion:

i. She begins her argument with a statement of her personal opinion.
ii. She states the question to which her personal opinion is the answer that she knows is the right answer.
iii. She selects some universal generalisations that support the different opinion that is already in the minds of the audience.

iv. She gives the Appearance of being guided by the question
v. She unsettles those universal generalisations by pointing to the differences left behind during their abstraction.
vi. She sets up a work of fiction with similarities to a real event.
vii. She uses the similarities in the case she imagined to support her own opinion.
viii. She ends her argument with the restatement of her personal opinion. She ends precisely where she began.

I repeat: - Thomson knew the answer to the question that she pretended to ask. That is emerging as a systematic regularity in the Action of Specialists in philosophy. She did not need to do more than introduce doubt about an alternative answer. However, she did need to do much more than that in order to avoid doing more harm than good. I will show you that she went much further in her Unsettling than seems to be recognised. While Thomson is busy manipulating many Meanings to apparently one purpose, a close look reveals that more than one purpose can be served by the argument she designed. Thomson does not draw Attention to the other possible conclusions, and may not have been aware of the Potential for Damage that her Art created. But the use of Reasoning would have brought awareness and that is the alternative Method to which I am drawing Attention by way of Contrast.

It is Certain that among those who read Thomson's paper, some at least will have been looking for material to serve different purposes and, being less pre-occupied with the issue of Abortion, they will have noticed the other possible uses of her argument. I propose that the Potential for harm that is in her argument was Realised. Thomson did not entirely rely on finding similarities for her argument in Real happenings. She also relied on constructing Meaning in Imagination and moving Meaning from Fiction into the Real World. It is a Fact that Meaning moves from Fantasy into Reality – Meaning is translated into Action in the external world. But it is a Fact that the figments of Imagination do not come from a Blank Slate, for the mind is not that. The tale of the violinist and the rule breaking but possibly well-intentioned music lovers who took advantage of you while you slept and now demand sacrifices from you is Not True in its Representation. However, the "parallel" Meaning, that struck the opposing position so effectively and is referred to as "analogy" is not a work of Fiction and is Meaning taken from the Context of the Real World and used to serve her particular purpose. The misuse is not the Action of Representing Meaning from the Real World in Fiction. The misuse comes from the False Association of that True Meaning with a doing and a happening that has a different type of Meaning in the Real World and the sending of the changed Meaning back to the Real World. Fiction is a Powerful Mechanism for change and it can be used well or misused.

The Fact that Fiction has the Power to affect Actions in the Real World is Systematically Denied. But Fiction Represents Meaning and Meaning is translated into Action. And Specialists in philosophy know that Fact. Thomson's argument is Evidence of her knowing that is a Fact. Evidence of knowing Facts is Not Evidence that they Know What they Do. Knowledge is Facts in their Intrinsic Relations. It is a Fact that the Denial that Fiction is a Powerful Stimulus for changing Cognitions, and thus for changing Actions, enables its use free from constraints. Those who use Formal logic make assumptions and it may be that they assume there is no malign intent. I am by profession a Forensic psychologist and I use Natural Deduction in searches for Truth; I do not use assumptions. Sometimes there is Evidence in Single Cases that: They Know Not What They Do. And in some Single Cases Evidence of malign intent does emerge. And the two things – knowing and intending - are not separate and are related. The number of Single Cases where there is Intent to Harm remains Unquantifiable while we cannot tell differences in Intention. The presumption in each Single Case is that there was No Intent to Harm. That is the presumption here in the Case of Thomson as the designer of the argument, and that presumption is Not Tested here. But the way her argument Functions in Social Dynamics is Testable and is Tested in this narrative.

Thomson's paper is a fine example of the Art of Persuasion. I will not dismiss her argument as "useless" fiction. Fiction is not useless in working with Meaning and does have Significant effects on Action in the Real World. Nor will I provide a different version of the same scenario to challenge the conclusion. In Fact, I will not challenge the conclusion on Abortion. I am not opposed to Abortion, nor am I Biased in its favour. Abortions are Natural happenings under certain conditions. Nature has its own Rules. The conditions in which particular Actions are Justifiable can be specified and Boundaries can be Marked. It is puzzling that Boundaries are Not Marked when they could be. Justification for Abortion under certain conditions can be found through Calculations of Meaning in Single Cases within the Process of Pure Science. There is no Justification for Asserting a Universal Artificial Rule on Abortion. Inductive Rules are never Justified. Induction suffers and benefits from the Insoluble Problem of Justification.

I will provide Evidence that the type of argument designed by Thomson is Not Reasoning and is Not Moral. The conclusion on Abortion is not the end of the conclusions to which her argument leads and I need to show you what the further conclusions are if my claim is to be Verified. In doing so, I will mount a serious challenge to the Method of Argument by Induction that has been traditional for Centuries. I propose that Specialists in philosophy have led Social Dynamics into Error and that proposition is Testable. The production of Error pays dividends for those of Strong Will with no Moral Constraints. The number of those who have Strong Will coupled with no Moral Constraints cannot be Quantified while there is No Qualitative Standard for the Judgment of Morality. We cannot distinguish between different Qualities of Meaning in things while peers with Authority agree that those differences should not be told.

Brown and Fehige state that Thomson distinguished between two concepts previously run together. That statement can be Tested for Justification. Thomson had an idea or thought that if she separated the terms "right to life" and "right to what is needed to sustain life" she could win the argument on Abortion. Thomson did not identify different Concepts. I propose that:

THOMSON TORE A SINGLE WHOLE CONCEPT APART.

In the present Calculation of the Meaning in Thomson's Actions in her paper, I will set aside the word Right for that is common to both terms. That leaves us with two terms: life and what is needed to sustain life. I have not removed a Concept by taking out a word that is common in the Equation. Human Rights are man-made Constructs. And I have just set that word aside for now. Constructs differ in their fit to any underlying Concepts. Concepts are Natural and flexible and Constructs must be too if there is to be alignment. A Universal Concept of Rights that is Natural and covers all Species could be developed and Constructs aligned to that could be developed. But that doing involves Complex Calculations of Meaning and is not the task I have set for my narratives. It is necessary to have a Method of Calculating Meaning before Meaning can be Calculated. However, I am looking for a Natural Concept that would unite the two parts that are left in this particular Calculation and could then be part of other Calculations.

Now, Concepts have more than one Meaning. In the debate on the topic of Abortion, it is clear that the word life has one particular Meaning. So, I will use the word life in the Context of Abortion with the particular Meaning of the type of State that is different to the State of Death. Taken in conjunction, life and what is needed to sustain life is a Reasonable definition of an Ecosystem and could constitute the core of the Natural Concept of Life on Earth.

An Ecosystem is Dynamic and parts within it are self-contained and related to other parts within the Whole System. A Whole Concept could have a Multiplicity of Meanings, but it is not the Quantity of Meanings that distinguishes a Concept from other ideas and it is the Qualities of Meaning that are distinctive of a Concept. Concepts can have just two types of Meaning as parts of the Whole. Concepts are Active on their parts and the Action is distinctive in type; Concepts Function in particular ways. All the different things on Earth are aligned by the Natural Concept of Life on Earth. Taking the view that the Meaning of Life and the Meaning of what is Necessary to Sustain Life are Actively Aligned in Intrinsic Relation to each other and are not separate, it follows that if the two parts of the Concept are separated by severing the relation between them then the Structure as a Whole would disappear. And this Natural Structure is the Ecosystem. So according to MOM, the parts 'life' and 'the means to sustain life' are joined together in conjunction as One Unified Point of Reference for the Natural System of Life on Earth. Nature designed that Concept. The Power for the Ecosystem comes from elsewhere. We do not know the Source of Natural Power. We do know the Sources of Artificial Power. We also know, or should know, that if there is no Natural Source of Power then there cannot be

Artificial Power. Quite simply, we cannot use or misuse what is not there.

Concepts with the Quality of Justification are Points of Reference that govern Meaning in arrangements through unifying the parts of which they are composed. Constructs do not Act in that way even if they are Justified. Meaning represented in the word 'life' and that represented in the term 'what is needed to sustain life' are Intrinsically Related and run so closely together that separation is not possible in the Natural World. The Evidence of Testing gathered through our Senses Verifies that. Only the Representation is different and that is what Specialists in philosophy tell us to pay Attention to. If we do what they tell us to do, we are led into Errors of Judgment. It is a Fact that life as a State of Being is dependent on that which sustains it. There is a Necessity in Intrinsic Relations. Thomson denies that Fact but she cannot actually be in Denial due to the overwhelming Evidence of our Senses. So, this aspect of her argument is puzzling and the agreement of her peers to also deny that Fact is even more puzzling. It is an Observable Fact that life extinguishes if there is nothing to sustain it and flourishes if there is something to sustain it. To deny the means to sustain life is to deny life itself. The Concept of Rights becomes Meaningless if those two Intrinsically Related terms are separated.

The Concept of Rights is a Moral Concept and we are told that Thomson's argument was in the moral realm. The question that should be asked is: What was Thomson actually doing when she stepped into the Moral Realm? And the answer is: She was doing exactly what she had learned to do. She stepped into that domain in order to Unsettle any Meaning that was Active there. Things that are Intrinsically Related have the same Meaning. Thomson paid Attention to the differences in Representation and ignored the Fact that their Meaning is the same. That practice is peer agreed in Specialist philosophy. The Mechanism of Attention is well understood and Operated for particular purposes. Paying Attention to the Meaning represented in the two terms and the relation between them leads to the Perception that Meaning is the same. Thomson could sever the relations between Intrinsically Related parts in the mind's eye though use of tools of the Art of Persuasion, but she could not sever the relation between those parts in the Real World. The

Appearance of Justification for Actions that are not and can never be Justified was being gifted in her argument.

BEWARE THE TROJAN HORSE!

Thomson Acted against the perception of a Natural Relation between two things. Why did she turn her Attention to the Human Right to Life? The Concept of a Right to Life stood in the way of acceptance of her Belief. Aristotle taught that the Structure of the Syllogism gives the Appearance of Justification to an Induction. In other words, Aristotle recognised the Power of Natural Deduction to discover True Meaning and also saw how to Selectively Disable that Power. Recall that the Structure of the Syllogism transfers Meaning from parts through relations to a conclusion. In Induction, the conclusion follows by rule from the Meaning Asserted in the Major and Minor Premises. The deductive Structure gives the Appearance of Justification for a conclusion in the absence of the Action of Justifying. The Inductive arguments on Abortion can be stated in the form of two syllogisms. The deductive syllogisms enable identification of the location from which to attack the established position. The major and minor statements specified here are not Facts or Propositions, they are Premises. Premises might or might not be True. We cannot Verify their Truth until and unless they have been Tested.

1. <u>The Anti-Abortion position in the structure of a syllogism</u>:

    Major Premise: All human beings have a right to life
    Minor Premise: - A foetus is a human being from the moment of conception
    Conclusion: A foetus has a right to life from the moment of conception

2. <u>The Pro-Abortion position before Thomson in the structure of a syllogism</u>:

    Major Premise: All human beings have a right to life

Minor Premise: - A foetus is not a human being from the moment of conception

Conclusion: A foetus does not have a right to life from the moment of conception

The Anti-Abortion position does not Appear to be fallacious and the Pro-Abortion position before Thomson does have that Appearance. The material difference between the positions is in the Minor Premise. In order to shift the conclusion that a foetus has a right to life from the moment of conception, Pro-Abortion advocates had tried to find a Quantitative point after conception at which a change of State from non-human to human could be said to occur. That search could not succeed. The Ancients knew that Fact, and a long-standing puzzle reveals the problem without giving a solution to that problem. That is part of the Insoluble Problem of Induction. And it is a Problem of Quantitative Induction. The Problem of Detecting and Marking the Boundaries of Changes of State is Not Insoluble when the Process of Pure Science is engaged. The solution is Qualitative.

The Minor Premises are Testable. Conception determines whether or not a human being will exist and that is the Significant Happening where parts Transition from one Qualitative State to another. The Meaning of things changes with a Change in State. Conception is a particular type of happening that makes a Real Difference in Meaning. Two small separate parts come together and Actively Relate to each other thus becoming Unified as one Whole thing with particular Meaning being drawn from each of the different parts. That is a Natural System and is a systematic regularity in Nature. The Meaning of a Whole is different to the Meaning of the parts and Energy produces the transformation. After Conception there is growth and development of the foetus but its Meaning does not change. Meaning is Qualitative. The Quantitative Construct does not deal with Qualitative Differences and cannot ever do that. A single type of thing is Qualitatively the same irrespective of its size. A measure of growth is just that and no more than that. Changes in type of Meaning can happen and the Quantitative Construct of size has nothing to do with that.

In short: Conception marks the Boundary at which the type of Meaning is different on either side. The parts that can relate to form a human being are in a different State before they relate than afterwards. Relations are Active. Action changes things. The relation adds Meaning to the parts. Abortion is also a happening that makes a Real Difference to the existence of a human being. Conception is Action that enables a human being to exist and Abortion is Action that enables a human being not to exist anymore. Abortion removes the foetus from that which sustains its life. Severing the life-giving relation leads to a change in State of the foetus from Life to Death.

The Minor Premises in the two syllogisms are different definitions of a foetus: in one a foetus is defined as being human and in the other a foetus is defined as not being human. So, we clearly see that different definitions lead to different conclusions and that the different definitions serve different purposes. These definitions have been Tested for Justification. The definition of a foetus as a human being is Justified. That definition should be accepted as Fact. The contradictory definition is Falsified and should be rejected. A definition is a foundation on which Meaning is built in arrangements. It is a Fact that definitions are Highly Significant to the conclusions that are drawn.

Thomson perceived that she could not change the Minor Premise of the Anti-Abortion position. She knew the Ancient Puzzle of Sorites. She needed to find a different location from which to attack the Belief she opposed. Thomson gives the Appearance of bowing to the inevitable. She states:

> "I am inclined to agree, however, that the prospects for "drawing a line" in the development of the fetus look dim. I am inclined to think also that we shall probably have to agree that the fetus has already become a human person well before birth".

The Minor Premise of the Anti-Abortion position cannot be Falsified, due to its Quality of Truth. She gives the Appearance of accepting that a foetus is a human being. That is a Truth to which she does not agree. And in Formal Contexts there has to be agreement on truth if there is to be truth - that is a strange and Unnatural position. So, Thomson does not change her view that a foetus is not a human being. She cannot Justify her own Belief and she cannot Falsify the Minor Premise. She is determined to Unsettle the conclusion that is the opposing Belief. She does not change the starting point for her argument to Conception, which would have had the Quality of Justification and would have changed the direction of the argument. And in the absence of Justification she resorts to an Assertion of the truth of her own Belief towards the end of her article. She ends precisely where she began:

> "....it should be remembered that we have only been pretending throughout that the fetus is a human being from the moment of conception. A very early abortion is surely not the killing of a person, and so is not dealt with by anything I have said here."

She is stating that she wrote Fiction. She used Fiction to give the Appearance of Justification to an argument that she could not Justify. In other words, she told Untruths in order to gift her Belief to other people. And in the absence of searching for Truth and Justification, her Attention had necessarily turned to the Major Premise. In order to Unsettle the conclusion on Abortion she focussed Attention on the Human Right to Life, which she proceeded to attack. Thomson had the Intuition that if she split the Right to Life from the Right to What Sustains Life then the Right to Life can still Appear to stand while actually losing all of its Meaning. Concepts are a type of Point of Reference and Specialists in philosophy systematically drop Points of Reference and thus do not reveal what their Attention is focussed on doing. Thomson's Intuition was based on the teachings of the fathers of Artificial philosophy and did not come from a Blank Slate. If you lose Meaning then you can insert in the gap any Meaning that you choose. Beliefs are characterised by the Feeling of Certainty and their truth is Assumed. Then Beliefs are Asserted. The Difference in Meaning between a Belief and a Premise is Trivial. Thomson created the False Appearance that she was Uncertain in order to argue against the opposing position and thus lead the audience to Certainty of the truth of her own personal Belief. She used theses and antitheses. The Fault lies in the failure of Formal logics to Test for True Meaning coupled with the Refusal to accept Evidence from Testing within the Process of Pure Science when it is produced.

So, let us again turn Attention to the Major Premise as we necessarily must do. The Concept of Human Rights is Complex. It relates to what people can do to other things, including members of their own species, and thus it relates to what can be done to them. The Meaning being transferred from the Major Premise to the foetus is a specific Meaning from the Concept; it is the right not to be killed by another human being. That specific Meaning was given in a different form as a Commandment in the Christian Religion – Thou Shalt Not Kill. A Commandment is a Rule that prohibits things. Her argument on Abortion is part of co-ordinated Action against Metaphysics. Thomson's argument aligns with the Actions of Hume, Popper, the members of the Vienna Circle and many others across the Centuries who aimed to take away the Power of all types of Metaphysics to Act on the Perception of the Meaning of doings and happenings. Religion is a type of Metaphysics. Pure Science is a different type of Metaphysics. All types of Metaphysics have some things in common and are distinguished by their differences.

If there is Error in the conclusion of the Anti-Abortion position where does the Error spring from? General Rules apply to groups of things. A group is a different Point of Reference to a Single Case. It is possible to transfer Meaning from a group to a group and for the transfer to be Justifiable under certain conditions. The transfer of Meaning from a group to a Single Case is Not Justifiable. That can lead to Error in Judgment. But, in the position of the Minor Premise in the Abortion Syllogisms where there can be a Single Case there is a group of things. The foetus does not have a name and a particular identity. The foetus here is a group – it is a class of many things formed by taking common Meaning from each unique part that has particular Meaning of its own. The Evidence that the word foetus represents a group of human beings at a stage of development is compelling. Meaning can be transferred from a group to a group with Justification and the conditions for Justification are met here – the foetus and a human being are one type of being in different forms. What applies to all human beings applies to them.

Religious Laws are man-made Laws and in common with other Laws of that type they should be interpreted in Single Cases. But that was not happening to foetuses. The Syllogism can be used to transfer Meaning Justifiably between groups, under the condition that the relations are Tested for Truth, but the Justification for that Action is lost when we do not perceive that it is groups we are dealing with and the idea that we are dealing instead with Single Cases takes hold. The term 'a foetus' suggests that a Single Case is the Focus of Attention for the transfer of Meaning and that suggestion is wrong for the group is the Point of Reference here. The Appearance of a Single Case is Deceptive. We are misled by that term. Differences must be taken into account when dealing with Single Cases. Differences are not taken into account when dealing with them in groups. Miss-taking the Meaning of the Group for the Meaning of the Single Case is a common Fault in Inductive arguments. And that is the problem to which Thomson directed her argument. But that is not the only problem here and all relevant problems must be addressed if an accurate solution is to be found.

The Anti-Abortion position does not take differences in the Meaning of Single Cases of a foetus into account. And that also applies to the Right to Life. It is not possible to Justify dealing with the same group of things in different ways when they are the same in type. If differences in Single Cases are to be taken into account in the Right to Life of foetuses, then they must be taken into account in the Right to Life of other human beings. Thomson did not argue that. She stated that:

> "it should turn out to be a truth that all persons have a right to life".

It is a Fact that the right not to be killed by another human being enables consequences to follow when killing happens and that consequences can reduce the risk of that happening. And it is necessary to reduce that risk. But Dynamic Systems are Complex Systems. It is always necessary to Calculate Meaning using different Points of Reference. Those who already know the answer do Not Speculate. That Fault leads to Errors of Judgment.

Narratives provide the Context to clarify particular Meaning from definitions and Concepts. It is Natural to tell stories. That which is Complex can be Simplified through telling stories. The Qualities of stories are variable. Stories deliver particular Meaning and the type of Meaning delivered is determined by the designer of the story. Thomson was a Specialist in philosophy and she knew that some of her peers deny Meaning that is actually True. But she accepted the Illusion that True Meaning is Trivial. She should have known the Fact that True Meaning is Significant. There was a choice and Facts had been shown. She also knew that some of her peers deny that stories are powerful in delivering Meaning and she did not accept that Denial. Thomson knew that she could use the Power of Narrative to her advantage. She harnessed the Power of Fiction to deliver her own Beliefs through the Structure of the Enthymeme.

Speculation is a Necessity if Truth and Justification are to be found. We have not reached the end of the particular story I am telling. If differences in Single Cases are to be taken into account in the Right to Life of foetuses, then they must be taken into account in the Right to Life of other human beings. But I will add a Universal Rider to that. Justification is an Issue in taking differences in Single Cases into account. The Judges in our Courts know that Fact. And I have shown you the problems they face when they search for Justification. If human beings use Formal logic to make decisions in any Context then those decisions will be in Error. And it is a Fact that Pure Science is Not in Operation in decision making in Social Dynamics. There will be an Unquantifiable number of Miscarriages of Justice in applying the Law, be it Moral or Legal Law, until the Method for reaching Judgments changes wherever conclusions are to be drawn. Lines must be drawn. Prohibitions are necessary. But Truth must be our guide and Justification must be an Issue.

Pure Deception is in Operation when an agent diverts Attention from the True Meaning of their Actions having recognised that if True Meaning were to be seen it would be clear that the type of Action is unacceptable. Pure Deception is Not Action in the Moral Realm. Thomson knew that it is necessary to hide True Meaning when Actions are not Justifiable. She knew that open Denial of the Human Right to Life would provoke a sharp response and defeat her purpose. But Thomson had learned Methods by which she can impose her Will on others when Justification is Impossible. And it is presumed she was full of Good Intentions. In order for her argument to succeed in Unsettling the opposition to Abortion, it was necessary to create the Illusion that she is not attacking the Right to Life. Illusions of Meaning are the Appearance of False Meaning. Meaning that is False is Not True Meaning. Now, Action follows the Meaning that is perceived. If the audience could be led to Believe that Thomson was not attacking the Human Right to Life, then she and her Actions would not be attacked by the audience and her conclusion on Abortion could be accepted by them. Clearly, she succeeded in creating that Illusion. It is definitely an Illusion. I have Verified that and it is a Fact. Our Judgments are in Error when we are misled by Illusions.

The outcome of Acting on the basis of False Meaning is not anticipated outside the Boundaries of Pure Science. But the effects of Meaning can be anticipated relatively easily and while not all of them are, some effects are anticipated. Truth is of no use at all to those who want to transfer their own Unjustifiable Opinion or Belief to another agent. Truth is of no use when there is No Truth in an Opinion and the Opinion serves a personal purpose. There is never a gap in terms of purpose, but it is difficult to see what Thomson's purpose was. It is not as difficult to see how the Actions Function. Thomson perceived that the lack of Justification for a Major Premise that No Human Has a Right to Life will be visible if she openly states that Premise. She denies that Premise is Active. In her article she states that to deny a general statement that human beings have a Right to Life would be "outrageous". She declares support for that Right. Thomson states:

> "But I would stress that I am not arguing that people do not have a right to life – quite to the contrary, it seems to me that the primary control we must place on the acceptability of an account of rights is that it should turn out to be a truth that all persons have a right to life."

In Formal logic truth is what peers agree it to be, so here Thomson is merely stating that peers should agree that all persons have a Right to Life. Truth is not an issue in peer agreements. She affirms her agreement to that statement and then Acts against that Right. Her Representation of Meaning in words did not fit her Actions. Deception is Complex Action and it is possible to mislead an audience while not actually telling a lie. I do not accuse Thomson of lying, nor do I suggest she had Malicious Intent. But it is necessary to look closely at her Actions and where they led. In order for her argument to succeed, Thomson had to produce the Illusion that only the Right to Life of the foetus is unsettled by her argument. She simply attached False Meaning to her Actions and Asserted the False Meaning was true. That is the Action of Premising.

It is Asserted that there is no alternative to Formal logic. We should not accept that Premise. It is Not Safe to give the Power to make Single Case Judgments to those who use Formal logic. While we do so there will be Miscarriages of Justice – the Innocent will be Convicted and the Guilty will go Free. Formal logics are interfering with Natural Deduction. Inductive science is interfering with Pure Science. I propose that Rights are a Natural Concept that is underdeveloped. Constructs have replaced Concepts. The Constructs of Human Rights are a product of "moral reasoning" using Induction. And reasoning using Induction is not actually Reasoning. How is moral reasoning defined within Specialist philosophy?

AN ILLUSION OF MORAL REASONING

As you would expect, there has been much discussion amongst the Specialists in philosophy and that hive of activity has resulted in the Appearance of Utter Confusion in Meaning, which is covered by Vagueness in Representation. There is No Doubt about that. Specialists in philosophy reach agreements on Representations and do not reach agreements on the basis of Meaning. From the perspective of Richardson (2013):

> "Moral reasoning is individual or collective practical reasoning about what, morally, one ought to do."

Thomson's argument fits that vague statement in which a definition of what is moral is absent. Thomson tells her audience what one ought to do and what one need not do. She Asserts her Opinion. Surely Moral Reasoning is Reasoning that is Justified to a Moral Standard in relation to the Natural World. That Standard would be a Concept. The Concept of Moral has not been developed in the literature of Formal philosophy.

Thomson understood very well the power of using Single Cases in an argument and of Representing Differences in Meaning. I quote from her paper. She stated:

> "For there are cases and cases and the details make a difference".

Details do indeed make differences in Meaning. Here Thomson makes visible that Specialists in philosophy know that Fact. But it Appears that Specialist philosophers do not understand the Power of Truth or how to access True Meaning and use that Power. In service of the Pro-Abortion argument, the parable of the Good Samaritan is unsettled by Thomson and a supposedly practical alternative is offered. I will deal with that practical reasoning and where it leads in the closing Chapter of my narrative, for the details of that part of her argument are Significant in terms of Functioning in Social Dynamics. For now, I draw your attention to the Fact that Thomson attempts to replace the Moral Standard of Conduct demonstrated in the Ancient Narrative of a Good Samaritan with the Construct of a Minimally Decent Samaritan. She states:

> "I have been arguing that no person is morally required to make large sacrifices to sustain the life of another who has no right to demand them, and this even when the sacrifices do not include life itself; we are not morally required to be Good Samaritans or anyway Very Good Samaritans to one another."

The sacrifice of one human life for another was abolished centuries ago. Or was it? You have read the first Chapter of this book, so you can bring that part to bear as you reflect on possible answers to that question. Slavery was abolished more recently. Or was it? Perhaps the Meaning represented in the words sacrifice and slavery stayed the same and was given a Different Appearance. The proposition that the Representation changed and the Meaning remained the same is viable and Testable. Thomson does not see Abortion as the sacrifice of a human being in the form of foetus and sees the failure to Abort as the sacrifice of the mother. A sacrifice is a loss. It is an interesting view but it Justifiable? I will in due course Test that view for Justification. The sacrifice of a foetus is the sacrifice of life itself. The type of sacrifice of the mother is not specified by Thomson.

In Thomson's argument we Observe, as we have done in other arguments on different topics that agents skilled in philosophical talk of a Formal type engage in unsettling the Beliefs of those who oppose the things that they themselves know for Certain. In conjunction with that, we have observed their argument that there is no such thing as Certainty. Now that is a puzzling discrepancy and they cannot account for that. And they are the philosophers who have agreed they cannot be held to account for anything that happens. That group is composed of a particular type of philosopher. I will Verify that there is a different type of philosopher who foresees what potentially could happen and takes that into account in advance. That is a Natural type of Philosophy and it requires development in order to realise its Potential.

Pure Science can solve puzzles of Meaning. That which Does Not Appear to Make Sense Does Make Sense when you take a different angle of view. And, that which makes Sense is not always Justifiable or Moral. From all that has gone before in this narrative and all that will come after, it follows that:

NATURAL REASONING TEMPERED BY EMOTION
LEADS TO
GOOD FUNCTIONING.

SO THAT IS WHAT I PROPOSE.

I will define Meaning in the terms of that statement in order to render the proposition Testable. Meaning is Relative. And that which it is relative to must be specified.

CHAPTER FIVE: GLIMPSING CAUSATION THROUGH EYES IN THE ANCIENT WORLD

PART ONE of Chapter five

WHAT IS WISDOM?

I will show the definition from the Oxford Pocket Dictionary:

wisdom: n. 1. experience and knowledge together with the power of applying them. 2. prudence, common sense....... [Old English: related to WISE]

When using words, Meaning must be specified in Context but it must be there in the parts if it is to be drawn from them. Looking at the definitions of the words experience and knowledge in a dictionary would not take you far towards understanding the Meaning of wisdom. But as the definition stands and if you grasp Meaning from your own assumptions, then the Meaning of wisdom looks clear. We think we know the Meaning of experience and knowledge, of prudence and common sense, and the Meaning of wisdom follows from that.

The definition of the adjective wise takes Meaning from the word wisdom and freely distributes that. The distribution is not random. The adjective wise is used to describe different types of thing.

wise: adj. 1. having, showing, or dictated by wisdom. 2. prudent, sensible. 3. having knowledge (often in comb.: streetwise; worldly-wise). 4. suggestive of wisdom. 5. US *colloq.* alert, crafty.................

Is that which is suggestive of wisdom an indication that wisdom is there? Is it prudent and sensible to be crafty? The direction of the drift of Meaning is clear. Meaning can flow from the parts of the definition of wise through the Whole that is the word wisdom, which is included within that definition, and on to those who are streetwise and crafty. The word crafty is defined as cunning, artful, wily. This definition is a catch-all definition. The type of Action on the Meaning of the word wisdom is clear. Different types of Meaning are

Confused in the definition and the Confusion does not become Apparent until we look elsewhere. Definitions are Assertions of truths that are mistaken as being Facts. We are misled into thinking that different types of thing with different types of Meaning are one type of thing when those differences are not picked up. It is a Fact that definitions are parts in Illusions of Meaning. The Ancients have many questions to answer in relation to their Actions. They cannot do so now. I doubt that their followers can do so or are willing to do so. That does not matter if Meaning can be Calculated. Many of the Actions across the span of Centuries were documented and are still visible.

I note that knowing the Truth has not been included in the definitions of wisdom and wise and that the word meaning is not there. The words experience and knowledge carry a suggestion that the Perception of True Meaning is a Factor in wisdom. Or so I presumed until I looked at those words in the dictionary:

<u>knowledge</u>: n. 1. (usu. foll. by *of*) awareness or familiarity (of or with a person or thing) (*have no knowledge of that*) 2. person's range of information. 3 a. (usu. foll. by of) understanding of a subject etc. (*good knowledge of Greek*) b. sum of what is known (*every branch of knowledge*)

The word truth is not there, nor is the word meaning. All parts of that definition are Constructs. I will draw Attention to part 3 b. in particular. That leans towards a Quantitative Construct of knowledge. A Concept of Knowledge awaits development and would be related to Truth and the Natural World. There will be an Association of these Constructs with the word truth somewhere along the line of Representation but that does not occur in the word experience:

<u>experience</u>: observation of or practical acquaintance with facts or events. 2. knowledge or skill resulting from this. 3. event or activity participated in or observed ….

The word fact provides a link to the word true. Now I will quote again from one of many famous Twentieth Century philosophers. Russell (1946, 2004) stated:

> "Science tells us what we can know, but what we can know is little, and if we forget how much we cannot know we become insensitive to many things of very great importance."

That Meaning is aligned to Meaning expressed by Plato and I will show you that in a moment. But first I will deal with Russell's version. I propose that the suggestion "what we can know is little" is not only a defeasible generalisation but also a misdirection of Attention. Acceptance of that Premise is limiting. We do not know what we can know until and unless we Test the Meaning of things for Truth. I would say that remembering how much we do not know in every Context is part of the Natural Philosophy of Pure Science. I would add in equal measure that if we forget how much we can know and how we can discover the Meaning of things, then we become insensitive to many things of very great importance. Pure Science is the route to Knowledge and Inductive science is the route to forgetting how much we know. Many things of very great importance are denied by men of science. Specialists in philosophy designed science. I will presume they did that without bothering to discover what type of Meaning Pure Science has and what type of Action it is.

Specialists in philosophy speak vaguely and in so doing they deliver Meaning precisely. Curiouser and Curiouser said I as I watched the performance. But, some of these Specialists are clever fellows with different and more Impressive Qualifications than I have. And they have Authority in a field that has the Appearance of being different to the field of psychology. Doubt entered my mind. If the Appearance of Impressive Qualifications is not designed to produce Uncertainty in those who do not have them, then at least that is how they Function. Given the Uncertainty I saw, I asked another question: Were the Actions I saw as foolish as they appeared to be to me? Is it foolish to search for True Meaning? Or is that what we should do?

In this Chapter I present parts of the Narrative Legacy from the Ancient World. That which people leave behind is not necessarily worth keeping. But if the Meaning they document and leave behind is in Error, I would not consign it to the flames for that which is left behind will be useful as a reminder of what was created and came into being. Comparative Analysis can lead to Causal Knowledge. But the passing of that which was created is Natural so we should not prevent that happening when its time is up. The memory should remain while we move on to a different place. Different Wholes are created in Nature and have Meaning that is different to the parts from which they emerged. I have proposed that Natural Processes in human beings have been disrupted and Artificial Meaning from the past has been installed. That which is installed in a System by human beings can be uninstalled and replaced with a different Operating System.

There is of course Bias in hindsight. What matters is the type of Bias. The Bias is in the direction of discovering Truth when the Process of Pure Science Rules the Action in the Arena. Now settle in your seats and observe some Action by the Ancients in conjunction with some Calculations of its Meaning.

THE ILLUSION OF KNOWLEDGE

I will begin with a quote from Plato's Apology (2002 translation by Grube). Plato (427-347 B.C.) put words into the mouth of Socrates to represent wisdom as follows:

> "what is probable, gentlemen, is that in fact the god is wise and that his oracular response meant that human wisdom is worth little or nothing, and that when he says this man, Socrates, he is using my name as an example, as if he said: This man among you, mortals, is the wisest who, like Socrates, understands that his wisdom is worthless"

I will turn later to the "god" and the words of the oracle to which Plato's Socrates refers. Many of you will be familiar with the narratives of Plato and interpretations of them while some of you may not be. I will show you my view. In the Apology, "Socrates" refers to the "wisdom" of those who are "reputed wise". Socrates is said to have tested samples of statements and narratives from the politicians, poets and craftsmen. Their wisdom is said to be "worthless" and their failure to recognise that as a fact is the difference that makes Socrates wise. If wisdom is worthless then it is not Knowledge but it could now be said that knowledge is worthless through use of the Mechanism of Association that is Active in definitions. The value was taken out of Knowledge. It could also be said that experience, prudence, and common sense are worthless for the things those words represent are part of the definition of wisdom. I disagree with the valuation of knowledge, experience, prudence, and common sense that Plato provided through his characterisation of Socrates. What are we left with that is of value by their standards? We can look for the answer by looking at the things to which the fathers of philosophy and their followers attached value. Their values are different to mine.

The term Reputed Wise is a useful one, for that leaves a question hanging in the air ready to be drawn down. The question changes a Premise into a Proposition for testing. Is their reputation for wisdom Justified? A reputation for having knowledge would be Justified if it were True. We do know that politicians, poets and craftsmen have experience in what they do. But we have no Concept of Knowledge to bring to bear on the question. In this part of the Apology we are not told what wisdom is except that Socrates is the wisest man among them because he knows he is not wise. That is Non-Sense. Can Sense be made of Non-Sense? The answer to that question is yes.

Plato's Socrates is not included among the Reputed Wise. The Appearance of being wise was tested in those who had that Appearance and the Appearance was found to be False. In other words, the Appearance of Knowledge presented by these professional men was shown to be an Illusion. We are shown that Appearances are Deceptive. We are also shown that if you give the Appearance of not being wise you are not Tested. So, I presume Plato understood that Meaning is Relative. The Appearance that Socrates does not have wisdom might also be Deceptive. It is said "Socrates, understands that his wisdom is worthless". But what Socrates understands is that their wisdom is worthless. His wisdom remains Untested for he makes no claim to knowledge. That is crafty and crafty is part of the definition of wise that we have been given by those who designed the definition of that word. Those who put Meaning into words through the Action of defining them are Specialists in philosophy. There were many philosophers before Plato and Socrates. Plato is one of the founding fathers of a particular philosophical position that spread and endured. In his narratives there are definitions in progress. And Plato attributes his sayings to Socrates. Attribution is an interesting Mechanism and a Powerful one.

Socrates was influential and part of a group. How do you give the Appearance that you are not wise while exerting influence? There is more than one way of doing that. We are shown in the Apology that you can give the Appearance that you are merely the instrument of a higher Authority. Gods are different to mortals. The message is that the god is wise. Socrates is wise for he recognises the knowledge and experience that he has is worth nothing relative to the knowledge and experience of the god. Being wisest is being defined as listening to the god and deferring to the god - "human wisdom is worth little or nothing". And, only those who can hear the god speak can pass on knowledge. Of course, the word god must also be defined and Specialists in Formal philosophy drop differences in type. The difference between a god and a human being was dropped in Ancient times. Oracles were human beings receiving messages from immortals. They drew Power through Association with the gods. In the Apology and other Essays, a definition of some particular human beings as gods was in progress.

Socrates, Plato and Aristotle are acknowledged as fathers of Formal philosophy. Children learn from their parents in various ways. Modelling Meaning through Action is a Powerful Mechanism for the Installation in us and in others of Meaning to guide Action. I have shown Evidence that the common definition of philosophy in our dictionaries is an Illusion of Meaning since the definition is restricted to Inductive philosophy. Plato's narratives enable us to see some of the Action involved in the Creation of that Illusion. Stories are Powerful Structures for the delivery of Meaning. Specialists in philosophy do not produce Knowledge, know nothing for Certain and they tell us nothing that we can pin down to them. They deliver conclusions surreptitiously through stories. The Structure of the narrative channels Meaning in particular directions towards particular conclusions. That type of Action in Formal logic is represented by the word Enthymeme. The Enthymeme channels Meaning in one direction towards a Single conclusion.

In the Apology, Plato did not give the details of the tests applied to the selected groups of specialists, but the term Socratic Method covers those details and they are given in other narratives by Plato. Socrates did not include philosophers among those whose wisdom he tested and found wanting. The suggestion is that he was testing himself by testing others. And in Plato's tales of happenings the Single Case of Socrates represents the group of philosophers. From all the suggestions in the Apology, it follows that as a group, philosophers are the wisest of men. And why is that so? Well that is so because they can show that the wisdom of other men is worthless. That is an Extraordinary Suggestion and yet that Meaning has been accepted! That is what Specialists in philosophy do.

I will repeat that suggestion while picking up a particular part of Meaning from the word wisdom. Specialists in philosophy spend their time showing that any knowledge gained from experience with our Senses and applied with prudence and common sense is worthless. I have shown you Evidence of that happening systematically in recent times. It is a Fact that the philosophers who are descendants of Plato do their best to show that the Reputations for Wisdom of all other men are worthless. They are fully occupied in doing that. Any sample large or small, of the writings of Specialists in Formal logic is testament to that. That leaves Meaning in an Unstable State. Knowledge must be found and cannot be found by showing where it is not to be found. Or can it? That is an interesting question. Bring into play the Fact that Meaning is Relative and you will grasp how that could work to advantage. But the Premise is: Socrates is the wisest relative to those who have a Reputation for being wise. And he apparently knows nothing. Thus, wisdom is not to be found there either. Yet he is not shown as being in a State of Uncertainty. Thus, he is in a State of Belief. Or, he may know some Facts he has not disclosed. Are you now thoroughly Confused? Meaning should be emerging through the mist. The clouds can be seen above you and sat there upon them are the Oracles that the Specialist philosophers are – they are contemplating the Meaning of Life on Earth and mediating between us mere mortals and gods.

Meaning can flow in one direction only, from the heavens through them to us. Clearly Meaning cannot flow in the other direction for human wisdom is worthless.

I propose that Knowledge cannot be found only by Testing the Meaning of existing cognitions and showing that Knowledge is not there. If Knowledge is not where it Appears to be then it is somewhere else and we do not know where it is. Thus, we must search in many places and we must also be able to recognise Knowledge when we find it. That brings us back to the point that the Concept of Knowledge must be clearly specified and has not been. Concepts have more than one Meaning but they are not vague. How do you find True Meaning when you do not know what that Meaning is? Well, that is what Pure Science is designed to do and Nature designed that Process. Pure Science is a Natural Process. Natural Processes are different to the Artificial Processes of human design. Pure Science produces Knowledge.

Socrates Appears to be a Single human being in Plato's stories but represents the group of philosophers. The Reputed Wise in Plato's narrative consisted of groups of people who have particular interests. I would include Specialists of all kinds, including philosophers, in the category of the "reputed wise" but I would restrict the term to the group. Groups of Specialists are "wise" only by definition and it is thus that they are the Reputed Wise. The definition includes those who are wise amongst those who are not wise, whichever Meaning you select for wise. General categories hide from view the differences in type that exist between Single Cases. Within the Boundaries of Pure Science differences in Meaning are identified. In Pure Science, Attention is focussed on Single Cases. The identification of True Meaning enables responding that fits the things we encounter. When we have Knowledge, the Meaning we perceive fits the Action that takes place – True Meaning fits doings and happenings. We know that Fact. But all we see is the Action. We cannot perceive Meaning directly and that is where the problem lies. The Capacity to Reason bridges the gap between Meaning and what we see and feel. When that Capacity is disabled, the Judgment is impaired.

Let us return to the suggestion left by Plato that the mortal who understands that his wisdom is worthless is the wisest. That is puzzling. I will specify two different Points of Reference. Some people are guided by what they think they know and some are guided by what they actually do know. But did Socrates actually know anything? That is a good question. The Premise is that he did not and in conjunction with that is the Premise that he did. We are left to draw our own conclusions from the rest of the material provided. That is the Action in an Enthymeme. But the conclusion is not our own. Plato Associated Socrates with truth. Meaning travels along channels. The Association of the word truth with Socrates travels to all those philosophers who Act in the same way. The Association of Formal philosophy with Truth is Not Justified. The common definition of Formal philosophy is a Premise. The proposition that Plato created an Illusion is Testable.

Plato had a habit of losing Points of Reference. When a Point of Reference is not given, particular Meaning is not given. But Enthymemes and the Socratic Method give particular Meaning and they do not give True Meaning. The Active Points of Reference for Inductive arguments are hidden. Plato was one of the designers of the Formal philosophical position and hiding Meaning is at the core of that. The Action of losing Meaning is threaded through every arrangement in the Context of Induction. Context clarifies Meaning and Formal logic clarifies that none should be specified while some should be given. And that which is given is particular Meaning. Only the Fact that the Meaning is given is hidden. And in order to demonstrate that the reputation for wisdom of all men in the professions was False, Socrates put Single Cases to the test. Then Meaning was generalised to the group of men. That Error has been systematically repeated across the Centuries. That which is characteristic of one man may be a common characteristic of a group of them, but it is differences that define Single Cases. The Meaning of each Single Case is different to the Meaning of the group as a Whole and vice versa.

The Title of Plato's Essay is translated as The Apology and it does not state in the title whose Apology that is. In Plato's time it was common practice to use a particular individual to represent Meaning of your own as it is given through a story. The poets did that. They used Single Cases in a particular way. Meaning was drawn from Plato's stories and from the stories of a small group of people associated with him, and as it travelled through time and space that Meaning was Represented in different forms by other people. The type of Meaning did not change. Systematic regularities of thinking across the Centuries have led to the position we are now in. There have been variations on the theme, but in the long run those variations in Appearance cancel out and we can see the core Meaning that holds them together and determines the trajectory. In our time we can see how the Artificial philosophy designed by the Ancient Greeks has Functioned in the long run. The outcome of continuing to persist with Formal thinking and the Structures that were built on those foundations has been accurately predicted by many people. But the Source of the problem was Not Verified. Causal Knowledge is needed to enable effective Action. Let us discover how much we can know.

SOCRATES AND THE SOCRATIC PROBLEM

Socrates was an Athenian living in the years between 469 BC, his birth and 399 BC, his trial and execution. He did not leave a written legacy. Sommerstein (1973), a Professor and an expert on Greek history, tells us:

> "There is no evidence that his contemporaries in general regarded him as a man of exceptional merit at all."

If a Quantitative Construct of merit is applied then there is no evidence that the contemporaries of Socrates regarded him as having exceptional merit. But a Qualitative Measure leads to a different conclusion. There is Evidence that he attracted the Attention of his contemporaries, and that there were different Opinions on his merit. Those differences were expressed in Qualitative Terms. Nails (2010), who is a Specialist philosopher, tells us:

> "Certainly he was impressive, so impressive that many others were moved to write about him, all of whom found him strange by the conventions of fifth-century Athens: in his appearance, personality, and behaviour, as well as in his views and methods."

It is a Fact that Socrates attracted Attention during the time of his being in this World. And he has done so ever since. One way or another Socrates acquired a reputation and his reputation has endured. The man who may or may not have had merit relative to other people became an Authority after his time. We have no Qualitative Standard to enable merit to be judged for we lack a Concept of Merit, but the Construct of Fame is there. Fame is judged by the number of people who have heard of you and there are different types of fame. People can be famous for doing good things or for doing bad things. One route to fame in fifth-century Athens was through the traditional Old Comedy. Being taken as the subject of Comedy is still a route to fame.

In 423 BC, Aristophanes wrote and presented a play in which Socrates was a major character. We are told that this play was not particularly successful at the time and yet it has achieved lasting fame. That is not unusual for works of Art. Sommerstein describes 'The Clouds' as tragic not comic and that description fits the narrative. When reading The Clouds, laughter is not triggered in response. According to Sommerstein, The Clouds is a Complex Political Commentary and "was apparently too dark for its original audience". Dark clouds are ominous and are portents of storms. Reading the play leads inexorably to the conclusion that it was Complex Political Commentary. But what determined that it was initially not particularly successful is a matter of supposition.

Sommerstein informs us that:

> "Aristophanes could have created, if he had wanted to, a fictitious composite of the leading 'sophists' of the day................ Many indeed would argue that that was precisely what he did. But the tradition of the Old Comedy was that major satirical targets should be given the identities, and usually the names, of actual contemporaries .... and Aristophanes gave his fictitious composite the identity and the name of Socrates."

That which is fictitious is Not True. Sommerstein accepts at the end of the paragraph that the character of Socrates was a "fictitious composite". But does that term fit? The name Socrates was the name of a Single Case. That was the name of a particular person in Athenian Society. When the name was used it was clear to the audience at that time which person was referred to and likely that they had seen him. Socrates would have been seen in the company of other people. It is clear that the tradition of the writers of plays at the time was for the identity of the person named to be a composite. That which is a composite is composed of more than one thing. The Whole is different to every part in the arrangement of which it is composed. Poets Represent Meaning in their narratives. It is Reasonable to accept that Aristophanes portrait of Socrates is Not a True Likeness of the man and represents Meaning the group had in common as Aristophanes saw that Meaning. The word Socrates is the Mark of a group. As a Single Case, the character was fictitious. But was the Meaning of the group fictitious or was the Meaning of the Whole group as it was portrayed by Aristophanes actually True? If so, then the term "fictitious composite" is in Error. And I note that Sommerstein puts Socrates in a group of Sophists.

Satire is Political Commentary and is an exaggeration of what actually happens. An exaggeration goes beyond the Meaning that is there and is in line with that. In other words, True Meaning is Represented in satire and Aristophanes is described as a writer of satire, which is defined as:

satire: n. 1. ridicule, irony, etc., used to expose folly or vice etc. 2. work using this.

So, it is recognised that Aristophanes used ridicule and irony to expose folly or vice in the doings of the group of philosophers of which Socrates was part. That was over Twenty-Five Centuries ago. If Aristophanes view was accurate, then its public expression had no noticeable effect. The Poets wrote Fiction. It can be accepted that the Representation on stage was not an accurate description of actual events in the Real World. It is a Fact that Meaning is different to its Representation and can be different in type as well as in form. And now we have some questions that are Highly Significant and to which we need answers. Facts can be Represented in fictional terms. Did Aristophanes Represent in his plays Meaning that was True? Was folly or vice exposed? If so, then why did that gift of Meaning fail to have a Significant Effect?

According to Plato, Aristophanes portrait of Socrates is Not a True Likeness of the Meaning of the group. Plato gives a different view of the doings of the group and that is useful. The Meaning of the group must be Tested and that involves Comparative Analysis in terms of doings and happenings. It is a Fact that in the Real World Socrates was not suspended in a basket above others in the cast of players there. Aristophanes did not suggest that that is what happened in Reality. The basket was symbolic of clouds. Aristophanes gave his interpretation of the Meaning of what he had Observed happening in the Real World. Specialists in philosophy have accepted that Aristophanes Representation was False. It is Plato's interpretation of the group and of his friend and mentor that has been accepted as true. The definition of the Meaning of Socrates that Plato created was passed on to future generations and the agreement on its truth enables a particular type of Action to continue. Plato set a Premise. A Premise is a Definition that has not been Tested for Truth. When the Process of Pure Science is engaged, Meaning can be Tested for Truth.

The Mechanism of Association was Active before human beings existed and long before there was Awareness of its Operation. The Power of Activating the Mechanism of Association became apparent once the Capacity for Awareness was Installed in things. Both Aristophanes and Plato used a named Single Person as a Mechanism to Transfer Meaning to an audience. That Mechanism is in frequent use today. The name was the Point of Reference for the Meaning and is a Simple vehicle that carries Meaning from one place to another. Particular Meaning is Associated with Single Cases. Aristophanes put Socrates in the category Sophist. Plato set Socrates apart from the Sophists. Plato Marks a difference in type of Action from that of the Sophists but there is Evidence that the difference he Marked was Not True.

MEANING THAT IS FICTITIOUS ACQUIRES THE APPEARANCE OF BEING TRUE THROUGH ASSOCIATION WITH THINGS THAT ARE FACTS

AND

MEANING THAT IS FACT ACQUIRES THE APPEARANCE OF BEING FALSE THROUGH ASSOCIATION WITH THINGS THAT ARE FICTITIOUS

So now I will recall the example of the Syllogism that Specialists in philosophy frequently use to demonstrate the deductive Action within Induction. I stated that this familiar demonstration of the Syllogism begins with Facts. The specified relations are True. The conclusion is True:

All men are mortal
Socrates was a man
Socrates was mortal

Being a man and mortal was part of the Meaning of Socrates. He was a human being. But now we see that the word Socrates as it is used does not represent the particular Meaning and Action of a Single human being and Represents a group of them. And it is a Fact that the Meaning that was Associated with the name of a man did not die when he did. Meaning is not mortal. Or is it? That Meaning has survived over 25 Centuries. Plato fought Aristophanes for the type of Meaning of Socrates and that was a battle to the death. Aristophanes Meaning survived, but it did not survive the battle to be perceived as True or Justified. But Meaning is not a human being and can be Resurrected. That is worth doing if the Meaning is True or Justified. I propose that this familiar and much-repeated example of the syllogism is part of the Action of Associating self-styled philosophers who engage in a particular type of Action with the different type of Action of Natural Deduction. The transfer of the Meaning and Action of Natural Deduction to the group of philosophers named Socrates is Not Justifiable.

Sommerstein gives us information and offers some views. He continues:

> "..... (if we are to trust Plato) Socrates, while constantly questioning the beliefs of others, generally avoided any explicit statement of his own beliefs. It could plausibly be argued that of all the intellectuals of the late fifth century, he would seem to the

> outsider to be the one who excelled in negative criticism but had least to offer in the way of constructive ideas".

It would not be in the spirit of Pure Science to trust Meaning given by Plato. There are two parts within the apparently plausible argument described by Sommerstein. There is Evidence that Plato's composite bearing the name Socrates constantly questioned the Beliefs of others and excelled in doing that - that part is negative criticism. Aristophanes and Plato agree on that part, though the agreement itself is not the Evidence of Fact. But is it True that the group named after Socrates had least to offer in the way of constructive ideas? Aristophanes and Plato are not in agreement on that part. There is Evidence that Plato's Socrates generally avoided any explicit statement of his own Beliefs, and that he did not always do so. The Evidence indicates that Plato wanted others to Believe that the Method that bears the name of Socrates was not constructive. That is Testable. Plato was not dismissive of poetry, acknowledged the Power of the poets to affect the perception of Meaning and used some of their tools. Aristotle was a pupil of Plato's. It is a Fact that Aristotle studied the techniques of the poets for delivering Meaning and laid out those tools for use. Those tools are constructive. Socrates, Plato and Aristotle are agreed to be the fathers of Formal philosophy. We have seen denial of any merit in Metaphysics by Hume and others. That is a difference from the fathers. But Aristotle and Plato's Socrates also used Mathematics, and Specialists in Artificial philosophy, including Hume, favoured that constructive part. The Quantitative Method is the constructive part of Specialist philosophy that is now dominant. Negative criticism of everything that is not Quantitative has led to the current position. The Quantitative Method is the Art of poets in code. I will need to prove that is a statement of Fact, so for now you can take that as a Proposition. That particular proposition will not be Tested in this narrative. But the Evidence presented here proves Beyond Reasonable Doubt that these clever fellows from the Fifth Century BC excelled in offering constructive ideas and in having them accepted by other people.

Formal logics are man-made Artificial logics that bear some resemblance to Natural Logic and are not the same as that. Specialist philosophers and mathematicians are closely aligned and that alignment was clear in Ancient times. I will borrow a phrase from Aristophanes play The Frogs, make a small adjustment, and put it as a Chorus:

Chorus: One Formal logic is so clever that you can't tell what is meant and the other is about as clear as the purest mud.

Quantitative logic is so clever that you can't tell what is meant. Statistics is the branch of the Quantitative tree that has grown in the Context of Social Dynamics. Statisticians are so clever that you can't tell what they mean and other Specialists in philosophy are about as clear as the purest mud even when they use Natural Language. Statisticians do not avoid explicit statements of their own Beliefs and are so closely tied to the disciples of Socrates, Plato, and Aristotle that the word philosopher can be applied to them. They have the same foundations. The Quantitative Method is both clever and muddied to the point where its Meaning cannot be seen, or perhaps to the point where the absence of Meaning cannot be seen. We just assume Meaning is there somewhere in the mud. Statisticians and Specialists in philosophy use codes. In order to give the rest of us the Meaning they want us to take, their Meaning must become visible. When they leave their codes and use Natural Language, as they must do if Meaning is to be delivered to the rest of the population, their weak spot is exposed. A Formal logic that is clever is clearly constructive and huts have been made of mud. Huts are nests for humans. The relevance of that statement will soon become clear.

Ideas form a basis for more ideas. Plato speaking for the group through Socrates had much to offer in the way of ideas. Plato became the Puppet Master pulling the strings. Socrates was cast on stage as a Leader. Those who pull the strings must be hidden if the fantasy is to have the Appearance of being Real. Plato was a young contemporary of Socrates. If he created a composite in the tradition of the Old Comedy then there is No Evidence that Plato regarded Socrates as a man of exceptional merit at all. The Single case was used to derive benefit for a particular group of people. Plato was the Power behind the Representation of Meaning for the group and he transformed the name Socrates into an Authority. But Plato did see exceptional merit in some of Socrates' Actions and there is Evidence for that. And Plato saw the Power of narrative and of using the Single Case as a vehicle for the delivery of Meaning.

Specialists in philosophy leave subtle hints that lead Meaning in particular directions towards inferences. Hints are constructive and have the Appearance of Not Being So. But Statistics and Formal philosophy are not helpful and positive, so they are not constructive in that sense of the word. There are explicit statements of Beliefs in Plato's dialogues alongside the hints. Speaking through Socrates, the Representation that is now the Authority, Plato did not leave the Perception of Meaning to chance.

I now quote from the translation of Plato's Apology by Grube and Cooper (2002). Grube was the translator and Cooper was a Specialist philosopher who amended that translation. It becomes hard to distinguish between Sources as one person Acts on the narratives of another person. I acknowledge the Sources and then Focus on the Meaning that is Represented and not on the people who Represented that.

So, Grube said that Plato said that Socrates said and, having noted that Meaning might have changed as the sayings were amended by Cooper as well as by each of the narrators, the end of that chain is as follows. Through the mouth of Socrates Meaning was delivered in the following words:

> "I do not know, men of Athens, how my accusers affected you; as for me, I was almost carried away in spite of myself, so persuasively did they speak. And yet, hardly anything of what they said is true. Of the many lies they told, one in particular surprised me, namely that you should be careful not to be deceived by an accomplished speaker like me. That they were not ashamed to be immediately proved wrong by the facts, when I show myself not to be an accomplished speaker at all, that I thought was most shameless on their part – unless indeed they call an accomplished speaker the man who speaks the truth"

Here, Plato's Socrates delivers Opinion without Evidence. The Opinion is that they are lying and the Premise is set that this man

Socrates speaks the truth. Note that here Plato recognises the Significance of Truth to Judgments and the significance of definitions to the perception of Meaning.

So, after a long life, the man named Socrates was put on Trial. In The Apology, which is Plato's account of the trial, there were many accusers and we are told that among the first accusers, whose influences are most feared, only Aristophanes was visible and could be named. There are signs that Plato had at least an inkling of what the Method of Pure Science requires for its Activation. I will show you that. The following words come forth from the mouth of Socrates in The Apology:

> "....the main body of these slanderers who from envy and malice have wrought upon you - and there are some of them who are convinced themselves, and impart their convictions to others - all these, I say, are most difficult to deal with; for I cannot have them up here, and examine them, and therefore I must simply fight with shadows in my own defence, and examine when there is no one who answers."

It is a Fact that Meaning cannot be challenged when particular Meaning is Not Represented and that it is necessary to access the details that lead to a conclusion in order to Unsettle the conclusion that has Appeared. When particular Meaning is Represented and the designer of an arrangement cannot be questioned, relevant detail cannot be elicited. Then what you see is what you get. That message was passed on through his narratives. You will all be familiar with that Strategy for it is often employed in Action.

Aristophanes play was briefly described in The Apology. Socrates trial and death were in 399 BC. The exact date of the death of Aristophanes does not seem to be known and is placed some years later than that of Socrates, between 386 and 380 BC. So, he was alive at the time of the Trial and it seems he was not called as a witness for questioning. The relevance of his plays to the Trial is questionable and hangs on suggestion. Perhaps The Clouds was not relevant to the Trial and perhaps it was. But, all those at the trial knew the Tradition of the Poets. Their plays did not even have the Appearance of being Evidence of the character of the person named. I presume it would have been understood that Aristophanes represented the Meaning of the group through the character named Socrates. But Plato knew the Power of Association and Suggestion. The message that Association and Suggestion are Powerful Factors in the Perception of Meaning was passed on through his narratives. Many messages have been passed through time and have been picked up.

It is a Fact that Meaning flows through channels. We all know that Fact and many other relevant Facts. But the Discovery of Intrinsic Relations between Facts has been neglected. That neglect stunted the development of Concepts that could Govern Action in Social Dynamics. Facts are related in arrangements but the Truth of the relations is frequently Not Tested. Specialist philosophers still portray Aristophanes as an enemy of Socrates who influenced the outcome of the Trial. Aristophanes attached a Negative Emotional Association to the name Socrates, which Represented a particular group of people. Plato attached a Negative Emotional Association to the name Aristophanes, who was a Single individual and may or may not have Represented the Opposition at the time. And Plato attached a positive Emotional Association to the name Socrates. But it was one type of Meaning and Action that was being dealt with by both authors and the different Emotions were being attached to that. Different types of Emotional Association have particular Meaning and lead to particular Action.

BEWARE THOSE WHO BEAR GIFTS OF MEANING COUPLED WITH EMOTION

AND

BEWARE THOSE WHO BEAR GIFTS OF MEANING THAT ARE NOT COUPLED WITH EMOTION

But it is not as Simple as that. The effects and the outcome depend on the type of Emotion evoked and where it is attached. In portraying Aristophanes as an Enemy of Socrates, Plato portrays him as an Enemy of Truth. In this Case the philosopher Plato supplied the Point of Reference. Socrates is said to Represent truth. So, is Truth actually the Point of Reference that governs the Action of that group of philosophers? There is peer agreement from the group of

Specialists in philosophy that Truth is Not the Point of Reference for them or for mortals on Earth. Truth is a Concept and Constructs Represent Concepts. But Constructs do not necessarily align with the Concepts to which they are attached. The Construct of truth by peer agreement is Dissociated from the Concept of Truth. Nails (2010) describes the Socratic Problem thus:

> "So thorny is the difficulty of distinguishing the historical Socrates from the Socrateses of the authors of the texts in which he appears and, moreover, from the Socrateses of scores of later interpreters, that the whole contested issue is generally referred to as the Socratic problem."

The Socratic problem is the problem of finding True Meaning. Hearsay is commonly accepted as evidence in Social Dynamics. Hearsay is recognised in the Forensic Context as a problem. Its acceptance as truth is a Fault that leads to Errors in Judgment. Nails continues:

> "The Socratic problem is a rat's nest of complexities arising from the fact that various people wrote about Socrates whose accounts differ in crucial respects, leaving us to wonder which, if any, are accurate representations of the historical Socrates."

Wondering about the truth of Representations can only result in a guess. Meaning must be Calculated. We will never know the person that bore the name Socrates in Ancient Athens and we do not need to know him in order to Track Meaning. The term "rats' nest of complexities" is an interesting one, but I do not see that as fitting. The representations are Complicated not Complex. Meaning at the foundations is Simple and its Representation is complicated by design. But given that the word "complexities" is in use, I would have thought the term nest of complexities would suffice. Since the term exceeds that which is sufficient, this gives me an opportunity to demonstrate that adding and/or rearranging words can change Points of Reference and in doing so will produce different Meaning. I will gratefully seize that opportunity.

Rats are animals. It is a Fact that a rat's nest is a nest for a rat and can accommodate more than one of them. The arrangement of words "a rat's nest of complexities" focusses Attention on the nest itself. Changing the arrangement, to a 'nest of complexities for rats', focusses Attention on the type of active agent that builds and occupies the nest. Humans are a species of animal. Humans build nests of complexities. The word rat has particular connotations when used to describe humans. Colloquially, that word means "unpleasant or treacherous person". So, there is a vague suggestion in the term used by Nails that unpleasant or treacherous people build and occupy nests of "complexities" and that suggestion is Associated with the foundations and development of Specialist philosophy. Of course, Specialist philosophers have told us they have difficulty using Natural Language, so it may be that the possible interpretation was not seen by Nails. Or, that suggestion may have been carefully placed. The Meaning is in line with that left by Aristotle as you will see in due course. Intention is not an issue in my narrative. Specialists in philosophy are clever, sophisticated people and I am not one of those. I will presume the Meaning I have just Tracked was not the Meaning intended. But the Meaning I have Represented is there for the taking if you follow the clue. I did not put the Meaning into the arrangement.

The Meaning of Socrates became a nest of "complexities". The Real person that was Socrates did not make the nest. The complications emerged from the narratives of different people. The scores of interpreters delivered Meaning as they perceived it to be and attributed their perception of Meaning to Socrates for that word represents an Authority. Attributions are not accidental. An Authority gives Power to the Meaning that others design.

PART TWO of Chapter five

A VIEW OF THE DEFENCE AT THE TRIAL OF SOCRATES

Plato (427-347 B.C.) was born 42 years after Socrates birth. So, Plato was influenced as a youth by Socrates during his later years. Socrates (469-399) was 70 when he died following his Trial. It is hard to tell the difference between what actually happened at the time and Plato's interpretation of that. Aristophanes (447-385 B.C.) was born 22 years after Socrates birth. Aristophanes was influenced as a youth by Socrates during his middle age. But the duration of an acquaintance is not its Meaning, nor even part of that. Aristophanes was influenced by Socrates in a different way to Plato and both were stimulated to write about a particular Approach to Interactions. We have only Plato's account of the Trial. But we have two different portraits of Socrates. And those are not portraits of a man. Meaning is Relative and we can only discover differences in Meaning when we have different Representations of something. What was written is not sufficient to determine which portrait of doings and happenings at the time is the better likeness and the original events whose portrait was drawn are not available to Act as a Standard. All we have left is their representations and Meaning can be Tracked through those. Other data from the time provide Context, and that helps as we try to Calculate Meaning. We also have recent examples of Actions based on the foundations set in early Centuries. All those parts can be taken in conjunction and Judged for Truth and Justification. There is sufficient data for Comparative Analysis and thus Meaning can be Calculated. Meaning is Judged in terms of the Action to which it Relates and in relation to a Standard.

The name Socrates represents a particular type of philosophy that was adjusted and developed by Plato and by those who followed the teachings. The word philosophy represents a type of Action and a State of Mind. Different types of doings and happenings lead to different States of Mind. And different States of Mind lead to different types of doings and happenings. A philosophy emerges as a Whole from an arrangement of ideas and is, in part and as a Whole, translated into Action upon the External World. Meaning must be Tracked through ideas as well as through doings and happenings when people are the subject of Analysis, and there is no shortage of ideas from which to take Meaning into Calculations. Narratives are Representations of Meaning. Meaning is made visible by the Action of Representation and then becomes Active in the absence of the agent who designed it. Meaning that was produced by the Ancients is still Active through those who accept the philosophy of the Ancients. The direction in which Meaning is taken is visible in narratives. We need to understand the direction in which Meaning is channelled and the Defence at Socrates Trial as described in The Apology is useful in doing that. I quote from Grube's 2002 translation of Plato's narrative "The Apology" which was revised by Cooper because that is the one that I bought. I did not even choose that from the range. I merely came across that version while browsing through a section of second hand books in a shop. But my purchase was not random. I knew that the works of Plato and of many others that I found without effort were relevant to the work I had in progress. In his introduction to that translation, Cooper says:

> "… in the revised form in which George Grube's distinguished translations appear here, they present Plato's wonderfully vivid and moving – as well as challenging – portrayal of Socrates, and of the philosophic life, in clear, contemporary down-to-earth English that nonetheless preserves and accurately conveys the nuances of Plato's and Socrates philosophical ideas."

Cooper does not know if that is True or not. And he would say that wouldn't he, for he is selling his version of a narrative that has passed through many hands before he got his hands on it. Each agent has to at least Appear to have done some work in order to profit from something that was done before by other people. Changing one small detail can change the Perception of the Whole Meaning, or it may not do that. Cooper was a Professor of Philosophy. He states that at a number of places he introduced "further revisions" in the translations. It is not clear to me what those revisions are. We are caught up in the Vagueness of the Representation of Meaning that is at the core of the Socratic Problem. But Meaning can be Tracked through any Representation and it is necessary to deal with each Representation as it Appears. I Focus on Meaning in the Single Case as it is told by the presenter and I Test the Meaning that the presenter Represents. Meaning is the same in type in different Specialist philosophical texts.

In Cooper's version of Grube's Representation, Plato's Socrates told the court:

> "I will refer you to a trustworthy source. I shall call upon the god at Delphi as witness to the existence and nature of my wisdom, if it be such."

Immediately we see an Error. A note explains that the god Apollo had a very famous shrine at Delphi, where his oracles were delivered through the mouth of a priestess, the "Pythian". Plato's Socrates suggests that there is a god at Delphi. Now that is confusing for the word oracle is used for the priestess and the god is different to that. The plural, oracles, is used for the words delivered by the oracle. The words delivered are said to have come from the god. Someone did not use the right Point of Reference. Plato associates the priestess with god and the words of the priestess with the words of god – the god is at Delphi. And at Delphi there was a human being. The Premise is that the words of the priestess are the words of the god. Thus, the Meaning of god is transferred to the oracle. The priestess who delivers the words becomes the god by the Mechanism of Association. Whichever way you look at this religious custom, the Power of the oracle is drawn from the god who cannot speak for himself and has no other physical manifestation. I could define Nature as god and while that will not get us far in understanding the Concept of God, it is a Fact that Nature speaks for itself. Nature is at least an Oracle. We do not recognise that as a Fact. We do not listen to Nature when it Acts. But Actions Represent Meaning. We should listen to Nature. But we should not listen to all oracles. Some oracles are False Idols.

In calling upon the "god" at Delphi as witness, Plato's Socrates calls on an Authority in a particular form we would not now recognise as such. But we recognise the Mechanism, for that is still Active. We do not even know if the consultation with the oracle by his friend Chaerephon actually took place as stated in The Apology, and we are told that at the time of the trial Chaerephon was dead. So, even the proposition that he consulted the "god at Delphi" was Not Testable. In The Apology, Chaerephon's brother was offered, if "proof" of a consultation was required. Even if Chaerephon had been alive Hearsay is all that is on offer. Hearsay is not Proof. But the words of the priestess would have been Believed at that time. So, if the audience can be convinced that the Source of the words is the Authority that is accepted at the time, then it was logical to assume that the Hearsay evidence would have been accepted as true. Here we see in Operation the Active Mechanism for the construction of the proofs that underlie the Premises in Formal Systems. Here you see Induction rising on the back of Authority in one form or another.

Then Plato demonstrates where Assumption working in conjunction with Assertion can lead. Having set up a Premise and presuming it was accepted that proof was there, Plato's Socrates hangs a question upon the words that are said to have issued forth from the oracle and are therefore true for their source is a god.

> "When I heard of this reply" (noted earlier here as: This man among you, mortals, is the wisest who, like Socrates, understands that his wisdom is worthless) "I asked myself: 'Whatever does the God mean? What is his riddle?"

Socrates gives the Appearance of being puzzled. The Premise and the question enable the offering of a Denial. That type of Action has a high frequency of occurrence and can be seen in parts of Thomson's narrative. Then Socrates states:

> "I am not wise at all."

There you see a Defence from Attack. And the Premise that the god has knowledge, leads to an effort to understand why the god, who clearly knows better than "I", would have an opinion that is different to mine. One Premise leads to another Premise: "I" am wrong for the god is right. The search is then for evidence and evidence is reasoning. But the reasoning shown is not Reasoning and is Induction. Meaning is moved from the Premise to the conclusion by Activating the Mechanism of Association. Socrates continues:

"For surely he does not lie; it is not legitimate for him to do so."

The word legitimate is in frequent use in the Context of Formal logic. Here the word legitimate is associated with truth. If the god does not lie then the god tells the truth. That interesting Association works two ways. That which is legitimate in Systems built on Formal logic is that which groups of human beings, called peers, agree is true. If it is not legitimate for the god to tell lies then the god is bound to do that which peers agree he should do. The power of Association is in harness. Gods become servants of men. Or rather, the idea of god then serves the purpose of men. Gods tell the truth according to man. That is clearly wrong. That is a Construct of God. The idea of God was Operationalised as Oracles, and those were human beings. That Construct is an Illusion. The Construct has taken various forms over the Centuries and stands where a Concept should be. The Concept of Power beyond Human Beings awaits further development. There is necessarily Power beyond Human Beings. And clearly some people do not accept that conclusion.

I am not of any Religious Persuasion and I accept it is a Fact that there is Power beyond Human Beings. The Evidence from Pure Science leads to that conclusion Beyond Reasonable Doubt. I have an underdeveloped Concept of that Power and No Construct to offer you, except perhaps Nature itself. And the majority of you have been turned against Nature. It is a Fact that the Earth is merely part of the Natural Universe that is a larger World, and humans are merely parts within the part that Earth is. Nature is treated as though it is there to serve the purpose of men. That is a False Premise. That Premise leads to abuse. The end of the world **as we know it** has become inevitable. But an ending always was inevitable, for things change Naturally. That Fact is being misused. The Significant Issue is what it will change to. And human doings lead to happenings that affect the type of change that occurs. Human doings also affect the timing of the changes. Parts of the Natural World have been damaged by ill treatment and some things have been destroyed. But other things will take their place. There is such a thing as Resurrection in terms of Meaning in the Natural World. That which is Resurrected is different to that which went before. And, Destruction reaches an end before there can be Resurrection. Selection is a Universal Natural

Mechanism. Resurrection is subject to Selection. I predict that the Human Species will Not Be Resurrected if they succeed in destroying the Ecosystem from which they emerged. That is a Testable proposition but we will not be able to evaluate the results of our Experiment in Destruction. No humans will be there to see the New World as it emerges. The prediction is Not Testable by us.

I will return from that short speculation to the story that Plato told. Plato's Socrates presents an Illusion of Uncertainty and indicates that Action is necessary in order to solve the puzzle he created. He says:

"I reluctantly turned to some such investigation as this".

You are shown that Plato's' Socrates is full of good intentions and he sets out to prove the god wrong, knowing of course that the god cannot be proved wrong. This is so close to the Actions of those Specialists in philosophy who came after Plato that it leads to a particular conclusion. Few people come up with new ideas, or with any ideas of their own. That conclusion took me by surprise for I had long been held under the influence of the Illusion that there are many New Inventions. Now there is Evidence that all Inventions are produced from different twists and turns of old ideas. Inventions do not emerge from a Blank Slate. New inventions are different to the parts of which they are composed and are novel only in that Sense.

So, the Premise is that gods cannot actually be wrong. It would be sacrilegious to say otherwise and a punitive Reaction would follow. A traditional Religious Premise is that gods deliver Retribution. The Premise that a god would deliver Retribution if he could do so and will do so when he can, is Not Testable. And, as we now clearly see, men deliver Retribution on behalf of the type of god that cannot deliver Retribution himself (or herself or itself). The oracle is a human being and the words of the oracle are translated into Action by the people who hear them. Presumably it was thought by those who constructed the Defence at the Trial of Socrates that the Judgment of the Court would not go against the words of a god. By definition, gods know and speak the truth. Clearly the argument of the Defence was not accepted at the time. Perhaps the Defence was perceived to be delivering False Meaning. Truth speaks for itself, so perhaps Truth was an Issue in the Court at that time and perhaps not. Perhaps the decision makers looked for Reasoning and perhaps not. All we can do now is Speculate. We learn from Speculation, so though that is all we can do, it is necessary to do that in order to grasp Meaning. As far as I can see, the Defence did not include Reasoning. The Defence was Not Justified. That can be Tested. Either Reasoning was there or it was not. And perhaps arrogance was detected, together with a lack of remorse. Either those Factors were there or they were not. We can look at the story and perceive Meaning as it emerges from Calculations.

The Defence began with Premises and built upon them. The investigation that was given a False Appearance of Necessity is described by Plato's Socrates:

> "I went to one of those reputed wise……. when I examined this man – there is no need for me to tell you his name, he was one of our public men – my experience was something like this: I thought that he appeared wise to many people and especially to himself, but he was not. As a result he came to dislike me and so did many of the bystanders. So I withdrew and thought to myself: 'I am wiser than this man; it is likely that neither of us knows anything worthwhile, but he thinks he knows something when he does not, whereas when I do not know, neither do I think I know; so I am likely to be wiser than he to this small extent, that I do not think I know what I do not know."

Here we see wisdom being defined as recognising that you know nothing worthwhile. That is not the commonly accepted Meaning of knowledge. But we can also see a different Meaning: - Wisdom is Not Believing we know more than we do know, and that is a different conclusion. That rules out Assertions of truth. So, we see here a contradiction of Meaning in The Apology. Plato and the group of philosophers aligned to that particular position use Premises and those are Assertions of truth. One type of Meaning was dropped or ignored at some point in time.

In the Defence at the Trial, Meaning is Not Tested for Truth. That has remained the custom among Specialists in philosophy to the present time as Evidence has shown. Plato's Socrates states he repeated the examination with another man and then "proceeded systematically". That gives the Appearance of collecting evidence. He did precisely the same thing time and time again. That is not how Evidence can be gathered. That type of Action merely demonstrates that an effect can be repeated. Here evidence is being defined in Quantitative Terms.

Attention is briefly turned to his rejection and to his feelings in relation to that:

> "I realized, to my sorrow and alarm, that I was getting unpopular, but I thought that I must attach the greatest importance to the god's oracle, so I must go to all those who had any reputation for knowledge to examine its meaning. And by the dog, men of Athens – for I must tell you the truth – I experienced something like this:

Socrates was not deterred from his appointed task by the responses of others. That is a familiar happening. To this day, those with skill in Artificial philosophy will not be deterred from their doings by anything that happens. That is Not Natural. In the version of The Apology from which I am quoting, it is said towards the end that:

> "If, on the other hand, death is a change from here to another place, and what we are told is true and all who have died are there, what greater blessing could there be, gentlemen of the jury? If anyone arriving in Hades will have escaped from those who call themselves jurymen here, and will find those true jurymen who are said to sit in judgment there, Minos and Rhadamunthus and Aeacus and Tiptolemus and the other demi-gods who have been upright in their own life, would that be a poor kind of change? Again, what would one of you give to keep company with Orpheus and Musaeus, Hesiod and Homer? I am willing to die many times if that is true. It would be a wonderful way for me to spend my time whenever I met Palamedes and Ajax, the son of Telemon, and any of the men of old who died through an unjust conviction, to compare my experience with theirs. I think it would be pleasant. Most important, I could spend my time testing and examining people there, as I do here, as to who among them is wise, and who thinks he is, but is not."

This is part of the Action that defines gods as men; here dead men sit in Hades. And Plato's Socrates shows that not even death could stop him doing what he enjoys doing. Given the opportunity, Plato's Socrates would do the same again and again. Perhaps the lack of remorse was a Factor in the decision by the Court Twenty-Five Centuries ago as it is in the present Century. Socrates recognises No Higher Authority than him, with the possible exception of the "god" at Delphi and then only as and when it is convenient to do so. The position of recognising No Higher Authority is one with which I am familiar in Social Dynamics and that Meaning and Action are not confined to one Context. The frequency of occurrence of that frame of mind indicates that there are determining Factors. Observation of Systematic Regularities in Meaning indicates that Attention should turn to the System in Operation. Systems are a matter of Process. But Specialists in Formal philosophy keep Attention focussed away from Meaning and thus away from the Processes that Determine what Happens. It is wrong to do that.

According to Plato, the group of philosophers that the word Socrates represents also represent Truth. The word Socrates Represents Plato, and also Represents all those who follow in that line. They acknowledge No Authority higher than them. Gods speak the truth and in seeming to speak for gods, oracles speak the truth and thus become gods. God is a Concept and the oracle is a Construct that stands in its place. All human oracles are False Idols. They are merely men. I use the word men in the generic sense to cover all human beings. Historically it is the males of the species who have Acted as gods in their own domain. That position is changing, but only slowly for there is Resistance. But if women Act as gods then they are no different to men and nothing actually changes. Women Appear to have accepted the Formal philosophical position. While I hope that Appearance is Deceptive, the Evidence so far indicates that the Appearance is True. After all, we are told there is no other way of gaining the Power we seek and need in order to change what happens. And hope may be the Most Evil of Evils and prolong our torment. We fail to recognise that it is not Single Human Beings who determine doings and happenings. That is the conclusion to which the Concept of Relativity leads. We need to see that that conclusion has the Quality of Truth. And we need to see that without a choice between different types of Meaning and Action the System remains the same.

I think it would be most unpleasant for there to be no change if we do move from here to a different Dimension. If in that other place we are destined to meet again the same people doing precisely what they did before death then that would be my definition of Hell in the Afterlife. Surely humans cannot design the Afterlife! Well they can if they get there and if there is no Force there that can restrain them. If so, then hopefully there is a fast route out of there. But also, I would not want to come back to a place that has not changed. Going Round and Round as a part for Recycling in a Dynamic System that repeats systematically would be movement for no purpose. That would not be Natural. People do not repeat systematically and are Not Invariable in form. Whether or not we come back to the Earth, those who are related to us will be here for a period of time. Whichever way you look at this and whatever you Believe it is necessary to change what happens in this World by changing the direction in which Meaning is moving.

I draw Attention to the word "dog" in The Apology. According to a note in the translation that is a "curious oath, occasionally used by Socrates, it appears in a longer form in Gorgias…as 'by the dog, the god of the Egyptians". In that sentence is the solution to the puzzle, so Curiosity is satisfied. Plato's Socrates is substituting the word dog for the word god. He is invoking a different god from a different culture and swearing an oath to that. The Representation changed and the Meaning remained the same. In doing so he again Associates his narrative with Truth and a higher Authority. Through the Action of these Associations, Plato gifts Ultimate Authority to the group of philosophers. Plato's Socrates continues:

> "In my investigation in service of the god I found that those who had the highest reputation were nearly the most deficient, while those who were thought to be inferior were most knowledgeable."

That firmly puts those who have specialised and gained Impressive Qualifications in their place – according to the fathers of Specialist philosophy, they are the most deficient among us, or nearly so. It is puzzling then that we continually strive to gain the Marks of Deficiency. That puzzle is easily solved. We are Forced to do so in order to work and earn a living in this Artificial World and we have reached a position where you cannot do that at all without a Qualification at some level. Specialists in philosophy aided by Specialists in psychology have had a hand in making that a necessity. Even those who sit in armchairs and ask questions from minds that are relatively uncontaminated by Data from the Senses are Marked, and that Mark serves the purpose of excluding others from the group. The Judgment of the relative value of the group of philosophers that the composite Socrates Represents is based on a Premise that the collective is actually working for the god. Specialists in Formal philosophy are intermediaries sitting aloft in the Clouds. Now that is what Aristophanes envisioned and Represented as a Construct! There is Evidence for his view. And Plato provided it himself.

Those who had the highest reputation included the politicians, the poets and the craftsmen. Socrates is set apart from them in Plato's narrative. Socrates sets the terms of the investigation, conducts the enquiry and delivers the verdict. He aimed to be the investigator, the prosecutor, the judge and the jury. The group of philosophers plays all parts except one; they are not members of the audience. That is a Significant exception for we can learn much when we take on that role. Socrates is portrayed as learning nothing from his own experiences over many years. He continued to Act in precisely the same way in the face of criticism and rejection. Politicians do that. That is Not a Natural Response. The argument at the Trial was doomed at the time. The Apology is not an Apology so why is that word used in the translation? Clearly the arrangement he presented did not Function well as a Defence. Whatever actually happened during the trial, Socrates was found guilty. As a Defence, the "Apology" is a Tragedy.

Accounts agree that Socrates could have escaped death and did not. So, there is another puzzle to solve. That which was Execution was also Suicide. Now it is an interesting thought that perhaps that is a hidden message. We will never know what really happened during the last days of Socrates the man. Plato drew a beautiful picture of the ending in words, and that image consolidates the Meaning of his other narratives. It is a Fact that:

THE EXECUTION OF THE MAN ENSURED THE SURVIVAL OF THE MYTH.

Now, some Myths have good effects and some Myths have bad effects. But there are Universal Principles of Myths that apply to both types. There is True Meaning and False Meaning in all Myths and we need to be able to tell that particular difference in the Quality of Meaning. Differences in the Qualities of Meaning are Significant and the Qualities can be seen in the detail of the happenings on which the Myth is constructed. Differences in the Qualities of Meaning lead to differences in Functioning.

It is Reasonable to conclude that Socrates and/or Plato did not Calculate Meaning. But they did Manipulate Meaning to their advantage. Or at least, Plato did. It is hard to see what advantage Socrates gained for we cannot see him in Action, and dying is not usually perceived to be an advantage, nor is unpopularity. But people do seek Attention and Socrates attracted Attention. Was the Manipulation of Meaning described at the Trial Justified? That is Testable. Let me show you the way the poets were dealt with in Plato's Tale of the Trial. The Apology continues:

> "After the politicians, I went to the poets, the writers of tragedies and dithyrambs and the others, intending in their case to catch myself being more ignorant than they. So I took up those poems with which they seemed to have taken most trouble and asked them what they meant, in order that I might at the same time learn something from them. I am ashamed to tell you the truth, gentlemen, but I must. Almost all the bystanders might have explained the poems better than their authors could".

Now Art is a Projective Technique. Artistic arrangements are designed for audiences to find Meaning as they wish. So, it is not surprising that many bystanders, who are in the audience, would find more Meanings in a poem than the poet would and could have explained the Meaning they found better than the poet. Different people have different Points of Reference. The number of angles from which a composition is viewed increases with the number of people viewing that, under certain conditions. The Evidence shows that Plato was well aware of that Fact. People see their own interpretation of an arrangement more clearly than they see any other. Aristotle also understood that Fact and misused it.

Plato's Socrates sets himself up as a Judge of Meaning. Plato's Socrates tells the Court that he can tell the difference between Qualitative Judgments of the poems in terms of their value. Value is a Judgment of weight. In The Frogs, Aristophanes portrayed a Contest of two Poets, Aeschylus and Euripides. In conversation, slaves describe the weighing on scales of lines from poems:

> "it's all got to be weighed and measured up proper, with rulers and yardsticks for the words, and compasses and wedges and I don't know what……. regular torture chamber yes, Euripides says he is going to put every line to the test."

PART THREE of Chapter five

THE SOCRATIC METHOD

The Socratic Method is an investigation that amounts to an inquisition. It is an intensive search of Meaning and its Representation as it is presented by another person during a dialogue through asking questions aimed at finding Justification for the statements that the "partner in dialogue" makes. The questions are necessarily constructed in Natural Language. Every line is put to the test. The Standard the fathers of Artificial philosophy set up for the weighing of Meaning is Quantitative as Aristophanes saw. The Quantitative Method is presented in many dialogues as a precise Method of reasoning without Error and the contrast is telling but it is an Illusion. In essence, the Socratic Method uses Natural Language to expose gaps in Reasoning. According to Ryan (2013):

> "Socrates was no transmitter of information that others were passively to receive, ...he helped others recognize on their own what is real, true, and good (Plato. Meno, Theaetetus)—a new, and thus suspect, approach to education. He was known for confusing, stinging and stunning his conversation partners into the unpleasant experience of realizing their own ignorance, a state sometimes superseded by genuine intellectual curiosity."

The initial statement is a Premise. The Premise that Socrates was no transmitter of information that others were passively to receive is True, and yet it is misleading. There is again the suggestion that this group of philosophers does not construct Meaning for the audience and that the "partners in dialogue" do the constructing. Within the Boundaries of Pure Science, it is True that others are helped to recognize on their own what is real, true, and good; they cannot do that in the absence of Evidence that provides Justification and Pure Science provides that type of evidence. Those who use Formal logic do not Act in that way. Plato's Socrates did not transmit information. He transmitted Meaning. Others were to actively receive Meaning that was transmitted. Their minds had to be Active for the transmission to have its effect. These specialists in philosophy manipulate Cognitive Structures in Human Minds. What do we call them? I would call them human beings, but they would not agree to being defined as ordinary mortals or even as mortals. They are not bound to the Natural World and float free in the clouds where they cannot be Observed by "the others" who receive their messages.

Socrates, the composite, helped people recognize that they were ignorant and here that effect is defined as "real, true and good". The Concept of what is Real, True and Good was replaced by a Construct that Functions in a particular way. Where there is Ignorance there are gaps in Knowledge and recognition of the gaps leaves people Open to the Reception of Meaning. The type of Meaning that they pick up determines whether or not that is a good thing to do. Now, the Method that bears the name Socrates is not said to lead to genuine intellectual curiosity, it is said that such curiosity was sometimes aroused. That suggests such arousal was accidental. Something that is accidental is not systematically determined. Curiosity can lead to the Discovery of True Meaning. Curiosity is the Most Blessed of the Blessings and that is confined. Its confinement is Not Accidental. These Facts have been known for Centuries, but they have been systematically Denied. But, perhaps the Significance of them when arranged as a Whole has not been known. That is a viable proposition. Either way, it is a Fact that the Action of Denial of Facts is a Problem in Social Dynamics.

Now, the purpose of drawing people into recognition that they are in a State of Ignorance is to create the conditions that enable them to be led from that State which is one of Disorder and Uncertainty into a State of Order and Certainty by giving them Meaning. The problem these clever fellows faced was that realizing their own ignorance did not systematically lead the "partners" in "conversation" to search for "knowledge". According to Plato's Socrates the realisation that they were ignorant sometimes led to Attacks on him, directly or indirectly. That was Not a Universal Problem. Some people will be easily led into the Certainty of Beliefs that others hold. And the problems that follow from both forms of Reaction are visible in the present day. There are different types of education. The type of Meaning and Action you are taught makes a difference to the way you Function. Other Factors also make a Significant Difference to Functioning as a

Whole. Different Factors work together in conjunction and the outcome in terms of Functioning is dependent on their combined effects.

Nails (2010) states that:

> "Socrates' irony consisted in his saying that he knew nothing of importance and wanted to listen to others, yet keeping the upper hand in every discussion."

The word irony is used by Nails and we all know what that means. I will include the definition. From the Pocket Oxford Dictionary:

<u>irony</u>: 1. expression of meaning, often humorous or sarcastic, using language of a different or opposite tendency. 2. apparent perversity of an event or circumstance in reversing human intentions. 3. theatre. use of language with one meaning for a privileged audience and another for those addressed or concerned. [Greek *eirōneia* pretended ignorance]

Formal philosophy is Art and part 3 of the definition is relevant to that. In his narratives, Plato included one Meaning for a privileged audience and another for those addressed or concerned. Specialists in philosophy are a privileged audience trained to see Meaning in a particular way. It might be True to say that Socrates had No Knowledge, but it is Not True to say that he knew nothing of importance. It is of course impossible to keep the upper hand in any conversation by knowing nothing of importance. The Appearance of knowing nothing of importance is Deceptive. Plato's Socrates Feigned Ignorance. Creating that Illusion enabled him to keep the upper hand in every discussion. That message has been widely received and has become Operational throughout the Centuries. I will express that colloquially: People play Dumb. People withhold information that could lead to the discovery of their Actions and its Meaning. But the Socratic Method has more parts than that. Plato's Socrates gave answers and often did so by eliciting the answer he wanted from the target. Meaning can be given while creating the Appearance that Meaning is not given. He asked questions and showed answers that progressively forced particular conclusions from the audience.

Within the Boundaries of Pure Science, Natural Deduction involves a search for the True Meaning of things in the Real World and the finding of that is Knowledge. Knowledge enables Survival. The direction in which Meaning moves using Natural Deduction is different to the direction in which Meaning is moved using the Socratic Method. The difference in direction was hidden. I will make Meaning clearer here. Outside the Boundaries of Pure Science, there is only the Appearance that the search is aimed at finding Truth through Justification. That Appearance is Deceptive. The search is actually aimed only at Verifying the absence of Justification for particular views. Gaps in Meaning are exposed. The Opponent's Beliefs and Opinions are weakened by the Socratic Method. And thus, the Opponent is weakened.

THE SOCRATIC METHOD AS A WEAPON

Plato's Socrates is said to have responded to the Rejection of his Defence through the verdict of the Court as follows:

> "Now I want to prophesy to those who convicted me, for I am at the point when men prophesy most, when they are about to die. I say, gentlemen, to those who voted to kill me, that vengeance will come upon you immediately after my death, a vengeance much harder to bear than that which you took in killing me. You did this in the belief that you would avoid giving an account of your life, but I maintain that quite the opposite will happen to you. There will be more people to test you, whom I now held back, but you did not notice it. They will be more difficult to deal with as they will be younger and you will resent them more. You are wrong if you believe that by killing people you will prevent anyone from reproaching you for not living in the right way. To escape such tests is neither possible nor good, but it is best and easiest not to discredit others but to prepare oneself to be as good as possible. With this prophesy to you who convicted me, I part from you."

That reminds me of a later saying I have heard: Vengeance Is Mine Sayeth the Lord. Meaning has flowed systematically since the dawn of time and been adapted to suit different purposes. The fathers of Formal philosophy were founders of a particular form of Ideology that has much in common with traditional Religions. Ideologies are a particular type of Metaphysics. All Ideologies have Meaning in common and differences between them. And, Plato's Socrates is Not Telling the Truth when he states that "nobody noticed it". I am not saying Plato is lying, perhaps he did not know the Truth. Perhaps he had not seen Aristophanes' play or read The Clouds. The Prophesy is acknowledgment that there is an army of younger people in reserve and that is in alignment with Meaning in Aristophanes' play. Plato's Socrates provided the Evidence at his Trial that this particular Meaning is True. Perhaps the Apology contains his Confession to knowing the Truth or perhaps Plato did not notice the fit in terms of Meaning. In The Clouds, Socrates and Chaerephon are shown teaching the youths of Athens. Aristophanes describes "the Thoughtery of wise souls" where "it seems they have two courses of reasoning, the true and the false, and that, thanks to the false, the worst law-suits can be gained". The Socratic Method of Unsettling Meaning for gain survived Socrates' death and it is clear that the course aimed at discovering truth was dropped from the curriculum at some point in time, if it was ever there. Aristophanes may have perceived a Potential in their doings that the group did not attempt to Realise.

Nails (2010) provides a quote from a collection of Benjamin Franklin Papers. This reflection from a politician shows that the Power of the Socratic Method is harnessed by those who use Induction in order to Unsettle the position of an opponent. Franklin wrote:

> "I found this [Socratic] method the safest for myself and very embarrassing to those against whom I used it; therefore, I took delight in it, practiced it continually, and grew very artful and expert in drawing people, even of superior knowledge, into concessions the consequences of which they did not foresee, entangling them in difficulties out of which they could not

extricate themselves, and so obtaining victory that neither myself nor my causes always deserved."

If that is what Socrates did, then it is no wonder he was unpopular. If he was unpopular then it is likely that is what he did.

Presumably Franklin found he could confuse, sting and stun those who held political views he did not subscribe to. He would have been unpopular with people who did hold those views. Franklin was not unpopular with audiences. Popularity is defined in Quantitative terms. He gained votes and was elected to positions of Power. He used the Socratic Method to serve that personal Purpose. How can the effect that Franklin described be produced? The answer is Simple when the Concept that Meaning is Relative is brought to bear. In Unsettling the arguments of the opposition, Franklin gave weight to his alternative Meaning. Only his opponent's Beliefs were Unsettled. Doing so in front of an audience clears the field. The enemy is routed. Franklin's own Beliefs remained untouched and could fill the gap that would otherwise have been there. He indicates that people did not enter a State of "genuine intellectual curiosity" through his skilful use of the Socratic Method. And he tells us he knew that Fact.

The Socratic Method consists of Questions carefully designed to fit particular targets and its Operation can only be shown through displaying the detail in Single Cases. Single Cases are Active as Models of the Method. Single Cases are Powerful in delivering Meaning to audiences. The Socratic Method Attacks the Beliefs of any who are prepared to defend views that have No Justification. The lack of Justification is the weak spot. If the Attack is properly conducted, the conductor of the enquiry cannot fail to win. The Socratic Method requires the target to find a Quality of Meaning that is not there in the Beliefs and Opinions to which Induction leads. It is impossible to find that which is not there. Quite simply, the problem of Justification is Insoluble for Induction and the Socratic Method uses that Fact. Hume merely highlighted a problem that had been known across the Centuries. But you have to engage the target in conversation. There are Rules in our Social Systems that Force such engagement and there is also informal pressure to engage in Interactions. The effects and the outcome of doing that are Variable and the differences in the Meaning and Action of individual people is not taken into account. A general Rule to engage in Dialogue is an Error of Judgment that can lead gentle folk into difficulties. Even those of Strong Will can get into difficulties under that Rule for they are not Equals in Cleverness. Franklin understood that Fact and he did not allow others to engage him in the same way. Yet we are social animals and there is a need for communication. Thus, the Rule has the False Appearance of Justification. And as Plato's Socrates clearly shows, the Rules in Social Dynamics do not apply to those who hold positions of Power and the Socratic Method shows how to prevent that happening. There is No Justification for the doings of these people. The happenings that followed their doings are Evidence of the Error to which this Ideological Distortion of Cognitive Structure leads. Peers have agreed that the Concept of Justification will not be developed or applied. That is a Fault. That Standard is necessary to enable Stable Functioning.

I repeat a Fact. As Plato's Socrates tells us, realizing their own ignorance did not always lead the "partner" in conversation to search for Knowledge. According to Plato's Socrates, the realisation that they were ignorant sometimes led to Attacks on him, directly or indirectly. That part is True. Those who are Reputed Wise do not easily accept that their Opinions and Beliefs have the Quality of Uncertainty in Relation to Truth and not all of them can be persuaded that their views do actually have that Quality. The story of The Apology suggests that Socrates had enlightened those who laid claim to wisdom and was unpopular for showing them the truth. That truth as he showed it deprived them of the State of Certainty they were previously in and were comfortable with. The Apology does not show how their Ignorance was proved. The Method is described in other essays and you can find it there, but demonstrating the Method does not provide Proof of its True Meaning. The outcome of Socratic Questioning is Uncertainty. Meaning that is Justified is Not Found. The fathers of Artificial logics showed how Uncertainty can be produced. They also showed how the Uncertainty that is produced can be used to Install new Beliefs. They did not show how to Calculate True Meaning and Test it for Justification. That is a Fault.

The Apology is a work of Art. Recall that Art is a Projective Technique and that Art is designed to hold different Meanings in one arrangement, any of which can be perceived by members of the audience. I repeat that Art is designed to do that. Plato designed the stories he told. What Meaning do you see in the term?

> "To escape such tests is neither possible nor good, but it is best and easiest not to discredit others but to prepare oneself to be as good as possible."

I will speculate

Do you interpret that as: It is impossible to escape tests such as the followers of Socrates apply and those to whom the tests are applied should not attempt to discredit those who test them? They should be as good as possible by not doing to those who apply the tests the same as they do to them.

I have seen that Interpretation in Operation as the following Construct:

Construct that Acts as a Rule: Those to whom the tests are applied are forbidden from testing those who test them and are forbidden or actively prevented from testing the tests they use. Thus, those who test others cannot be discredited.

I note that access to the details of tests is often restricted. Now whether or not that is Justifiable depends on a number of variables, the Meaning of which must be Calculated. I will not do that here. I will show an alternative interpretation of the term I am dealing with.

Or does it Mean? If you prepare yourself to be as good as possible, then you will not be discredited by the tests when they are applied to you as they necessarily will be.

And do both Meanings work in conjunction? As a Whole the message is to be prepared in case the Rule fails and the designer is tested. And what is your definition of the word good. Does that mean having good Defences for Opinions that are Not Justifiable or having good Opinions that are Tested in advance for Justification and are Justifiable? Those who use the Method of Induction for which

Justification is impossible can only prepare good Defences. They were told to do that by their fathers. And recall that Confrontation is used as a Defence. The Socratic Method is a form of Confrontation that has the False Appearance of Reasoning.

The descendants of Plato have prepared themselves to be as good as possible in the sense that they will not be discredited by the tests when they are applied as they necessarily will be. They designed the Action of testing to remove its Power to affect them. That was good enough up to the present time and is not good enough for the future. There is a new Method and a formidable foe setting out to gather together an army that will be extremely difficult to deal with. The acolytes of Formal logic will resent them. And it is clear they were taught to seek vengeance. It is wrong to do that. Perhaps given time they will come to see that they too will be saved if they change their ways. But now there is little time left to give them. The War to End All Wars must be fought soon and it must be won. But if it is not won, then Nature will move on without us. Destruction is Part of Creation.

The Socratic Method is very effective, so what is the Source of its Power? Causation is a Matter of Process. I will demonstrate that Fact. Recall that Franklin's own Beliefs remained untouched and could fill the gap that would otherwise have been there. The Socratic Method has the False Appearance of being a dialogue. That is an Illusion. One of the participants presents as being a Blank Slate. Given that I know nothing and you know something then Attention can only Focus on that which is there. You claim to have wisdom and I claim to have none, so Meaning can only flow in one direction. That direction is from you to me. That is clever. When asked to do so, who among you can resist giving me Meaning? Few people can resist that. But I am Resistant to all Meaning that you cannot Justify. That position is Reasonable. You begin to see the need for Justification and your temperature rises as you struggle to provide that in your attempt to transfer to me Meaning that is Unjustifiable. Your cognitions become disordered as your temperature rises. Many different Factors Act in Conjunction and determine the direction in which the Energy produced by your failure to fulfil my incessant demand for Justification is channelled. I can ensure that the Energy produced is channelled elsewhere. Or I can just leave you to give vent on me to your frustration in front of an audience. Then you have demonstrated that you are Unreasonable. I remain calm and that adds to the False Appearance that I am Reasonable. People are often roused to anger by any challenge to their Authority and their Beliefs. I note that Heat leads to Disorder and Disorder leads to Heat. There is a Chain Reaction. And there is an Ending of the Chain where there is the Beginning of a New State of Order.

The Socratic Method places the burden of proof on the person who is questioned and the questions are aimed at producing Justification for each Claim to Knowledge. That is what Popper did. That is an Ancient Strategy. Popper did not clearly reveal his Sources. Justification is an Insoluble Problem for those who use Induction. The Claim to Knowledge of those who Assert their Inductions is False. When the questions are designed to elicit the vague theories on which Opinions reached by Induction are based, there can only be one outcome to the conversation. But let us return to the Denial that Specialists in philosophy are constructive. The Meaning given by you to me will be Unsettled by my questioning and as you can see, I give you none to replace that for I am merely a Blank Slate seeking enlightenment. The questioning will show that there is different Meaning to be had, but if you take that Meaning then I have not given it to you. As you can plainly see, the Meaning is simply there for the taking. The Method is very effective in evading responsibility. The Socratic Method is used in Enthymemes. The Illusion of Evidence is Active in all forms of Inductive arguments. The Premise that the audience determines the Meaning they take from the Structure of the Enthymeme is False. The Method of Induction is Not Justifiable, and all its forms are Unjustifiable in part and as a Whole.

The Projective Technique is very effective in evading responsibility for any Meaning that the audience picks up and thus for any Actions those in the audience take on the basis of that. Enthymemes are a form of Projective Technique that leads audiences to a single conclusion, and the single conclusion depends on the Point of Reference of the audience. The Enthymeme is structured to set the Point of Reference for the audience. In other words, the designer does that which you think you do. It is a Fact that Specialists in philosophy Select the Focus of Attention for an audience. Plato wrote plays with expressions of irony in the Context of theatre; he used language with one Meaning for a privileged audience and another for those addressed or concerned. Hume did not suggest those texts were cast into the flames. Hume was a follower of the Quantitative Ideology and aimed to protect his Beliefs. I will pick up in the next Chapter a Significant Difference between traditional Religion and the Quantitative Ideology.

The Socratic Method moves from a particular conclusion to the Abstracted Meaning on which it is based and moves back from that to the Sense data from which the Asserted truths were abstracted. When the gear of Inductive reasoning is put into reverse and the machine moves backwards, the Socratic Method cannot fail to find Meaning that was not taken into account. The Socratic Method strengthens the questioner's own position by default, as Franklin gleefully observed. His glee is not Justified and he knew that. Franklin did not aim to engage the Capacity of the audience to Reason. He knew that the failure to do that was wrong and led to Errors in Judgment. But it was hugely rewarding to do that. The Temptation was there and he succumbed willingly. Did he have a choice? There was no other Method in view. Or was there no other Method in view that would suit the purpose of rising to a position of Power on the back of Unjustifiable Meaning. Franklin makes it clear that he knew that if the tables were turned and his Opinions were targeted, he would suffer the same fate. He did not allow those to whom he applied the tests to discredit him by doing the same thing to him. As it is exercised, Politics is a form of Art but it does not have to be that way. Pure Scientists are intrepid Explorers in pursuit of Knowledge who venture into Unknown Territory. Franklin was not engaged in the pursuit of Knowledge when he engaged in Action on human beings. But he was in other Contexts. He would have had an inkling of what made the difference.

Those who specialise in Formal philosophy know the importance of questions in eliciting details. And Thomson stated explicitly that "there are cases and cases and the details make a difference". If they do provide detail to support their Opinions, Specialist philosophers fail to provide accurate detail, knowing that the absence of Justification for their argument would be exposed if they did that. The demonstrations of the Socratic Method in Plato's Meno and elsewhere show the Structure of an Enthymeme. Plato's Socrates gave answers and often did so by eliciting the answer from the target. He showed evidence that progressively forced particular conclusions from the audience. Thomson did that in her Enthymeme. Sometimes conclusions are stated by the designer to help Meaning to flow in the right direction and sometimes the gap is left for the audience to fill. Psychological Mechanisms are involved in achieving the effects. The Structure of the Enthymeme resembles the Method of Natural Deduction and is not the same as that. The differences in Action and Meaning are Significant.

Ignorance does not necessarily lead to Knowledge, nor does it necessarily create a need for Knowledge. I would not define Ignorance as an Absence of Meaning. Ignorance is the presence of False Meaning in relation to something. Gaps in Meaning create the need for Meaning and are necessarily filled. In other words, gaps in Meaning are Unstable. Action is Intrinsically Related to Meaning and to its Qualities. Where there is a gap in Meaning, Action is Unstable. That which is Unstable does not endure in a Natural System. But human Social Systems are Artificial Systems in which Instability is maintained. It follows that the Appearance of Uncertainty is Not True.

INSTABILITY IN SOCIAL DYNAMICS IS MAINTAINED BY FALSE CERTAINTIES.

It is a Fact that there never is a Blank Slate. Such a thing cannot be envisioned by us and if it can be envisioned at all there is something that does that. Space is not empty. Human minds are not empty. There is always suggestion threaded through the Art of Persuasion. Questions are Actions with Meaning and lead to responses that are Actions with Meaning. There is only the Appearance that when they do not know, neither do they think they know. That Appearance is Certainly False. Plato's Socrates was not a Blank Slate. The Single Case of Socrates was used to deliver Meaning from the gods. The difference in Meaning of the word god was dropped and the messenger, who we recall is a group of people, became a god to whom others could refer. Specialists in philosophy are False Gods.

UNSETTLING AND SETTLING; ORDER AND DISORDER

The Socratic Method is used to discredit others and it is also used for different purposes which are better than that. But it is not used to access True Meaning and is not sufficient to do that. The Method that bears the name Socrates was not invented by him. It is Natural to ask questions and it is Natural to search for True Meaning using them. Socrates had a knack for asking questions and developed the skill to a high level relative to his contemporaries. But the name Socrates stands for a group of philosophers who did not and do not search for True Meaning. The Difference that a Focus of Attention on Truth and Justification makes is Significant.

Premises are the foundations for Judgments when Induction is in Operation. In The Apology there is reference to a "trustworthy source" which comes in three parts. At the core we have god. The word god represents the Intelligence that is assumed to exist as an entity that Knows All Things. The Oracle is the Representative of that highest Authority and speaks the answers for us to hear. The Premise that the god who Knows All Things gives the answers to questions, leads to the Premise that the words of the Oracle are truths. And those truths become Premises to guide our Actions. Premises are False Certainties. We assume he or she who personifies the god asked the god questions and received answers. The third source is the friend who hears those answers from the Oracle and passes the Premises on to us. The Sources are the evidence of truth of the answers. That is the Model for Inductive science where the proofs are sayings of gods and are not Proof of Truth of the Meaning that is delivered.

A Philosophical Approach is a Psychological Approach and a Political Approach and vice versa. All Active Agents are guided by a philosophy and use psychological techniques in order to exercise Power over other agents. The Formal philosophical Approach is not the only possible Cognitive Approach to the World. Yet that is the Illusion that was created by human beings and still stands.

The Socratic Method does not just confuse, sting and stun the conversation partners into the unpleasant experience of realizing their own Ignorance. Their Ignorance is often shown to a larger audience. Thus, a backlash of one type or another is inevitable. The Socratic

Method is not a Force for Stability. It is possible that the death of Socrates was taken to heart and a lesson learned by his followers. They did not learn to change direction. It is a Fact that those who use Induction took their Meaning under cover. Even death did not stop their Actions on the World.

THE ANCIENTS RULE THE WORLD EVEN NOW.

That is extraordinary.

BUT IT IS NOT SUPERNATURAL.

THAT IS UNNATURAL.

The Socratic Method resembles part of the Action of Reasoning within the Process of Pure Science. The Socratic Method cannot work without a partner in conversation. The target must be engaged. There are two Methods of producing engagement. The tools of the Art of Persuasion can be brought to bear and those have the False Appearance of No Apparent Force. When those tools fail then Force must make its Appearance one way or another. Interactions can be Forced by those who know the Power of Dialogue. But the effects of the Dialogue depend on the Point of Reference that is applied. I presume those in positions of Power know that, but if they do then the game changes for Intention becomes a Factor. So, I will remove the presumption and set up the alternative proposition. Propositions must be Tested. My experience has shown that what seems obvious to me is not obvious to other people. On occasions I have Observed their Surprise in response. And none of us perceive True Meaning until Testing intervenes. But it is clear that in terms of effects and outcome as a Whole:

THE SOCRATIC METHOD HAS A FATAL FLAW IN ITS OPERATION

The flaw is the small and Highly Significant detail of type of Governance. The Socratic Method is not governed by the search for Truth. But that difference in detail is only a flaw if you want to discover True Meaning. If you want to strengthen your own Beliefs for transfer to an audience and gain Power, then the Socratic Method is flawless once it is in your hands. The type of use depends on the Point of Reference. Structures are open to misuse when the Point of Reference is left open to be filled by the user. And the gap is not left fully open, for in Social Dynamics it is closed to Truth. Abuse has a high frequency of occurrence for it goes unchecked and without consequences. The Power of the Socratic Method to Unsettle Meaning through the False Appearance of Dialogue is undeniable.

The Power of Rhetoric to Settle Meaning through monologue is undeniable. Rhetoric is openly persuasive. Those who use Rhetoric do not Feign Ignorance. They Feign Knowledge. In Rhetoric evidence is presented as proofs. But the evidence in the Art of Rhetoric has only the False Appearance of Evidence; there is some resemblance to a degree that is not specified. Those who use Rhetoric are Reputed Wise and the Structure of Rhetoric keeps them safe from Unsettling by the Socratic Method. The Structure of Rhetoric prevents their being engaged in dialogue. But the difference between Rhetoric and Dialogue in the Context of Induction is Trivial. The involvement of additional Factors increases the Power of Rhetoric to deliver Meaning successfully to the audience. The audience is entranced. A measure of dialogue allowed at the conclusion of Rhetoric does not have the Power to Act on that which has been settled by the Illusion of Evidence.

When an argument is in progress, the audience that surrounds the combatants is Not Active in the play. The audience is Not a Context for the Action, for a Context clarifies Meaning. The audience Observes and picks up Meaning from the play and is only Active on Meaning as it is given to them. That is Theatre, pure and simple. The writer of the play sets the Context and thus is the provider of the Meaning that flows through the Structure to all who see the play. Designers of Representations gift Meaning for Perception. Meaning leads to Action. We should be careful which gifts we accept and we should Test each gift for its True Meaning. The position of designer was not mentioned by the Fathers of Formal logics. They filled that position automatically without showing it was there for the taking or showing there is an alternative Method.

Meaning must be Unsettled before it can be Settled. Designers escape Attention and their Inductions remain in place in their minds as they Design the Structures through which Meaning passes to others. The Beliefs of Designers govern their Action in Designing. The Appearance is given that Designers do not determine the Meaning that enters and leaves the Structure. That Appearance is an Illusion. The Disciples of Socrates, Plato and Aristotle have had Centuries to Tie Meaning in Knots. But:

WE ARE WHERE WE ARE AND WE HAVE TO DEAL WITH THAT.

That can be done. I will now state a Fact:

WHEN A POINT OF REFERENCE IS CHANGED, MEANING AS A WHOLE CHANGES SUDDENLY AND AUTOMATICALLY.

Actions change in line with the type of Meaning. Thus, the type of Action can change suddenly and automatically. Truth is the Point of Reference that governs Pure Science. Truth is not the Point of Reference that governs Induction but it can be brought to bear on Inductions. It is a Fact that the Socratic Method is disabling and that thus it is enabling. Disabling and enabling are different types of Action. Two different types of thing do not simultaneously occur in the same time and space. The Socratic Method disables one participant in a conversation and thus enables the other to Act. I propose that if Attention is focussed on ideas and not on the people who have them and if the search is for Truth, then False Meaning will be Unsettled wherever it lies. Once discovered, True Meaning must be dealt with.

I will repeat some Meaning that I want to Install in you. The Illusion of Tabula Rasa is a Core Illusion of the Formal philosophical position. That Illusion is Active in the Socratic Method. The idea that Meaning comes from nowhere at all is False. The Appearance is given that I am the audience for your wisdom. I take on the Appearance of a Blank Slate upon which you can write. That opens up Temptation. I show that I am earnest in my attempts to improve my knowledge. You cannot fail to be pleased and impressed by that. You succumb to Temptation. You provide me with Meaning. I then merely ask you to explain your Meaning in detail in order for me to pick that up precisely. The problem you have is that your Meaning is Not Precise and your Representation is Not Precise and worst of all there is No Justification for it. All those Factors Acting in conjunction are now your problem. I gift the Trojan Horse to you. You have been led into opening your gift to me outside the gates of the city. That is crafty. But under different conditions that is the Intelligent thing to do.

When I am crafty, I examine for Errors the detail you provide and draw the Errors to your Attention. The Errors are puzzles you have given me. You now have the need to solve the puzzles that you have created. I do not have to do that. Attention is now focussed entirely on you and is not focussed on me. Here we see that is what Popper did. He had read Plato's Manual. He had also read Aristotle. If he had read Aristophanes, he had ignored that Source of Meaning. You cannot solve the puzzles and you have now become the Blank Slate that I Appeared to be. But the Truth is that I am not the Blank Slate that I Appeared to be. That Appearance was Deceptive. I am full of Meaning. And the Meaning of which I am full is Untested Beliefs. And now you are the Blank Slate and that is not a False Appearance. You are in a State of Confusion where no Meaning is held firm. I can write whatever I choose upon the Blank Slate of your mind. Or putting that Simply: your Perception of Meaning is now in a State of Disorder and I can lead you into a new State of Order. The Intelligent thing to do is to Test Beliefs in both locations for Justification.

CHOOSE YOUR LEADERS WITH GREAT CARE

Before you can do that, differences in type of Meaning and Action must be disclosed and the Structure must allow Selection of the Fittest in the Sense that they use Pure Science to enable preservation of the Natural World and the Species within it. And fitness comes with exercising the muscles for Reasoning.

PURE SCIENCE TAKES PRECEDENCE OF NECESSITY

But that Process was Deposed. The Construct of Precedence by Rank Order is Artificial and Unjustifiable. Can the pretender be unseated? Can the Natural Concept of Precedence by Necessity be resurrected? Only Experiments can give us answers to questions.

The Intelligent thing to do is to solve puzzles ourselves using Reasoning. The Concept of Preservation must include the Concept of Development. Concepts are Wholes that become parts in other Wholes. Pure Science is based on Common-Sense so do not be afraid of that and understand that nobody will be excluded and all must play a part. We just cannot all be designing the same thing at the same time in the same space. But we can align harmoniously if we use the right Method for Judgment and ensure that Controls are in place to keep all parts on Track towards Truth in relation to the Natural World on which our existence depends. That is not happening now.

We actually know a great many Facts. Facts can be used well but they are not used well in the Context of Social Dynamics for Pure Science is banned from that Realm. Truth is not part of the governance there, and that Fault leads to irregular, systematic tremors of great magnitude that frequently occur. I predict that those tremors will continue to increase in frequency and magnitude unless the Fault is corrected. Those tremors will end in Destruction. That is a pretty safe prediction and I cannot tell you when the end will come for humans. But the end will inevitably come if the Formal philosophical position remains standing. And after Centuries of misdirection, the possibility that we can change the outcome is now small. But there is a chance and while we should seize that opportunity, it is likely that we will not do that. And change takes time and time is running out for many of the life forms that now exist on earth and are Intrinsically Related to our species. If we allow movement to continue in the current direction human beings will disappear from the World. The ending of human existence will be a new beginning and it will be the beginning of something different. There will be new and different life forms. It is a Fact that Nature is not dependent on us.

IF THE HUMAN SPECIES IS TO BE SAVED, HUMANS HAVE TO BECOME WORTH SAVING.

Does either Potential actually exist? In Qualitative Terms we have not come far in Twenty-Five Centuries. Once I have reduced you to a State of Confusion, I can just leave you disabled and defenceless in the wilderness. Only a Moral Standard will keep me from doing that.

AND PERHAPS I CAN TAKE THE AUDIENCE
WHEREVER I CHOOSE TO GO IN TERMS OF MEANING.

OR PERHAPS NOT.

CHAPTER SIX: A TOPIC CANNOT DEFINE MORAL REASONING

PART ONE of Chapter six

FINDING CAUSAL KNOWLEDGE ENABLES MORAL REASONING

The Art of Persuasion is aimed at producing agreement. I do not propose that reliance on reaching agreement between people is a Fault in Social Dynamics. I propose that reliance on peer agreement alone is a Major Fault in Social Dynamics. Peer agreement is mistaken for a Standard. Humans relate to each other and how they relate makes a difference to what happens. Many Factors should be taken into account in reaching agreements between peers and True Meaning in relation to Functioning in the Natural World should be the Point of Reference. The Quantitative Construct of Reaching Agreement by Numbers stands where a Concept of Reaching Agreement with Justification should be. The Concept of Truth Related to Justification can be developed and become Operational as a Standard through the Action of Pure Science. And it must be taken into account that some people will not agree to things that are Reasonable. That Factor must be dealt with.

In the absence of agreement there is argument. Humans fight battles in words and in deeds. Peer Agreement is a necessary Factor in producing Stability in Social Dynamics and is not sufficient to do that. The type of agreement is a Significant Factor in determining the Qualities of a State. It is a Fact that some types of agreement lead to Unstable States. The Art of Persuasion can be used to draw people into agreement to fight. Under certain conditions humans can be persuaded in large numbers to agree to anything at all and once they have done so, agreements are enforced. The question we need to answer is: How can Meaning be Independently Tested for Justification before agreement is reached? A Bias towards Truth is the Mark of Independence. Those who engage in the Process of Pure Science are acting Independently of other types of Bias. But they are Biased and that is a Significant part of the definition of Pure Science. Pure Science is Biased towards the Perception of True Meaning and its relation to the Natural World. Inductive science is reputed to be free from Bias and since there is the Appearance of only one type of science the Premise that all science is free from Bias has been accepted. That Premise is False. Inductive science is Biased in a different direction towards a different purpose. Pure Science and Inductive science have Different Objectives – they seek or aim for different things.

The solution to the problem of Calculating Meaning is a Natural one that has always been available. So, why do we fail to use Natural Deduction to Test Meaning for Truth? The answer is that we have been persuaded to accept Meaning as it is Represented. The Premise that the Appearance of things, in part and as a Whole, fits their True Meaning was set. That Premise survives despite repeated Falsifications in Single Cases within our own experience and repeated Falsifications through our Observations of what happens to other people and other things. Illusions are Powerful and some of them are Stable and Resistant to Unsettling. It is a Fact that what you see is not necessarily the Meaning of what you see. Facts are True and Fiction is False but the words True and False have no particular Meaning until they are assigned to something particular. Then their Meaning is in relation to that particular thing. It should not be hard to tell Fact from Fiction, but dropping the difference between Meaning and its Representation Interferes with those Qualitative Judgments.

Within the Boundaries of Pure Science, Analogies are Models of particular Meaning. A Model is a Representation. The same Meaning is represented in different ways through Analogies. Meaning and its Representation can perfectly align and Testing enables the fit to be adjusted until alignment is reached. A Representation can be Fiction in all its parts while the Meaning Represented can be Fact. Meaning cannot be True and False at the same time, nor can its Representation. But Meaning is a different type of thing to its Representation. Different types of thing have different Qualities and many different types of thing co-exist in harmony in Nature.

Designers of narratives know many Facts and they Certainly know how to deliver particular Meaning. That is not the problem that leads to Instability in Social Dynamics. What leads to Instability is the type of Meaning that is delivered to audiences, large and small, and how that Meaning relates to things that are already in their minds. The

Fact that audiences do Not Believe that Fiction is a Mechanism for the delivery of True Meaning is a problem. One Fact relates to another Fact. Audiences also do Not Believe the Fact that when False Meaning is delivered it can be translated into Action in the Real World. Meaning that is developed as Fiction can become Fact. The Denials of Facts by men of Inductive science have been Asserted and accepted. People can be disturbed by Fiction when they see it but Action against its showing is disarmed by Premises. After all, Fiction is a single category and all things within a single category have one type of Meaning. And since the Premise that nothing is Verifiable was accepted, the category Fact has been removed and all Meaning can be categorised as Fiction to varying degrees. The Illusion that categorical differences are differences in degree within one dimension is a problem that leads to Errors in Judgment. That Illusion was created to accommodate the Fact that Quantitative differences cannot Represent Qualitative differences. Differences in type are categorical and have been hidden to gift the Power to Act Universally to the Quantitative Method. That is a Fault. While under the influence of the Quantitative Method, we fail to see changes in the type of Meaning as they occur and those changes are Highly Significant to Functioning.

I will express that Meaning in a different way and develop it a little more. Fiction is Not Fact by definition. The Premise that Fiction is Fantasy and the Premise that the Products of Imagination are Independent of the Real World are Active in conjunction with each other. Those are False Premises. The Premise that the World of Fantasy is Dissociated from What Actually Happens leads to the Belief that all human beings can safely roam the World of Fantasy without restrictions. What happens to other people and other things in the Real World is also Dissociated from what happens to us. Dissociation is the disconnection of Emotion from particular Meaning. Association is the connection of Emotion to particular Meaning. Association and Dissociation are Powerful Emotional Mechanisms that produce different effects and Operate in conjunction. Association is a matter of positioning, and proximity enables certain things to happen; the effects of proximity are variable. Happenings depend on position relative to other things and doings change positions.

Physical proximity is one Factor among many that work together in channelling Meaning. I will speculate with an Analogy to demonstrate that particular Meaning. Physical objects can be physically connected but are not Associated until they are also connected to a Power Source and the connection becomes live. Association is Action that switches the Power of Emotion on. Dissociation is Action that switches the Power of Emotion off. When there is Dissociation the connection may still be there but it is Not Active. Parts that are Associated are in a relation and Parts that are Dissociated are not in a relation. So, these Mechanisms are more Complex than they Appear to be. But we do know Facts that enable their Activation and that is Denied. And we know the effects of these Mechanisms so Prediction is possible. If Association is defined in that way, positioning is only part of the Mechanism and the Principles of Association are Principles of Emotion expressed in terms of the effects of positioning.

I have provided Evidence that Formal philosophy is an Ideology that gained Power in the way that other Ideologies have done. I will summarise the Reasoning. Formal philosophy is a matter of Belief. Spanning the Centuries BC, the Belief was that oracles showed, through passing on messages from gods, how people should Act in the Real World in order to gain an advantage. After his life, Socrates became one of the gods and before that, during the time he was on Earth, he was perhaps an oracle who was consulted by some individuals, though in the stories about him he personally made no claim to being one. Oracles are Reputed Wise. An oracle picks up Meaning from a Source and passes it on to others who translate that into Action. After his passing and transformation by his followers into a god, the followers became oracles through which Socrates spoke. The Premise that Specialists in philosophy can discover and tell the Truth was metaphorically written in stone. Plato and Aristotle joined the gods when they passed from life on Earth and living beings speak for them. The words of an oracle are the guide for those who Believe in the god. That is just a rough sketch, but you will recognise the description as a Process through which a Religion can develop. Religions are philosophies. This particular Religion is thriving while others are failing. It has no name other than Formal logic and that term gives the Ideology the False Appearance of being merely the

Action of Reasoning. That which was written by Disciples became a Manual for Action in and on the World. The word Theocracy represents that happening.

All Religions have Formal aspects. Religions are Ideologies rooted in customs and Beliefs. But there is a Significant Difference between this particular Ideology and other Religions. It is usual for Religions to have a Moral Thread and Formal philosophy does not have that. There are Ideologies in which the founding fathers made progress in the development of Moral Concepts. The founding fathers of Formal logics made progress in undermining Moral Concepts. Formal philosophy has been virulent in its Action against traditional Religions. I am surprised that I have come to this view and am writing in this way. I have throughout my life been Agnostic. Two Factors in conjunction led to that. I saw the misuse of Religion by people and I am by nature Non-Conformist. I have a Natural tendency to ask questions. The existence of God as an entity is Not Testable and I have not changed my view that Testing Meaning for Truth and Justification is vital to enable good Functioning. But the Functioning of Religious Beliefs in Social Dynamics is Testable and that changes the relation between me and Religion. I remain an Agnostic but I now see that Religion is a topic to which Pure Science can be applied, and probably has been applied more than we realise. Thus, I could Justify paying more Attention to that topic. In Fact, it turns out that paying Attention to a form of religion is precisely what I have done, here and in the past. I now see that I was subject to an Illusion of Science.

I was not taught Formal logic and did not learn in a professional Context how to do that, but I Observed the Action informally. The fathers of Specialist philosophy showed how some human beings can be highly effective in Forcing their Will on other human beings. Religions can be well used and misused but if there is no Moral Dimension within a System then misuse is inevitable. The Ideology of Plato and Aristotle teaches people how to make War not Peace. Pure Science is a Process and is Metaphysical but Pure Science is not an Ideology. I propose that Pure Science invariably takes us into a Moral Dimension while we are engaged with that Process.

I will repeat a question at this point: How do you find particular Meaning when you do not know what that Meaning is? The Ancients struggled with that question and though they did not provide an answer, they gave the Appearance of having grasped parts of an answer. Some turned away from Metaphysics and towards Physics. They turned from the Abstract to the Physical. They turned from Poetry to Mathematics. They turned from Concepts to Constructs. They turned from Marking Boundaries to removing Boundaries. Plato and Aristotle secretly embraced Sophistry and Illusion and gave it a different Appearance. That Appearance is Deceptive. All those turnings led in one direction. They turned away from Nature and from the World that had emerged from that Force. They failed to recognise that the differences between things are necessary if there is to be any Meaning at all. And they distanced themselves from the happenings to which their doings led.

In general, Meaning is drawn from the parts of a composition by a Point of Reference and then comes together as a Point of Reference. The mistake is to perceive the Point of Reference that draws Meaning from the parts as being the same as the Point of Reference to which the Meaning that is drawn leads. That is a fundamental flaw in the Method of Induction. It is Not Natural for the Meaning at the end of the chain to be the same as the Meaning at the start of the chain. It is Natural for differences to develop. Those who do not know that the conclusion in Formal logics is pre-determined assume that the parts of the chain are where the answer came from. If they have found the parts and designed the arrangement then they succumb to the Illusion that they calculated the concluding Meaning.

Concepts can draw different types of Meaning together, but they cannot draw all types of Meaning together for some Meaning works against other Meaning. Concepts draw together the different types of Meaning that can be Reconciled. That is not a difficult Action to comprehend. Those types of Meaning that Interfere with Reconciliation must be removed if the happening of Reconciliation is to be enabled. It is difficult to remove False Meaning once it has been seen and Installed, but Truth with Justification is a Powerful Force. That Force has yet to be unleashed.

IS THERE A FATHER OF PURE SCIENCE?

Probably not! Aristophanes is a treasure trove of Meaning that I recently found and I have drawn some from his legacy, but only to describe Meaning that I found through Testing Meaning elsewhere. Art can give Power to the expression of particular Meaning but it does not Test for Truth and Justification the Meaning to which it gives Power. Nor does it Test the effects and the outcome of the Meaning delivered. All types of Artist draw on the Emotions. The fathers of the Ideology named logic drew forth Emotions and taught their followers how to do that.

Aristophanes was an Artist and he was a human being. Natural Deduction is available to all human beings. He represented his Perception of Meaning in the Real World through Art. Different types of Meaning are represented in Art. Aristophanes was a prominent Political Player. Socrates, Plato and Aristotle were also Artists and prominent Political Players. During the course of this narrative, Testing has shown that some Truths were captured in Aristophanes' plays. But the Proof of Truth did not lie in his plays. Aristophanes did not Verify the Meaning he perceived. The Frogs is my favourite on brief acquaintance with his work. I will use another part from The Frogs. I invite you all to join in this chorus. But you can choose whether you do so or not.

<u>Chorus</u>:

"If you look at the stuff that is written today
And the stupid things our statesmen say,
You would think, people had utterly lost the knack
Of telling the white from the utterly black –
They've thrown away all caution!

They sit at the feet of Socrates
Til they can't distinguish the wood from the trees,
And tragedy goes to POT;
They don't care whether their plays are art
But only whether the words are smart;
They waste our time with quibbles and quarrels,

Destroying our patience as well as our morals,
And making us all talk ROT."

Plato sat at the feet of Socrates and picked up some Meaning there. Aristotle sat at the feet of Plato picking up Meaning from him. They passed the Meaning on having Represented that Meaning in different forms and added their own amendments. Aristophanes was an Observer of Social Dynamics in his time. This part of his play Represents True Meaning. That can be Verified by Evidence and cannot be Falsified. The Fact that his description precisely fits the Action that can be Observed in our time is part of the Evidence of a systematic regularity and should trigger Alarm Bells together with a search for that which has Determined the playing out of a long-running Tragedy in the Real World. The Whole Tragedy has many parts in which we are all caught up one way or another. The Meaning I have just quoted from The Frogs fits my experiences in the Twentieth and Twenty-first Centuries. I am not alone in seeing the alignment. Many translators and interpreters of Aristophanes' work have seen the alignment and have referred to the fit as surprising. Systematic regularities are not due to Chance, something determines their happening. I have provided Evidence that specifies some determining Factors of those regularities and the direction of their combined Force.

True Meaning and where that leads has been shown by many practitioners of Pure Art in their stories. But proving Meaning is True requires a different type of Action to perceiving it. Meaning is Propositional until and unless it is Verified. The Actions of Perceiving Meaning and of Calculating Meaning are different in type and Intrinsically Related. Only focussing Attention on Truth aligns Meaning to that particular Point of Reference. Verification and Falsification require engagement in the Process of Pure Science. It is a Fact that Untested philosophical ideas are played out in all forms of Art. It is a Fact that Action follows the Perception of Meaning. Action clarifies the Meaning that is there. Observing the Action that is related to philosophical ideas enables their Meaning to be Calculated. Great care should be taken when putting ideas into Minds and great care is taken in order to produce particular effects. While

Attention is focussed on the effects, it is not focussed on the outcome as a Whole. Thus, the outcome can take us by surprise.

Meaning depends on the Point of Reference. Humans take care to do what they want to do and to achieve the goals they want to achieve. They can do that through putting ideas into the Minds of others. Puppet Masters watch the performance of their Puppets. That might seem a harsh way to represent what happens in the World. But, if True Meaning is not delivered through being Powerfully Represented, then nothing will change. We can all say that we have been both a Puppet and a Puppet Master at different times in different plays. That produces Confusion in our minds for we cannot tell the difference between the roles we are playing after a while. It is a Fact that Power can shift imperceptibly from one thing to another.

People who have fallen under the Spell of the Illusion of Formal logic do not tell the Truth. They do not aim to do so. The Truth has been told by many people who have not succumbed to that Illusion, but True Meaning is not recognised as such. One thing can Act against another thing. When True Meaning is placed in conjunction with False Meaning and the difference in Quality is dropped, then Truth loses its Power to Act. Either there is Interference with the Calculation of True Meaning or a lack of Knowledge about how to do that. Those alternative propositions can be Tested once the difference between Pure Science and pseudo-science is shown. And only once the Truth is known does it become possible to decide how to deal with it.

POWER AND WILL, BUT JUST BRIEFLY

I will quote a line from the Hobbit (well it is either from there or from the Lord of the Rings and I will not waste time checking that Trivial detail):

> "The Wild is not a place for gentle folk who can neither fight nor defend themselves."

That is open to interpretation. The angle from which we view arrangements makes a difference to the Meaning that we see. We think we Know what the words wild, gentle, fight and defend mean and we are given definitions of those words in our dictionaries should we feel a need to check. I look there for the Meaning held in some particular words, but I look in order to find Meaning during Testing and I do not pick it up as a gift. I have shown you that when we look for guidance, we will find it. Accepting guidance can mislead us. It is again time for some definitions.

The word wilderness is defined as:

wilderness: n. desert; uncultivated region or garden area. 2. confused assemblage.

That definition is in Error.

Uncultivated regions are not confused assemblages. They just Appear To Be So.

Confused assemblages are not uncultivated regions. They just Appear To Be So.

cultivate: v. 1. prepare and use (soil etc.) for crops or gardening. 2 a. raise (crops). b. culture (bacteria etc.). 3 a. (often as cultivated adj.) improve (the mind, manners, etc.). b. nurture (a friend, friendship etc.)

It is a Fact that a culture can produce different types of thing. The common definition is Biased towards the Perception that the types of thing produced by a culture are good things. The particular examples show that. The etcetera's hint otherwise. As a verb, culture has the Meaning:

culture: maintain (bacteria etc.) in suitable growth conditions. [Latin: related to CULTIVATE]

All things, those that are good and those that are bad, require suitable conditions in which to grow.

cultivator: n. 1. mechanical implement for breaking up the ground etc. 2. person or thing that cultivates.

Formal logic is a mechanical implement for breaking up the ground of Meaning ready for cultivation. Natural logic is an implement that is not mechanical and can break up the soil ready for planting, but does not always do the planting. Sometimes the ground is left fallow for Meaning to fall Naturally into place and then grow. There are many things in Nature that break up the soil. Trees do that with their roots and ants do that; it is not the size of a thing that makes a difference to what it can do and it is the type of Action that makes the difference. When setting the conditions for growth, great care should be taken to understand the type of thing that those conditions favour. The things that grow in Natural conditions can be different in type to those that grow in Formal conditions or not, but whatever their type, it is Certain that there is a difference in the Quantities produced once conditions change. Conditions have particular Meaning. Conditions are sets of Qualitative Variables. Different types of Qualitative Variable Interact.

Natural Language is a Dynamic System. Round and Round we go in Dynamic Systems but the System that is Circular does not stay in one place unless it is held there by a Force exerted upon it. I have demonstrated that:

IT IS INSTRUCTIVE TO OBSERVE IN DETAIL THAT WHICH IS CIRCULATING.

As I Observed more and more happenings in different Contexts, I found the absence of Reasoning puzzling. The Wild is the Natural World and the Natural World is Not Unruly. Gentle things do survive and flourish there. Gentle things have difficulty surviving in the Artificial World of human design. The Wild is the Natural Environment and is a place where gentle folk can defend themselves using Reasoning and can enable others to do the same. The Formal System is constructed by human agents and Acts against the Operation of the Capacity to Reason. The defences of gentle folk are not warranted in the Artificial World of human design. In other words, there is a long-standing agreement to remove the tools that gentle folk use to defend themselves. Removal of the Power of Natural Deduction by hiding that tool so that few will find it leaves gentle folk defenceless against attack. That is extraordinary. That Is Outrageous. We Certainly understand the Power of Agreements. But do we actually understand that there is an agreement of that type.

Natural Deduction cannot be removed entirely for it is a gift from Nature but when Truth is found it is disarmed. Truth is disarmed by requiring peer agreement on the Qualities of Meaning. Social Structures are currently designed to enable human agents of Strong Will to Act on everything in service of personal purpose without being held to account. That is where the disregard for Truth and the Natural World leaves us. Specialists in philosophy redefined Science as Inductive science. Pure Science was left unnamed and undefined. That is a Fault. The failure to show differences in Meaning gives the Appearance of Justification to that which is Unjustifiable. It also gives the Appearance of Being Unjustifiable to that which is Justifiable. Pure Science does not leave us Defenceless in the Wild. Things that Function well are Selected by Nature and survive. So, there is the Appearance of a contradiction between what happened to Reasoning and the Principle of Natural Selection. Or at least there is so far, but since Natural Selection takes a long time to Operate it makes Sense that the Apparent Contradiction will be resolved.

I have, with increasing frequency, Tested Meaning during my lifetime and particular Meaning has consistently Emerged. It is clear that peers have reached agreement that you can copy their Actions and join them in displaying Strong Will and attacking the defenceless, or you can join the ranks of the defenceless and accept the consequences of the Actions of those with Strong Will. But it does not have that Appearance. The conclusion is not stated, but it is there to be drawn. Illusions are designer products. Under the current philosophical regime choices are limited to different forms of that which is Unjustifiable. There are a few defenders of the defenceless and they are rendered relatively impotent by the System. The choice is limited by design for there is a viable alternative. Pure Science is a different route to a different destination. The problem of the current philosophical position has been seen by many but the alternative does not Appear to have been seen.

The lack of limits on human Actions has often been described as a problem in general terms. Or rather, it has been vaguely described as a problem. Vagueness is all that is left when differences in Meaning have been dropped. For example, Russell (1946) wrote:

> "To formulate any satisfactory modern ethic of human relationships, it will be essential to recognise the necessary limitations of men's power over the non-human environment, and the desirable limitations of their power over each other."

That has the Appearance of a Nice Sentiment. It seems that Specialists in philosophy are looking for a solution to the problems that are related to Human Actions. Is there now Reasonable Doubt about that? Or is there Reasonable Doubt that people are looking for a solution to the same problem? When Points of Reference are not accurately specified Meaning is open for misuse. Many agents have succeeded in overcoming the Will of others and in limiting the Actions of other agents while avoiding any limits being placed on their own Actions. It is only desirable to limit the Power of other agents over each other when their Actions harm others and the non-human environment. If there is no harm in relation to those things and there is benefit, then it is not desirable to limit their Actions. But in order to tell the difference in effects and outcome it is necessary to be able to Calculate Meaning. Benefit is Relative and Desires are Variable.

It is a Fact that the Formal philosophical Bias has led to Harm and to No Benefit as a Whole and that it will continue to do so while it is in place. There is a Natural alternative. But it is a Fact that Evidence alone will not produce a change in Beliefs. I need to bring Positive Emotions into the Equation and relate them to Pure Science. It is a Fact that Reasoning and Emotions Acting in conjunction are more Powerful than either Factor alone.

EMOTION IS NECESSARY TO MAKE A CONNECTION LIVE

PART TWO of Chapter six

THOMSON AND THE GOOD SAMARITAN

The strength of the Process of Pure Science lies in Testing Meaning from all possible Points of Reference. Truth is not Known until you do that. Only one Meaning will align to Truth. Meaning must be discovered and dealt with. Meaning that is True does not always Function well. The Selection of the Standard for Judgment of Meaning and Action is related to the type of Functioning in the World. The Artificial World of human design is different in type to the World designed by Nature. The standards of Artificial logic are different to the Standards of Natural Logic. Now, I am going to Appear to digress initially, but do not be deceived by that Appearance. Narratives progressively position parts in the field of view and enable the development of relations between them leading to the Emergence of Whole Meaning from a particular Point of Reference.

I have always loved stories and most forms of narrative. I say most because my Attention has always been focussed on Meaning. Some narratives carry little Meaning or hide Meaning in Confusion, and I do not enjoy those. In my youth I strongly resisted attempts to send me to Church, but as a very young child, I was sent to Sunday school. Religion did not engage me, but some of the narratives did. I had not thought about them over the years since then, but their memory lingers on in the depths of my mind and can be brought to the surface. Many parts in this narrative drew those memories to the fore, but no part of this narrative is based on Belief in the Truth of the detail of those stories as they were told. I took Meaning from the details and some of that I judged to be True. I remember the tales of Miracles. Children's Attention is always caught by tales of magic doings and happenings. Now, a few Whole ideas can be broken into an ever-increasing quantity of parts that are also ideas and those can be distributed among a large audience hungry for Meaning. That could be viewed as a form of Miracle. Perhaps it was shown in the Parable of Loaves and Fishes that ideas can divide and multiply when distributed, and that ideas keep body and soul together. And perhaps that was not an anticipated conclusion. Speculation is necessary to find Meaning and the stories triggered that. Are parts of the Bible Analogies delivered after Calculating True Meaning? It would be strange indeed if some of them were not, for Deduction is Natural and it is through the spoken and the written word that Meaning is passed on.

But Inductive analogies are Not Analogies of Meaning and there will be some of those too. Myths were created from the Actions of the original people around whom some Meaning crystallised. Stories are a large part of the Process of Channelling Meaning in particular directions. Truth is a particular Quality of Meaning. Narratives cannot transform one type of Meaning into a different type of Meaning but they can give a False Appearance in relation to its Qualities. Religions are related to the Real World in Complex ways. There is Art in Religion and Illusion is part of that. Different types of Illusion Function in different ways. It is a Fact that the Meaning captured in any Religion can be used well and can be misused. Misuse of the Power drawn by people from their Association with traditional Religions has been a Factor in their decline and, as we have seen, other Factors have been Active in conjunction with that. Remember that if you have not been as good as possible, you are open to attack. And detecting weakness in those who oppose you is a philosophical skill. Now, the Whole of Religion is different to the human parts that engage with it and yet the Disclosure of Evidence of Abuse by human beings who were Associated with Religions has played a part in Unsettling the Perception of Meaning in the Whole. Some humans defined as oracles crossed the line drawn in the Criminal Law. Sometimes it is not possible to adjust the Meaning of the Whole to accommodate Falsifications of Meaning in Single parts. And the type of transgression makes a Significant Difference to the Potential to accommodate them. But since the Whole is different to the parts, we could look again at some of the ideas in Religions once we have a greater understanding of the type of Action that characterises Pure Science. Ideologies are Powerful determinants of human Actions. So, it is strange that some of those well versed in Religion have Acted in a way that is Not Aligned to the teachings. There is also a need to understand that happening.

Narratives have the Power to determine Actions through the Meaning they deliver. That Power is recognised and is Acted against, but that Power is also harnessed by those who would deny its Power to others. I will provide Evidence of that Fact from more than one Source in more than one narrative. Thomson's Action in her famous argument was an Attack on particular Meaning using Formal logic Represented through narrative. She told a story. There was the Appearance of a Single Target. I have shown that Appearance to be Deceptive. It cannot be Denied that she saw some of the other Meanings in the line of fire. Some of the many particular Meanings she Acted against were Concepts that had been a Force for Good. She did not ignore such Concepts, she Unsettled their Meaning. We should all be puzzled that she did that. Perhaps that Action was not noticed while the Attention of the audience was kept focussed on Abortion. But it is puzzling if nobody noticed. I scanned so many things as I searched for Meaning that I cannot precisely remember what drew my Attention to her article. I was not attracted by the topic. I had a different Focus of Attention to that of Thomson when I read her article. I will again draw her Action Against Concepts to your Attention. It took effort to Calculate the Meaning and the type of Action involved. To me, the result is worth the Effort it took. This detail of Action to remove a particular Standard is enlightening, in the Sense that I use that word.

Thomson's argument was in line with the teachings of Plato and Aristotle, and was designed to have an impact on the Emotions of the audience. The Art of Persuasion is effective in doing that and I am not immune from the Arts. I have an Active Capacity for Emotion. When I read Thomson's argument in the Context of Abortion it struck several discordant notes in me. I found her treatment of the Parable of the Good Samaritan particularly striking. It would not have occurred to me to include any parts from traditional Religions within my own narrative. I do so now because she did and because I recognised the Significance of her Action. My Point of Reference is not Religion and I do not argue in that Sense. Thomson did not argue in a traditional Religious Sense either, but she did argue from an Ideological position while I do not. Beliefs are not my guide. The Formal philosophical position is not categorised as a Religion. Religions are part of the Metaphysics that the Vienna Circle and some other Specialist philosophers across the Centuries have forcefully dismissed. That is a Fault. We can learn from differences in all types of thing including differences in Beliefs. But those who have one set of Beliefs frequently and systematically Act against those who have a different set of Beliefs and do not aim to learn from them.

Part of Thomson's purpose was clearly to unsettle Meaning in a story from the Christian Religion and I presume she saw that Ancient narrative as part of the support for the Anti-Abortion position she attacked. If so, her judgment was in Error. It was not necessary to Attack Concepts in order to Verify the Absence of Justification for Beliefs on Abortion. Both sides of the argument suffered from the Insoluble Problem of Justification that is Intrinsic to Induction. She must have known that Fact. Verification is impossible using the Method of Induction. I will show two other Facts that she should have known.

THE ENDS DO NOT JUSTIFY THE MEANS.

THE MEANS MUST JUSTIFY THE ENDS.

That is an old saying. I will Represent the same Meaning in a different form: The Whole Meaning does not Justify the parts and their Meaning – the parts and their Meaning must Justify the Whole Meaning. I will continue my Analysis of her position using the Point of Reference that governs the Action in Pure Science. In this case the Process of Pure Science provides Justification for an Ancient Standard in a Religious Context. The word Moral can accurately be applied to this Standard. I did not Intend to do that. The definition of the word Christian is firmly associated with a particular Religion. I am not a Christian. So do not Miss-Take the Meaning of this Analysis. I presume there are Moral Standards in other Religions that could also be Justified. I have no Religious Beliefs but I do not Act against any form of Metaphysics that has a Moral Dimension. Morality is something to be developed in Social Systems, not eradicated. Moral Standards are not confined to those who have Religious Beliefs. And …. dare I say it? Well yes, I must do so in order to cover the range of possibilities. Those who have Religious Beliefs are not confined by Moral Standards. Unless we have Standards that are Moral, we cannot Judge who meets them and who does not.

In Part Three of Chapter Four of this narrative, I briefly showed that in service of the Pro-Abortion argument Thomson attempts to replace the Standard of Conduct set up in the parable of a Good Samaritan with an alternative standard, that of a Minimally Decent Samaritan. That Action merits closer examination. I would describe the Parable of the Good Samaritan as a Simple Model of particular Meaning that is not a specific prescription for Action in the Real World and is a Sign pointing in a general direction as a guide to enable a destination in the vicinity to be reached. The so-called problem of Universals is Not a Problem. Flexibility is Natural and enables good Functioning. The story of the Good Samaritan delivers a Moral message. The Meaning of the Sign was clear. Representations of Meaning do not have to be True in order to serve as a Model of Meaning and of the Action that relates to it. That is an important Fact. Irrespective of the Truth of a Representation, active agents learn through the Perception of Meaning shown in Models. The Representation is Symbolic. Standards are Active in Focussing Attention. Their Meaning must be Tested for Truth and Justification. It is Reasonable to presume that the Meaning in this Parable was Tested using Natural Deduction by the Ancients who designed it.

A Single human being or a group of Ancients born several hundred years after the founding fathers of Artificial philosophy had their being, set up a particular Point of Reference. Thomson set out to change that Point of Reference. She does not align her work with the particular Ancients whose philosophy had a Moral Dimension. Thomson aligns with the philosophy of Ancients who predate them and did not include that particular Dimension in their teachings. And I would define a dimension in Qualitative Terms, not in the Quantitative Terms of Artificial logic. I have already drawn Attention to her Unsettling of the Aspiration embodied in the related Concept of a Right to Life. That these Unsettling Goals were set at all is of particular interest. Thomson's article was set within the broad Context of Public Policy. It is a Fact that those who design Public Policy affect each and every one of us by the Meaning that is given through Structures they design. It is necessary that human agents set the Points of Reference to govern Meaning and Action in Social Structures. Designers are necessary, but great care should be taken in their Selection. Great care is taken! Choice is restricted when Selections are made. Only one type of designer is put up for the final Selection. What you get at the end of the Action is fixed by type at the start of the Action. We succumb to the Illusion of Choice for the same type of thing comes in different forms, but if we do not succumb that does not matter for that Illusion is enshrined in a Structure that was built to Resist Disturbance. The Structure for the Distribution of Power was built to last, has lasted for Centuries and suffers from the Problem of Justification. That Problem is said to be Insoluble. I propose that there is a Solution and that it is not a difficult one. The Solution is actively prevented from happening. And perhaps the Solution is not seen. Only Experiments carefully designed for the purpose could tell us whether there is Intention to create and maintain problems or not.

Taken as a Whole, Thomson's arrangement of Meaning, which is defined by peers as a "moral argument", left the State of the audience in a poor condition. I will show you more Evidence to Verify that statement. Thomson aimed to change a Qualitative Standard to a Quantitative standard. That is a change from a Concept to a particular type of Construct. There was Method in that madness. An attempt to remove the Standard altogether would have shown that a change in the type of conduct was being advocated. She took care to avoid making the change in type visible by using the Formal Strategy of introducing the notion that one type of thing can be split into Fractions. Once the Premise that only one type of thing is involved takes hold, that enables Significant changes to be perceived as being Trivial in Meaning. That Premise is part of the Quantitative Action that enables the happening of changes to Meaning as a Whole. Her tiptoeing around the True Meaning of her Actions is clearly visible in the data pattern she designed. The softly, softly approach is very effective in the Practice of Deception. Of course, I am not implying there was any Intention to Deceive on her part. It would be Outrageous to do that without Proof of Intention and I am not even searching for Evidence of that. I am merely stating that the Action Functioned as Deception. She was doing what she had been trained to do. An Illusion was created in a Time-Honoured Tradition. Deception is not a Tradition I honour.

The Meaning of small changes is hard to pin down. When changes are small the possibility of a successful attack on the thing that is moving is substantially reduced for the movement is not noticed or its Meaning is dismissed as Trivial. Where there is no movement, nothing is happening. Where there is the Appearance of no movement it is assumed that nothing is happening. When Attention is drawn to a small movement it is Asserted that nothing different is happening in terms of Meaning. Thomson was an Expert in Formal philosophy and must have read the teachings of the Fathers and many interpretations of those. The direction of the movement of Meaning towards and within the Quantitative standard is clear and is strategically placed to co-ordinate with other Nudging's in her argument. I need to show you the Action in detail and the way it Functions. I can do that through telling stories in Natural Language. That too is a Time-Honoured Tradition and it is a Natural Tradition that is part of Pure Art and is also part of Pure Science. I honour that Tradition. Pure Science as a Whole is a type of Metaphysics that is Not Art. Pure Science is aligned to Pure Art and is antagonistic to the Art of Deception.

Like dictionaries, encyclopaedias are compiled to give us Meaning and do so at length – the difference between them is Quantitative and is Not Qualitative. It is stated in the entry on Thomson in Wikipedia:

> "A Defense of Abortion" is a moral philosophy paper by Judith Jarvis Thomson first published in 1971. ...... Her argument has many critics on both sides of the abortion debate, yet continues to receive defense. Thomson's imaginative examples and controversial conclusions have made "A Defense of Abortion" perhaps "the most widely reprinted essay in all of contemporary philosophy".

Perhaps others did notice the Meaning I noticed. I have not read the criticisms of other people. But the False Representation of her paper as moral is repeated in the entry in Wikipedia so the critics did not Falsify that description. To describe her narrative as a "moral" paper is a peer agreed philosophical Trick of Representation. Her article is a philosophy paper using Formal logic and there is no possibility of Justification for that. That which is Moral is Justified.

In order to perceive the problems that her argument created you have to move away from her Focus of Attention and Track the

Meaning of her narrative in relation to other things. I have noted that she ensured her success by fixing the Attention of her audience to Abortion. Thus, the Action on other Issues may not have attracted much Attention. Now, observe closely her treatment of The Good Samaritan. I will now, and temporarily, fix Attention on that. Thomson quotes as follows from Luke 10: 30-35. I have not checked in the Bible for accuracy. It is not relevant to do that and I have not read the Bible. But this fits my memory of the story from the time I was sent to Sunday school to listen to the ministers there.

THE PARABLE OF THE GOOD SAMARITAN

> "A certain man went down from Jerusalem to Jericho and fell among thieves, who stripped him of his raiment, and wounded him, and departed, leaving him half dead. And by chance there came down a certain priest that way; and when he saw him, he passed by on the other side. And likewise a Levite, when he was at the place, came and looked on him, and passed by on the other side. But a certain Samaritan, as he journeyed, came where he was; and when he saw him he had compassion on him. And went to him, and bound up his wounds, pouring in oil and wine, and set him on his own beast, and brought him to an inn and took care of him. And on the morrow, when he departed, he took out two pence, and gave them to the host, and said unto him, 'Take care of him; and whatsoever thou spendest more, when I come again, I will repay thee'."

That is, quite simply, a nice story with a Moral. How could anyone object to that? Well I have shown you how, so maybe a better question is: Why would anyone object to that? What purpose could rejection of that Moral Standard serve? I cannot answer questions of Intent but I can show you the Structure of the Manipulation, the origins of that Structure and how the change in Meaning Functions in terms of the Action to which it leads. I can provide a body of Evidence through Reasoning that Concepts can hold parts with different Meanings together. I propose that the Emergence of a Concept can produce Stability and remove Confusion. I can show that Removal of a Concept can produce Instability. My next narrative gives more of the relevant details and has already been written; that merely awaits proof reading before I can show it to you. For now, I will use Thomson's argument to demonstrate the effect of removing a Concept in the Context of the Real World. Her argument can be turned to the advantage of a few and to the disadvantage of the many.

According to Wikepedia:

> "Jesus is described as telling the parable in response to a question regarding the identity of the "neighbour", who Leviticus 19:18 says should be loved."

I will not divert to particular questions of what love is or how lovable neighbours are or are not. People are necessarily Judged as Single Cases in terms of the Action in which they engage. The parable is not concerned with such questions and is concerned with how we conduct ourselves in relation to other people in general and with the setting of a Qualitative Standard for Judgment of our conduct. The Meaning of all things must be Judged against a Qualitative Standard through their Actions in relation to the things they encounter. The Parable is an example of Good Conduct and Acts as an Aspirational Standard. The story is Not Fact, it is Fiction. Fiction is based on Observations of happenings in the Real World and draws particular Meaning from those. Fiction is not a precise or accurate description of Single happenings. Meaning modelled in Imagination can Represent happenings that have not actually taken place in the Real World but could have done. Imaginary events are not Real but the Meaning they Represent can be True or False. The difficulty lies in distinguishing Meaning that is True from Meaning that is False. Only Testing can make the distinction and writers of Fiction do not do that. But that can be done.

Fiction can be translated into Action in Reality under certain conditions. The Parable was a description of a Fictional happening in which particular Meaning was Represented for transfer to the Real World. The story would not have caught Attention in the way that it did if the Meaning Represented had been what systematically happened in the Real World at that time. A Contrast was shown. The parable was a Qualitative Formula designed to change the Perception of Meaning by an audience and thus to change Action in Social Dynamics. The parable was designed to show a difference between good and bad conduct. The audience was shown that there is a choice and what the choice Means. In order to show the Qualitative difference in Meaning the story showed the difference in Action. That is not Operationism but there is some resemblance. Narratives are Not Dissociated from happenings in the Real World. This Parable is not Formal philosophy and is Natural Philosophy. In this example of Natural Philosophy, a Moral is taught. Moral teachings are aimed at changing and improving the Condition of the Audience. The narrative in Wikipedia surrounding the parable continues thus:

> “Portraying a Samaritan in a positive light would have come as a shock to Jesus's audience. It is typical of his provocative speech in which conventional expectations are inverted. Some Christians, such as Augustine, have interpreted the parable allegorically, with the Samaritan representing Jesus Christ, who saves the sinful soul. Others, however, discount this allegory as unrelated to the parable's original meaning and see the parable as exemplifying the ethics of Jesus.”

As a very small child I made a simple, literal interpretation: - helping others when you have the opportunity and the Power to do so is a good thing and there are good people in the world who help others. I knew even then that there are bad people in the world or at least that there are people who do bad things. It seemed clear to me that all I had to do was find the ones who are good. I had no idea at that time precisely how difficult it is to make an accurate distinction.

My experiences in my youth were within the range of normal experiences and have been since then. The many stories, parables and fables that I read in my youth were like hot soup on an icy cold day. Hot soup revives and sustains you against the cold. The stories took me away from Reality and soothed my Senses following those normal experiences in the Real World that were difficult to handle. I now see that the stories were far too effective in soothing me. Some things do not go away, and being soothed in relation to those is not entirely helpful. True Meaning must be found and dealt with. But soothing is necessary to enable the Action of Reasoning. And that soothing effect and the Context from which it came must have stuck with me because when I read Thomson's Tale on the Topic of Abortion, I could not ignore the implications. She threw the hot soup away. I had survived in the Real World. I saw that others would not be sustained as I had been. But since I was at an older age, I saw more than that, and I am in the course of showing you what more I then saw.

So, I had liked the niceness of the original story and what I saw in her adaptation was not nice at all. Fundamentally, her position advocates leaving individuals alone, hurt and vulnerable to experience the icy cold environment without succour and comfort. That is not a Reasonable position.

THAT IS NOT A MORAL POSITION.

Yet Thomson seems comfortable with that, for she gives the Meaning that she substitutes for the Original Meaning an Appearance of Justification. That Appearance was accepted by her peers. They should have known better and I presume they did not. Frankly there is no possibility of Justification for her position. Thomson must have known that there is No Justification for her argument by Induction. She is a Specialist philosopher well versed in the Formal Approach. It is quite some time since Hume pinned that information into position on the noticeboard in the training academies of philosophy and that Fact was known long before Hume's time in this world. As we have Observed, they do not look for Justification for their Actions. But she also knew that an Appearance of Justification is necessary to have an argument accepted. In her narrative she tells us she knew that.

While the Parable is a part of Christian religious teaching, I have glimpsed the same Meaning Represented differently in other religious teachings even though I have merely peeped and did not look far. In attacking Meaning in one Representation, Thomson attacked that Meaning in whatever form it was represented. The influence of the original story had extended beyond the Context of Religion. The word Samaritan and the Meaning Associated with that became enshrined in the common dictionary. From the Pocket Oxford Dictionary:

samaritan: 1. (usually Good Samaritan) a charitable or helpful person (with reference to Luke 10.33) … 2. a member of a people inhabiting Samaria in biblical times, or of the modern community claiming descent from them, adhering to a form of Judaism accepting only its own ancient version of the Pentateuch as Scripture.

The words good and samaritan were firmly and lastingly Associated by the Ancient Parable. Any charitable or helpful person is represented by the word samaritan. This definition is Evidence that Natural Language is a Powerful Method of delivering Meaning. In this case, the story, which is Fiction, Justifies the definition. The word Samaritan became a Symbol of Goodness. The story is underpinned by a Concept of Good and Bad Conduct and the Parable is a Model of the Concept. But the story is not a Construct that replaces the Concept of Good and Bad. The story teller made Meaning Operational and clarified some Meaning in the Concept. That is Not Operationism. That is Modelling Meaning. That is the Delivery of Meaning from a Concept through the Mechanism of Analogy.

Yet, according to Thomson, we:

> ".. have in fact to distinguish between two kinds of Samaritan: the Good Samaritan and what we might call the Minimally Decent Samaritan."

Here we do not have a Fact; we have a False Assertion of Truth. This is an attempt to set a Premise. That is a version of a very different Scripture. That was news to me! Here you see Evidence that Specialist philosophers Act on Definitions and Disassemble Concepts. Here you have a Trick within the Art of Deception. Thomson's Action is Operationism – she replaces the Concept with a Construct that is impoverished in its Meaning. Now, according to a narrative and a definition that has Stood Firm for Centuries, someone who is Minimally Decent is not charitable or helpful and cannot be called a Samaritan. If you change the definition of the word Samaritan you change the Parable for that is where the Meaning came from. If you change the Parable you change the Standard of Conduct.

Let me show you Thomson's new standard in her "moral" paper. The new "standard" is actually No Standard at all. There is only one possible conclusion to this part of her argument. Thomson did not give you the conclusion in a straightforward way. She delivered a particular conclusion by subterfuge. I will be straightforward in delivering the Meaning she gifted. If that had been done by Thomson then the audience would not have been able to accept the Representation and it would not have been popular or so widely read. Possibly it would not have been accepted for publication, but I am not sure about that. Hints are Powerful in delivering Meaning, for you do not see precisely what is being delivered or its Source. In reading the story of the Samaritan who is Not Good and is Minimally Decent you will find that this Meaning is Too Hot to Handle. The new story is not a parable for the lesson is Not a Moral One. The new story is longer than the original and runs like this:

THE NARRATIVE OF THE MINIMALLY DECENT SAMARITAN

A GIFT OF MEANING INSPIRED BY THOMSON

A certain man went down from Jerusalem to Jericho and fell among thieves, who stripped him of his raiment, and wounded him, and departed, leaving him half dead. By chance, a certain priest came down that way and, when he saw him, he passed by on the other side. And likewise a Levite, when he was at the place, came and looked on him, and passed by on the other side. They did not know the man. They left him to his fate. His fate is quite simply no concern of theirs. The priest and the Levite had other things to do, and after all time is valuable and limited. It was perfectly reasonable to just walk on by. There was no obligation to help. Along comes a Samaritan who stops and helps the injured man. The man was a stranger to him. He did not need to stop and help. Why do that? There is no reason to do that! If the Samaritan could have helped without any inconvenience to himself, then just possibly helping might have some merit but he could not be faulted if he failed to assist. The law does not require people to help each other. Surely the law would require that if it were reasonable! Pay attention and listen

> to me. Others have passed by and left the man to his fate. You have no obligation here and there would be a cost to you if you did not pass by. There is no cost to you if you continue on your way. You need to earn your living. To leave people to their fate is the normal thing to do. Even the priest who earns his living by preaching standards of conduct does not stop. The priest is a moral leader. Learn by his example.

That is not a nice story and is the Natural Logical conclusion of the Enthymeme that Thomson designed. That rearrangement shows clearly that the Moral Concept was removed. I perceive the gift of her Meaning and I am not persuaded to accept that. I can Predict Precisely in Qualitative Terms how that Functions in Social Dynamics as a Whole. And so can you. Surely, we can all predict the way that narrative Functions. And surely that Prediction should have been made around the time the article was being published. And it is a Fact that I have merely specified the Meaning that Thomson gave in her argument which was defined as a "moral" argument. But, while the Moral Standard that was a Concept has been removed through her Actions, the Appearance of a Standard has been retained in her article. A Quantitative Construct was substituted for the Concept. That is Operationism. Now can you see the Fault in Operationism? In Formal Operations, differences in type are dropped in order to form a Single type of thing that is then split into fractions. At best, splitting things into fractions weakens those things and at worst it kills them. Certainly, it kills them if they are living things. And Meaning is Active; it is a Living System. Meaning can be weakened and extinguished. I have shown Evidence that proves Operationism is wrong and that people know that type of Action is wrong. The Appearance of Standards in Social Dynamics is an Illusion. Ethics are Constructs that stand where the Moral Concepts should be. It is a Fact that Meaning is translated into Action. It is necessary to strengthen and develop Concepts in order to prevent the continuing erosion of Moral Standards. Let us look at the common Meaning given in our dictionaries to the word moral:

moral: adj. 1a. concerned with goodness or badness of human character or behaviour, or with the distinction between right and wrong. b. concerned with accepted rules or standards of human behaviour 2a. virtuous in general conduct. b. capable of moral action **3. (of rights or duties etc.) founded on moral not actual law** [Latin *mos mor* - custom]

Specialists have inserted at part 1b. Meaning that is different in type to other parts. But the words good, bad and virtuous are still associated with the word moral in the common definition. Perhaps they find it useful to keep that Meaning there while they misuse the Representation. Or perhaps they simply haven't got a clue what they are doing with Meaning. All is not well with this definition. In this case Confusion lies at the Latin roots, and that Confusion is carried into the definition. Custom is habit; it is a particular, established way of behaving. The type of behaving is not specified at the root. The word "accepted" in part 1.b. points to peer agreement. The type of agreement is not specified. I have highlighted part 3 for that part picks up a difference of importance which Thomson dropped. She argued that the moral standard is or should be the same as the standard set by law. The definition still shows a difference. It is the tradition of Specialists in Artificial philosophy to tread Meaning and Natural Language underfoot in a stampede to get what they want.

Thomson's argument was concerned with goodness or badness of human character or behaviour, and with the distinction between right and wrong. But her Bias would be unexpected and not seen. The direction of her argument was away from goodness and towards badness, away from what is right and towards what is wrong. Her argument interfered with Qualitative distinctions in the Moral Realm. Yet, her argument was welcomed. Specialist philosophers did not Unsettle that argument. I have to ask a question: Why not? It is their custom to Unsettle Meaning given to them. Clearly, they do not Unsettle all Meaning given to them. The answer to the question is clear. Specialist philosophers have long aimed to Unsettle the distinction between good and bad, right and wrong. Thomson's argument is in line with previous Action. Specialist philosophers do not pick up and distinguish differences in type of Meaning and they have No Concepts to guide them. But that does not mean that nobody can distinguish one type of Meaning from another and that nobody has Concepts to guide them. I have provided a wealth of Evidence that the type of philosophy in place in Social Dynamics is Inductive and governs all the Action that takes place there. It does that because those who follow that type of philosophy rise to Power using the Structures put in place by the Ancients to support their rising. The teaching of Artificial philosophy as a Specialism helps maintain the Unstable Systems that type of philosophy creates. A change to Natural Philosophy and Recovery of the Process of Pure Science is long overdue. Perhaps you now agree with that statement and perhaps you do not. Beliefs are resistant to change.

A gift is a present and a present is something that is presented in a particular Context. It is necessary to recognise that something presented to our Senses may contain danger irrespective of the Context. Appearances are Superficial and cannot be relied upon for Meaning. When we bear in mind the possibility that the Appearance of Meaning may be False, we Act accordingly and Test for the type of Meaning hidden within. Aristotelian Rhetoric is strongly persuasive. The success of the Persuasion depends on keeping hidden the Meaning that is delivered through the argument. I have shown you precisely where Thomson's Enthymeme leads us. You can Test whether I have exaggerated where this part of her argument leads by reading the original article. Of course, the responsibility for that interpretation lies with me alone. Or does it? I would not have written such a story if I had not read her argument. No other conclusion is possible. And I have made it clear that I do not accept that conclusion or her Method of argument. I showed you a difference in Meaning that existed. I do not accept Artificial philosophy. Truth is Not an Issue for those who align with the Formal philosophical position. That which is credible takes time to produce and will be produced by those who wish to keep Truth at bay. We should not accept that which is merely credible. Truth is an Issue for those who align with a Natural Philosophical position and is Not a Problem for that type of Philosophy. I do not know how many people align with Natural Philosophy for their number cannot be counted until the difference is made visible.

Meaning in other parts of Thomson's argument was aligned with the notion of a Minimally Decent Samaritan, and had similar effects. Particular effects align to produce particular outcomes. Now, while Wholes are different to the parts of an arrangement, they are not completely different. Bad effects will not align to produce a good outcome. It is over 40 years since her article was published. The outcome is already upon us. I do not suggest that her article alone produced the current state of Social Dynamics. I have tracked the Fault back to its Source. I suggest only that it was a part of a Whole position that led us in one direction to the present State. I have used this Single Case to demonstrate the type of regularity that is recurring systematically. There have been many Disciples of Plato and Aristotle before Thomson and since. I do not single her out as a person. She was trained to do that which she did. I do not hold her responsible for picking up Meaning that was delivered to her. But without examining her argument in detail I cannot show you where that Approach leads.

It is difficult to analyse Single Cases and make the results public. The Law Acts against doing that and there must be safeguards. You will gather that I am in favour of more safeguards not fewer, and that I take a different position on the type of thing that must be safeguarded. But the Law does not Act against Analysing Single Cases when there is Justification for such Analysis and when the Single Cases taken for Analysis have been put in the Public Domain by the people that designed them. Thomson's article was published. Many others have critically appraised the detail and attempted to Unsettle the argument while many have been friendly and supported her argument. The detail was put there in public for the taking. There is a licence to Analyse that Representation. I have merely used the Licence that is there.

Carefully leaving the conclusion out of view in an Enthymeme protects the designers from responsibility for what happens. But Enthymemes are constructed with precision and the Meaning that will be perceived is Certain. So, I will repeat a question: Do designers have responsibility for the effects and outcomes of the Structures they design and put in position in Social Dynamics or are the members of the audience responsible when they perceive Meaning as it is given, accept it and Act in alignment with that?

Action follows Meaning that is hidden more often than it follows the same Meaning when brought into awareness. Standards are particular Points of Reference. In attacking the Parable of the Good Samaritan, Thomson aimed to lower a general Standard for Conduct in Social Dynamics. That is a Fact. Thomson did not hide that Fact. She worked hard to produce that effect. Thomson designed what might be "the most widely reprinted essay in all of contemporary philosophy". Meaning that was and is Unjustifiable has been spread far and wide and Action follows Meaning. To lower Standards is Unjustifiable and Unconscionable. The word Infection is used for the Installation in an audience of Beliefs that lead to Action that harms them and other people. I leave you to reflect on that. Failures of care in hospitals and nursing homes, abuse in our institutions, our homes and our workplaces are frequent happenings. Meaning and Action that is repeated is Systematic and is determined by Factors Operating in Conjunction. The young, the old and those of all ages in between are suffering abuse on a large scale. There has been no stopping point and the frequency has increased at an alarming rate. Is it all incomprehensible? Are we never to progress towards solutions?

I propose that Causal Knowledge is within our grasp. We can change the State of our Societies if we turn our Attention to Meaning and if we create and enforce stopping points based on Calculations of Meaning. That has long been known and the Law is there do that, but is that actually what it does? And the Law is applied too late to prevent Disorder happening. The Force of the Law is applied after Disorder has occurred. The solution must come before that point and the development of Moral Concepts is at least part of the solution. Deliveries of Moral Concepts can Install Internal Stopping Points in the Minds of human beings.

It is a Fact that the effect of lowering one Standard is Significant in lowering all other Related Standards. Thomson relied on producing that effect. She lowered many Standards in order to enable acceptance of Abortion. That is Outrageous. The effects and the outcome of the Qualitative changes she managed to achieve by giving the Appearance that the changes were Quantitative would be Significant. But it would be impossible to prove that is a Fact by engaging in the type of research that is peer approved by Specialists. Causation is ignored under Quantitative Rule. Only Uncertainty is produced by Induction covered by the term “scientific research”. And the frequency of Research that is actually Pure Science is Unquantifiable while we cannot distinguish Pure Science from Inductive science. There is Intention behind that and we cannot know whose Intention that was but we can know that they were selfish and narrow minded for they did not look past their own horizons. The Evidence points only to that conclusion. I have not merely expressed an Opinion and given vent to Emotion at this point. The Formal philosophical position encourages us all to be selfish and narrow minded. Where else could The Narrative of the Minimally Decent Samaritan lead? There are many more tales in the published literature of the type that Thomson produced. Seek and you will find them.

The answers to Calculations of Meaning using Natural Deduction and taking many angles of view in a search for Truth are Justifiable; they cannot be Unsettled. The lack of Causal Knowledge and of good Qualitative Theoretical Models is directly related to the absence of Pure Science in the Context of Studies of Human Action. It is possible to access Causal Knowledge through Analysing details in Single Cases and Calculating their True Meaning. That has been made very difficult to do.

Thomson did not Act only on Representations as Specialists in philosophy tell us they do. She did not even create that Appearance. She Acted on Meaning and she did so in Natural Language. Effort is applied to creating the Appearance of Justification. After showing you the Parable of the Good Samaritan, Thomson stated:

> "The Good Samaritan went out of his way, at some cost to himself, to help one in need of it. We are not told what the options were, that is, whether or not the priest and the Levite could have helped by doing less than the Good Samaritan did, but assuming they could have, then the fact that they did nothing at all shows they were not even Minimally Decent Samaritans, not because they were not Samaritans, but because they were not even minimally decent."

By now, you will be aware that your Sixth Sense is being Manipulated by her argument and that the True Meaning, which I have shown, was hidden. Several things are going on here in terms of Shaping Meaning and Projecting it from her mind into the minds of her audience. Those things are vague in form, which makes it difficult to Track Meaning, but the Action is precise in Functioning and that is easier to Track. The first sentence in the small part just quoted is a statement of Fact from the original Ancient narrative. The second sentence is a small narrative in itself. The second sentence suggests there was a middle road, separates the word Samaritan from the word Good and replaces the word Good with the term Minimally Decent. That is coupled with the suggestion that there is an extreme below that and the suggestion that the extreme is unacceptable. Towards the end of the second sentence she contrasts the term minimally decent with the word Samaritan, thus avoiding the contrast between the term minimally decent and the word good. That is a precise manoeuvre by a Specialist in philosophy delivering a change in Meaning using Natural Language. Thomson fails to point out that minimal is not far above nothing at all. Her audience is pushed in the direction of a disregard for others. That is a Fact. And yet we are told by Specialists in Artificial languages that Natural Language is not a tool that can deliver particular Meaning precisely! They are wrong. But do they know that Fact? The Fact does seem to be obvious, but if it now seems obvious to you then perhaps that is because of what you have read so far. It is Reasonable to presume that Meaning has not previously been Calculated until and unless the type of Intention can be Verified.

I repeat: Minimal is a short step away from nothing and I would add that the Boundary is not Marked. Decency slips easily away to nothing when there is No Mark to stop its fall. The fall is given the backing of Authority. That is a Truly Strange thing to support. That is Not an Intelligent thing to do. Something has gone seriously wrong. We all know from experience that is a Fact. I am showing you what went wrong and how to correct the Fault. I am giving you my Opinion on Causal Processes and the Evidence on which the Quality of my Opinion can be Judged. The question now is: What Standard will be applied to Judge my Opinion and the Evidence that emerges from Single Cases? An Inductive standard is Not Fit for That Purpose.

The Concepts of the Good Samaritan and the Right to Life are Moral Principles. A Moral Standard is an Abstract Model of a presumed Ideal that does not exist in the Real World. An Analogy springs to mind that might clarify the Complex Meaning I am trying to express. A star in the firmament can be seen shining brightly. Vessels set their sights by that star in order to get to a destination that is not in the firmament and is on firm ground in this world. They are not aiming to get to the star. The star guides them to somewhere that is different to that and is different to where they began their journey. But they could not reach their new destination without setting their sights by the star. Moral Principles are there to Enlighten and guide us along the way through Life. Without clear Direction our Actions are Disordered.

Thomson knew that the word Samaritan represents Meaning in the word good. The word good represents a particular Quality. She avoided using the word good between the words minimal and Samaritan. She used the word decent. So why did she do that? I will draw the common definition into view:

<u>decent</u>: adj. 1a. conforming with standards of decency. b. avoiding obscenity. 2. respectable. 3. **acceptable, good enough**. 4. kind, obliging. [Latin *decet* is fitting]

Thomson chose her words carefully to fit the Meaning she wanted to deliver. She avoided being highly offensive through using the Mechanism of Understatement. She aimed to give details of a standard of decency that is good enough but did not show what it was good enough for. The Fact that Meaning can be delivered precisely in Natural Language is Denied by Specialists in Formal logic while they Selectively do precisely that.

It is perfectly possible that Specialists in philosophy and their allies are in such a State of Confusion about Judgments of right and wrong, good and bad, that they do not understand the Meaning of any of their Actions. They tell us they do not look where their Representations come from or where they go in terms of Meaning, and I am perfectly happy to accept that until that statement can be Tested. That is not done in this narrative and that can only be done once Meaning is shown in full and their Reactions to that are Observed. Those who have accepted the Artificial philosophy shaped by Plato and Aristotle and risen to positions of Power see no further than the tunnels in which they live. Only when the ground shakes do they see the need to provide additional support for their dwellings. And perhaps they will emerge and look around when it is clear that an earthquake of large magnitude is imminent.

MISUSE OF SINGLE CASES

Thomson removed the Quality of Goodness from the Parable of the Samaritan and then she put the notion of fractions of decency in the field of view. The Quantitative notion of goodness at one end of a single dimension and badness at the other end was left in the minds of the audience. That idea was backed up in several ways. Representing the same Meaning in different forms has the effect of embedding that Meaning in the minds of the audience. The Principle of One-Many Relations is a Principle of Meaning and clearly this Specialist in philosophy knew that Principle. Thomson goes on to refer to the murder, in 1964 in New York, of Kitty Genovese and she does so in a way that serves her purpose. Throughout her argument, Thomson uses particular details and generalisations to add or lose Meaning to fit her conclusion. The details in Inductive analogies do not need to fit happenings in the Real World. There merely needs to be some Meaning that fits to some unspecified degree the conclusion being delivered. She did not invent the technique of Adjusting Meaning to fit a conclusion; she learned how to use a particular strategy. Truth is not an Issue for those who use Induction. We are all taught in Formal education and informally in every Context to use Induction and we are drawn to an understanding that Truth is Not an Issue. The philosophy of Plato and Aristotle proved to be virulent and is still potent.

It is hard to keep Track of Meaning as Thomson moves it for this is Artificial Logic, which is Not Logical. We are caught up in a tide and swept in one direction. We necessarily let some things go if we are swept along by the Force of her narrative. Having stealthily switched Attention from the word good to the word decent in the Ancient Parable, Thomson asserts that the Qualities of decency and therefore of goodness are a matter of degree. She uses False detail to support her argument:

> "These things are a matter of degree, of course, but there is a difference, and it comes out most clearly in the story of Kitty Genovese, who, as you will remember, was murdered while thirty-eight people watched or listened, and did nothing at all to help her."

As I read her paper, I wondered how it could be that "thirty-eight people watched or listened, and did nothing at all to help her". There are particular conditions under which that can happen, but one has to identify the conditions that pertain at a particular time in order to understand the Dynamics of a particular happening. We are not actually told in her paper the detail of the watching and listening despite Thomson telling us that it is the detail that matters. Apparently only some Selected Details are the ones that matter and she is the designer who selected the parts for her argument. Selection is Natural, but there is Justification for Natural Selections. There is no Justification for the Artificial Selections in the arguments of Artificial logic.

Observations are necessarily Selected but their Qualities should be Tested. For True Meaning to Emerge, it is necessary to start with more than one Observation from more than one Source. It is also necessary to abandon any notion of trusting what you are told. In my experience, a sample size of 38 people should be sufficient to catch at least one or two people who would be proactive in the presence of an attack on another individual. Many different types of proactive response are possible and some are Reasonable while others are not. Even No Response at all can be Reasonable but I classify Non-Response as a type of Response. According to MOM, there is no such thing as a Blank Slate in terms of Action. The Action of Non-Response when a Stimulus is Detectable has particular Meaning. All types of Response must be Justified in the Context in which they occur. Action is Relative.

Thomson's description conveys the idea of people standing around or passing by, observing the attack as it occurred and doing nothing. A certain woman was attacked, wounded and left to die while 38 people saw her wounded and passed by on the other side. We now have a large number of people supposedly modelling in the Real World the Action she favours. That Meaning is easily Tested for Truth. In order to do so it is necessary to access other accounts for comparison. I have done so and you can do that too. You will find that Thomson repeated exaggerated reports in newspaper accounts in the USA around the time of the murder. Many accounts of the murder were not subsequently confirmed by other accounts and at Trial. Thomson's description was part of her creation of an Illusion.

But whether or not the Trial had taken place before she wrote her article, where there was Uncertainty that Quality should have been Marked. It seems that Thomson did not check the details, or alternatively she ignored them. She was not alone in doing so. Some social psychology textbooks also used the exaggeration to support an Opinion that the actual details did not fit. It is a Fact that the details of what actually happened in Reality lacked the type of Meaning Thomson required to Unsettle the case against Abortion. At best she exaggerated, but in the Simplest and Truest of Terms, She Did Not Tell the Truth. Truth is not a matter of degree. Perhaps she did not Know the Truth. For Specialists, Truth is Not an Issue. They have agreed to Ignore that Quality when Representing Meaning, so her activities in selecting Meaning without Testing for Truth are not surprising. They are Not Justifiable either. But she has a Licence and is certified by peers to tell Tales that are Not True. The False Meaning fits her argument and Meaning Tested for Truth does not fit. I do not see her paper as an Argument on Abortion for it is far more than that.

SIGNIFICANT DIFFERENCES SHOULD BE PICKED UP

Specialists in Formal logics drop Significant Differences, and divert Attention towards Trivial Differences. That which Appears to make No Sense does make Sense when you disclose the Point of Reference. Specialists in Formal logics do not disclose their Real Points of Reference and since the Meaning they Assert is completely Unjustifiable they usually leave a gap where the Point of Reference should be. They cannot find a Point of Reference that would have even the Appearance of a Fit to a Moral Standard. That failure to disclose particular Meaning prevents the Perception that the Meaning bears No Relation to Truth and that the Action does not fit Natural Expectations. A Reaction can be delayed or prevented when Meaning is not clear. Thomson wrote:

> "I should think, myself, that Minimally Decent Samaritan laws would be one thing, Good Samaritan laws quite another, and in fact highly improper."

The thrust of Meaning in that sentence is: - It is improper for the Law to insist on a person being charitable and helpful and therefore it follows that it is improper for there to be any obligation at all to be Good. That is fallacious. That is Induction. That does not even have the False Appearance of Deduction to give it the Appearance of Justification. Yet her work was accepted for publication, applauded and widely distributed. That is a Curiosity. She is muddling, or should I say mixing for consumption by an audience, the Contexts of Law and Morality. That is not accidental. In keeping with her role as a Persuader in the Context of Public Policy she aims to change the condition of her audience. She may or may not have aimed intentionally to worsen the condition of her audience but the parts of her argument and the Whole Meaning that Emerges from her paper leads inevitably to that effect.

Thomson did not see any Significant Difference in Meaning and Action between that which is Moral and that which is Legal. Thomson wrote:

> "Indeed with one rather striking class of exceptions, no one in any country in the world is legally required to do anywhere near as much as this for anyone else. The class of exceptions is obvious. My main concern here is not the state of the law in respect to abortion, but it is worth drawing attention to the fact that in no state in this country is any man compelled by law to be even a Minimally Decent Samaritan to any person; there is no law under which charges could be brought against the thirty-eight who stood by while Kitty Genovese died. By contrast, in most states in this country women are compelled by law to be not merely Minimally Decent Samaritans, but Good Samaritans to unborn persons inside them ……. it does show that there is great injustice in the existing state of the law. And it does show also that the groups currently working against the liberalization of abortion laws, in fact working toward having it declared unconstitutional for a state to permit abortion, had better start working for the adoption of Good Samaritan laws generally, or earn the charge that they are acting in bad faith."

In order to use Reasoning in a Moral argument, it is necessary to know what type of Meaning is represented in the Concept of Moral and be able to distinguish Action that is Moral from Action that is not. Thomson was contributing to an argument on the Law in relation to a Moral Issue, so understanding the difference between the two Contexts was an additional requirement. But Thomson was Not Using Reasoning.

Could Thomson be described as a Tin Woman to whom the Wizard gave an Artificial Heart to display to those who might question her Morals? I merely ask the question. I leave you free to answer that, but read her paper first. Or do I and should I leave the conclusion unstated? Reasoning tells me that when Meaning is delivered it should be clearly stated. If I state the conclusion you have a potentially Testable Proposition. And since particular conclusions are Testable when given, we have a possible motive for Not Stating Meaning when it is delivered. Intention is Not an Issue in this narrative but Functioning is, and that is how Vagueness Functions. The Power of the Enthymeme lies in the Absence of Testability of the Meaning that is delivered.

I will specify a conclusion. I do not in this case regard that character from the Wizard of Oz as an Analogy for Thomson. The Meaning Represented by the Tin Man is not the same as that displayed in the Action on Abortion (by agents on either side of the division). The story of the Wizard of Oz led to the conclusion that the Tin Man was a lovely, kind Man who simply lacked a visible sign that he had a heart. The visible sign was mistaken by others as proof and since it was mistaken as proof it was desirable - that is the Meaning the fictional character Represents. Thomson presented the visible sign of a heart and her Actions in her argument Did Not Have the Meaning the Sign Represents. Perhaps it is True that those who have "the highest reputation" for skill in the Context of Formal logic are "the most deficient or nearly so" among us. Perhaps some of them are still in free fall to the bottom of the Moral Dimension and will necessarily reach the bottom and exit that Dimension.

I find no Evidence of feeling for other human beings in Thomson's work of Fiction. The 'this' to which she refers is carrying a baby to term. Thomson is Dissembling. She is a Specialist and Licensed to give her Opinion. I am also a Specialist Licensed to give Opinion.

We both have the required Marks of Deficiency to do so. Philosophy and Psychology are Intrinsically Related. We can both do that which we are Licensed to do, and disagreement is permitted. But is the Telling of Truth permitted? I increasingly had doubts about that as I explored the literature and the possibilities for publication of my narrative. It is not against the Law to tell the Truth, in Fact the Rules are that you do that in Court. But the Law does not determine the Selections for publication. A report to Court is the work of the expert presenting it and should not be worked upon by other people for if that happens, the Opinion is no longer the Opinion of the Witness. I saw as I explored the territory for publication that agents and publishers insist on reviewing work in advance and on amending it. And they are not experts in my field. I present my own Opinion. Only then can Testing take place. In Fact, there are no experts on the Meaning and Action of Pure Science, so there is not even a particular group of peers who could review this work. This is a new field. Though that is surprising, that is a Fact. So, I have Justified not sending this book out for review in advance of publication. Now my work can be reviewed by all who wish to do so.

Perhaps I should conclude my narrative with a statement similar to the one that Thomson used. I could say I have only been pretending that Inductive science is pseudo-science and is not Pure Science, and that I am only pretending I can tell the difference between them in type of Meaning and Action. But if I said that I would be lying. So, I will not say that. I am not pretending. If there is no such thing as Verification you cannot say my narrative is True. If there is no such thing as Falsification either then you cannot say my narrative is Not True. And I cannot do that either. In the Courts the difficulty is resolved by Selectively taking Opinion as evidence for testing in Court and, sometimes, by putting witnesses in a hotter tub than that. That is only safe if testing the Opinion fits the purpose of finding Truth and Justification. But under the condition that the evidence is Inductive science, Truth and Justification is not there and thus only Uncertainty will be found if the Socratic Method is used with great skill. I claim that the Evidence I present is a product of Pure Science and if that is actually so, then the Socratic Method cannot touch me or my Evidence or the Opinion I express. I have already subjected my views to Comparative Analysis in a search for Justification and applied the requirement that my Standard of Truth is met. I give my Reasoning for my Opinion. My Reasoning and the Opinion are Testable. I show you the Method for doing that.

Thomson gave the False Appearance of Reasoning for her Opinion and that gives the False Appearance of Science. Thomson's argument is an example of Inductive science and my Analysis clearly demonstrates the harm that can do. Thomson states that her main concern is not the state of the Law and then she says there is great injustice in the existing state of the law. The Law on Abortion was the Context for her paper and the ultimate target of the Attack. Some types of offence are Illegal, while for the type of offence that is not entirely against the law a stopping point in terms of Quantities is Marked in the Law. Numbers are not necessary for the Representation of Meaning and can interfere with its Perception. Quantities are often expressed in Qualitative Terms. The Law deals with Meaning. Only Qualitative Terms have particular Meaning. The Law marks stopping points that are applied where Moral constraints have failed to be Active. If you change the Law then you can change or remove the stopping points on Action that the Law Marks. Alternately and depending on the type of change, you can add stopping points by the same Action of changing the Law.

There are particular Systems for changing Law in democracies and soundings of Opinion are taken, so it is puzzling that she attempts to divert Attention from her goal when it is clearly threaded through her narrative. The Law is drawn up by Politicians and is part of decision making on Public Policy. Thomson did not have to Justify making a contribution to a debate on Moral Issues and the Law that relates to those, for a contribution is allowed. So perhaps hiding her purpose is just an acquired habit and too strong to break. Those who use Induction habitually hide the True Points of Reference for their Actions. I have shown that her argument has many effects that could serve different purposes. I do not know her purpose. I only know the purpose she stated. Intention is not a concern in this narrative and Functioning is. Her lack of Attention to True Meaning in parts of her argument is clearly Observable. While she states that details are important, she does not allow True Meaning in the detail to affect the conclusion. The detail is important to her in a different Sense.

Thomson's paper had the Appearance of being a Dialogue in which she took both parts. Thomson flutters around the central Issue setting scenes so that Meaning can be transferred from them to Abortion. We are not primed by Nature to reject our own progeny. Unborn persons inside women are by definition not strangers; they are physically connected and related at that stage. The connection is Active – it is a Live Connection through which the means to sustain life flows. The word production is not used to describe pregnancy and childbirth, and the word reproduction is used. The foetus that becomes a baby is part of a woman and is a different Whole Being at a particular stage of development. At one point in her argument Thomson says:

> "There you are, you were kidnapped, and nine years in bed with that violinist lie ahead of you."

That of course is Not True. But if that were True you know precisely how you would feel in relation to that happening. Her Inductive analogies were celebrated. Let us look at this one in particular. Thomson's arrangement evokes Emotion through stimulating imagination in the style of other authors of Fiction. She uses exaggeration at many points in her article, usually taking care that the exaggeration is not extreme enough to trigger effective resistance. While she begins with 9 months in bed with that violinist, she ends in an exaggerated 9 years. The Quantitative change is sizeable yet barely noticeable. The number remains the same and we recall that any similarity will suffice to transfer Meaning using the Art of Induction. This scene is set up to trigger a Negative Emotional Response in order to transfer that feeling to pregnancy. If you do that then you will necessarily increase the demand for Abortions. And as the demand increases, the doing becomes normal by definition of Formal logic for normal is defined in Quantitative terms. Aristotle told his audience that, "the success of the persuasive efforts depends on the emotional dispositions of the audience" and there are many tools for changing those, which he displayed. The tool does what it is directed to do by an Active agent.

You have not in Reality been kidnapped and you are not trapped in bed with a strange adult for nine years, though both those happenings can occur in the Real World and the timescale is variable. In Thomson's Tale, the different Contexts do not have the same Meaning and her time scale is relatively exaggerated. But the story she tells contains particular Meaning. If you pick up that Meaning and map that onto the Context of Abortion, and if you fail to Test the Meaning and its transfer for Justification, then you have been cornered and her opinions will prevail. So, if you go along with the transfers of Meaning and accept the Associated Emotion without Testing them for fit in the Context of Pregnancy, then you have metaphorically been kidnapped. The kidnapper is Thomson, her Opinions are the violinist and the bed is Cognition.

Thomson states:

> "There may well be cases in which carrying the child to term requires only minimally Decent Samaritan-ism of the mother, and this is a standard we must not fall below. I am inclined to think it is a merit of my account that it does not give a general yes or a general no."

I can find No Merit in her account. An Appearance of Strength in the argument is given by the suggestion that we move away from generalisation to Single Cases and possible exceptions. But she did not focus on that strength as she properly could have done. She did not Calculate Meaning and that should have been done.

Total bans and total freedoms leave no room for adjustment to the different circumstances of Single Cases. That type of Action is Unnatural. Rules to cover many cases are necessary for they provide guidance. Allowing a few conditions as exceptions to a Rule does not permit sufficient movement for good Judgments in Single Cases. There must be Boundaries for Meaning and Action. Boundaries are Qualitative and have the flexibility that is required to enable good Functioning. Judging Meaning is a Complex Action. The Ancients knew that Fact.

Carrying a child to term is not the end of the responsibilities of a parent and 9 years does not cover those. Minimally Decent Samaritan-ism is not sufficient while the children are growing up. They need to be loved, cherished and provided with the means to sustain life. The failure of a parent to provide the means to sustain life for a child is against the Law. Do we "in fact" have to distinguish between two kinds of parent - the Good Parent and what we might call the Minimally Decent Parent? And how do you define that which is minimal in caring for a child. Why would you even want to do that? Well, recall that Thomson is a designer of Meaning in the Context of Public Policy and children who are neglected and abused end up in the care of the Authorities. The notion of providing Minimally Decent care has the Appearance of being useful there. But Appearances are Deceptive. Those who are brought up in care leave care at a fixed age. Then they interact with other people in a way to which their experiences have led them. Experiences change things and different types of experience change things in different ways.

Reasoning Verifies that Thomson moved blindly through a tunnel towards a single target, and let us presume she did not understand what she was doing. Either she did not see or she did not care about the potential of her arrangement for Collateral Damage. Clearly, she excelled in doing that which others have done for the audience of her peers rose from their armchairs and applauded her. Did they come from the Real World? Do they visit the Real World? I have lived there a long time. I do not applaud her argument. But I am grateful I discovered that. I have shown you in detail a practical example of Inductive science and precisely what it means to be Objective in that Context. When you are Objective you pursue a particular Objective and you do so relentlessly. The type of Objective has particular Meaning.

Those who use Inductive logic rely on deriving Power through the Mechanisms of Association and Dissociation worked in conjunction. That which is Natural has become Unnatural by Design. There would have been No Collateral Damage if the Process of Pure Science had been engaged, for the Faults within an arrangement and the Errors to which the arrangement leads would have been detected and made visible. Within the Boundaries of Pure Science, it would have been possible to construct an argument with Justification in line with True Meaning in the Natural World. Pure Science is Biased towards finding Truth. The finding of Truth is an Objective when Pure Science is engaged. Those who engage in that Process engage their

Emotions in conjunction with their Capacity to Reason. Natural Induction is part of Pure Science.

Thomson's argument was Stable; it was effective and resistant to attempts to Unsettle it. But her argument did not Unify the different views and there is no such thing as Certainty when the Method used is Induction. Her argument did not change the Beliefs of those who oppose Abortion and on occasions the controversy rages still. Arguing is Not Reasoning. Arguing on a Moral Issue does not make an argument Moral. But there is no need for Confusion. Meaning can be Calculated and Moral Standards can be established. Unless True Meaning is threaded through an argument, a conclusion is not Stable and Unified. Truth gives Strength to an arrangement.

Beliefs are rooted in Emotions. Human beings have the Capacity for Reason. The Capacities for Reason and Emotion are Intrinsically Related. When one part is brought to bear on the other part, the Whole that emerges is different to both. The Difference that Emerges is Dependent on the Type of Relation that is Active between them. When the Capacity for Reason is reduced or disabled, Emotion rules the Actions of human beings. Aristotle knew that Fact and he knew that Emotions can be manipulated to gain advantage. Thomson knew that too.

Before I leave the Topic of Abortion, I will add some words in conclusion. The Major Statement common to the Syllogisms that I identified can lie undisturbed by argument on that topic. The Minor Statement can be changed in a way that would move us towards the Moral High Ground. A foetus is a human being after the point of Conception and is not a human being before Conception. A human being does not have a Right to be Conceived. We cannot kill that which does not have life. Attention should shift to the Topic of Conception, but only within the Context of Pure Science. The Judgments of those who use Formal logics are likely to be in Error. The Topic of Conception could occupy Attention for quite a while since many things relate to that. And Abortion would be Judged on a Single Case basis governed by Standards to which we aspire. Moral Judgments are Complex Judgments and we have yet to Calculate Moral Concepts to Act as Standards. There is a Necessity to do that. And I would add that while Intention has been set aside in this narrative that is actually a Significant Variable for which we lack a

Concept and Operational measures. Once we are armed with Causal Knowledge, Attention can turn to that variable. Intention is Intrinsically Related to Moral Standards, and those can be Installed.

That which is Artificial can be aligned to that which is Natural. It is Natural for there to be stopping points. Only the Artificial Construct of Quantities extends one type of thing without limits to infinity. And those who use that Construct do meet the need to impose stopping points. Stopping points imposed when the Quantitative Method is in Action are Not Justifiable.

PART THREE of Chapter six

ARISTOTLE AND THE CONDITION OF THE AUDIENCE

I have arrived very close to the ending of this narrative with barely a mention of Aristotle (384-322 BC). Aristotle was a pupil of Plato's. The type of philosophy they designed over Twenty-Five centuries ago has determined the course of Social Dynamics across the Centuries. Their philosophy did not arise from a Blank Slate and was a development and adaption of ideas and techniques that had previously existed. The Socratic Method involves Dialogue. Now I will focus on Rhetoric. As the number of people in the audience increases, there comes a point at which Dialogue does not work. Rhetoric is Monologue. But the difference in Meaning and Action is Trivial. It is the Point of Reference that is the Significant Variable and the words dialogue and rhetoric are used in the Context of Formal logic. Rapp (2010) states that Aristotle taught his pupils that:

> "Rhetoric must take into account that its target group has only restricted intellectual resources, whereas such concerns are totally absent from dialectic."

When Formal logic is in Action that which has the Appearance of Fact is a Premise. I have shown Evidence that the intellectual resources of the audience are a Factor in Dialogue. The Premise that such concerns are totally absent from dialectic is False. It is suggested that the condition of the audience is poor before Rhetoric operates. Here we are not told why it is necessarily so that the target group for Rhetoric has "restricted intellectual resources" only that the Factor must be taken into account in designing Rhetoric. There is a difference in Meaning between the Whole audience that is the group of individuals and each Single part of which it is composed. The Whole is Qualitatively Different to the sum of the parts but the term restricted intellectual resources would not fit that difference. The range of propositions is restricted in Inductive arguments. It is of course possible that the intellectual resources of the target group are restricted by the Structure of Rhetoric, but that is not mentioned as a possibility. Expressing Meaning colloquially, Aristotle might have meant that the target group is stupid or can be stupefied. The idea that stupefying the audience is a necessary Factor for the creation of Illusions and must be a Focus of Attention in Rhetoric is there, and is hidden from view. The effect of Structures on audiences is routinely denied, yet an effect can be enduring when systematically evoked.

The Meaning represented in the term "restricted intellectual resources" is not specified but the attachment of the word intellectual to the word resources indicates that what is restricted is the intellect. That might refer to the intelligence of the audience, though it may refer to their cleverness. The common definition of the word intellect is:

intellect: n. 1 a. faculty of reasoning, knowing, and thinking. b. understanding. 2. clever or knowledgeable person. [Latin: related to INTELLIGENT]

The type of relation between intellect and intelligence is not specified at the root. As I would define the two words, intelligence is a different type of thing to cleverness. Both types are covered by the word intellect. Different types of thing have different Meaning. Meaning is expressed in Action. And different types of thing are related in Dynamic Systems. In my definition of Intelligence, Common-sense is included as a part – those two things are in alignment. The relation between Intelligence and Common-sense is harmonious. Cleverness Acts Against Common-sense in the audience. That relation is antagonistic. The word intellectual is used to describe Specialists in Formal logics. It is a Fact that Formal logics are not Common-Sense. Those who do not use Common-Sense, Interfere with the use of Common-Sense by others. That emerges from many of my Observations during a life-time of experiences and from the Calculations you have seen in previous parts of this narrative. The effects of Common-Sense are largely disabled in Specialist Contexts. I leave you to find your own examples of that. You will find that you are spoilt for choice. So, we have two points to take forward into the present Calculation - Rhetoric is produced using Formal logic and Formal logic does not align with Common-sense.

It is a Not a Fact that Intelligence is whatever peers agree that it is but it is a Fact that peers agree on the definitions of words. Operationism took a firm grip on Specialist psychology. Intelligence is not distinguished from Cleverness. The difference in type was dropped. The definition of intelligence in the Dictionary of Psychology begins:

> "Few concepts in psychology have received more devoted attention and few have resisted clarification so thoroughly".

Yet as you will all know, there are Quantitative Measures of Intelligence that have the False Appearance of Precision. The Premise that we can measure precisely something that we cannot pin down in terms of Meaning enables acceptance of Operationism. Round and Round they went as they worked on Meaning in our dictionaries. When we follow each direction we necessarily arrive at Vagueness in definitions, but only because Vagueness has been put there by designers and it Functions in particular ways. I will presume that the Quality of Flexibility in Natural Language was misinterpreted by Specialists in philosophy as Vagueness.

Stupid is an adjective that describes a lack of Intelligence. Since we do not know what Intelligence is, we do not know what Stupidity is. But we think we know and we Certainly know it is not good to be Stupid and it is good to be Intelligent – different Emotions have been attached to those words. We strive to gain the Appearance of Intelligence. Cleverness is precisely that. Cleverness is defined as the use of Formal logic and its products and is not defined as the use of Natural Deduction. I will pick up that difference, leaving the definition of clever as it stands and defining intelligence in part as the use of Reasoning, which is the use of Natural Deduction. Then it emerges that Cleverness is the False Appearance of Intelligence. Those who are Acting in a Clever way create the Illusion that they are Acting in an Intelligent way. Those who are clever are Reputed Wise. Cleverness is Marked by Impressive Qualifications and Intelligence is Not Marked at all. But we all have Natural Intelligence, and there is a difference between using that ability or not and in the level of skill in Reasoning that a person develops.

Aristotle was familiar with the Socratic Method by which deficiencies in the Judgments of the audience can be disclosed. The "partner" in Socratic dialogue is an audience drawn into the play. Plato's Socrates had proved through repeated sampling of Single Cases from groups of men who claimed knowledge in particular areas, that their Opinions and Beliefs are deficient. That is Not Evidence that all those who are Reputed Wise are deficient, for the Meaning of Single Cases cannot be transferred to the Meaning of the group and vice versa. But the evidence gathered using the Socratic Method was taken as proof of deficiency for all those within the groups. That is a Fault and a Fault leads to Errors in Judgment. It is clear that in general all those who claim to know things are the target of philosophers, for whom there is no evidence of deficiency. So, the Premise that the audience is intellectually deficient is supported by evidence from Socratic questioning. There is no evidence that the group of philosophers is intellectually deficient for they are not tested. And they are not in the audience either. The group of philosophers is not the target of Rhetoric. All those who are not Specialist philosophers are targets of Rhetoric. Specialist philosophers and those aligned to them know the answer to all questions, and words do not penetrate their armour. But they do not give directly the answers they know.

Those who use Rhetoric have developed skill in Formal logic and accept the Premise that the audience is stupid. They also accept the Premise that they, who are the agents that stand in front of an audience and deliver speeches, are not stupid and that their Rhetoric is a sign of that. Those Premises are False. It is a Fact that audiences have a wide range of Intellectual Resources. To be fair I must add that those who deliver speeches also have a wide range of Intellectual Resources, and I would add that their Intellectual Resources have been restricted too. I have shown you Evidence of that. They were an audience once upon a time. While Meaning is Complex in its details, it is Simple as a Whole. So, if Intelligence is Natural and the operation of the "intellectual resources" of the audience is restricted, then something restricts the use by the audience of Resources they actually have.

Rhetoric is used to deliver Meaning to an audience through the Structure of the Enthymeme. Rhetoric is not designed to prevent an audience from Perceiving Meaning in the arrangement. Rhetoric is designed to deliver particular Meaning precisely using Natural Language. So, Rhetoric is designed to Act in a way that the Formal philosophical position Asserts is impossible to do using Natural Language. The Premise that Natural Language is an inadequate tool for the Representation of Meaning is False. It is a Fact that the Premise is False. Specialists in philosophy know that Fact. In order to restrict the Action of Intellectual Resources it is necessary to block Tests of Meaning using Reasoning, for Reasoning is at the core of our Intelligence. The effect of Rhetoric depends on the audience being Stupid when they are Not Stupid. I will rephrase that and adapt its Meaning in the light of the Evidence we have seen. I propose that the Power of Rhetoric depends on the audience having been Stupefied. I will use the common definition to clarify the Meaning of that word:

stupefy: v. 1. make stupid or insensible. 2. astonish or amaze.... [French from Latin *stupeo* be amazed]

The Evidence that an audience can be Stupefied is overwhelming. There is No Room for Doubt about that. Many Factors Act in Conjunction to produce that effect. When we have been rendered Insensible, we are ready to be amazed by that which Appears to be Sensible and ready to accept any Meaning that is given. We are aware that Art is Deceptive. Rhetoric is the type of Art that deceives us Below the Threshold of Awareness of Deception. We are not entirely conscious that Rhetoric is a particular type of Art. In Fact, we are not entirely conscious during the hearing of Rhetoric, in the Sense of being aware of what is happening to us. Rhetoric is mesmerising – we are spellbound by Illusions.

It is a Fact that ideas can be Unsettled by Reasoning and replaced by ideas that have the Quality of Justification, but the definition of the word rhetoric does not cover that happening. Rhetoric changes cognitions and that word is restricted to the Context of Induction. In that Context, the Art of Rhetoric does not involve Reasoning on either side and the ideas delivered to the audience are Not Justifiable. The acceptance by an audience of Unjustifiable Meaning is dependent on the Absence of Justification being Invisible. That which is Invisible can be made Visible. Rhetoric is designed to disengage the Capacity for Reason and engage the Capacity of the audience for Emotion. Rapp states that:

> "..... though emotions are one of the most important topics in the Aristotelian ethics, he nowhere offers such an illuminating account of single emotions as in the Rhetoric. Finally, it is the Rhetoric, too, that informs us about the cognitive features of language and style".

Rhetoric is designed to play on the Emotions. The Speaker is the Conductor of the arrangement and finely tunes the instruments as they are played. Aristotle cast light on how to play on Emotions precisely using Natural Language in conjunction with non-verbal tools. The manual and hymn sheet for this religion has long been in existence and they were not cast into the flames. Rapp comments thus:

> "Aristotle stresses that rhetoric is closely related to dialectic .... Now, if rhetoric is nothing but the counterpart to dialectic in the domain of public speech, it must be grounded in an investigation of what is persuasive and what is not, and this, in turn, qualifies rhetoric as an art."

Rhetoric is designed to persuade an audience to accept particular Meaning. The designer engages in an investigation of what is persuasive prior to designing the speech. There is no Qualitative difference between rhetoric and dialectic. The Dynamics of the persuasive process are the same irrespective of the size of the audience. My analysis of the Socratic Method aligns with the Action in Rhetoric. The success of the Socratic Method depends on there being deficiencies in the Opinions of the target. I picked up Meaning someone quoted from Kant, though I cannot recall the paper from which I picked this up. This represents well some Meaning for which I produce Evidence:

> "According to Kant, the ancient Greeks used the word "dialectic" to signify the logic of false appearance or semblance. To the ancients, "it was nothing but the logic of illusion. It was a sophistic art of giving to one's ignorance, indeed even to one's

> intentional tricks, the outward appearance of truth, by imitating the thorough, accurate method which logic always requires, and by using its topic as a cloak for every empty assertion."

Thus, Specialists in philosophy understand that Rhetoric is also the logic of Illusion. So surely, they understand that Illusion is Not Logic. But they might not for, as we must bear in mind, they do not use Natural Deduction. Illusions are Powerful. Natural Deduction is a core ingredient of Natural Logic which is a Thorough, Accurate Method of Calculating Meaning. Rhetoric and Dialectic Imitate Reasoning. The False Appearance of Deduction is a Trick that serves the purposes of those who make False Assertions. Inductive logic is nothing but Illusion. Sophistic Art is Induction. Induction is Sophistic Art. Formal logics are a sophistic art of giving to one's ignorance the outward appearance of truth. The Ancients taught people how to do that. It was understood that the outward Appearance of Truth must be there for Meaning will not be accepted by an audience if it is recognised as being False. A resemblance to Natural Deduction is all that is required and the Structure of Artificial Induction provides that Appearance.

I also picked up from the Source I cannot now rediscover, the following, often repeated, description of dialectic:

> "In classical philosophy, dialectic … is a form of reasoning based upon dialogue of arguments and counter-arguments, advocating propositions (theses) and counter-propositions (antitheses)."

Thomson's argument on Abortion fits that description. Dialectic has the Appearance of a Dialogue between two people for it sets out opposing positions, and it is designed and delivered by one person. Plato did that. Poets in the Tradition of the Old Comedy did that. The word rhetoric can be used for that type of argument and is usually reserved for speeches. Following that description, there was a jump to a statement that cannot be Justified:

> "The dialectical method is discourse between two or more people holding different points of view about a subject, who wish to establish the truth of the matter guided by reasoned arguments."

But the dialectical method is not discussion between two or more people who wish to establish the Truth of the matter, if it is nothing but Illusion, giving to one's ignorance, indeed even to one's intentional tricks, the outward appearance of truth. Those are different types of discussion. The Point of Reference is different for each type. The definition of dialectic in the Pocket Oxford Dictionary has the same Meaning as that quoted:

dialectic: n. 1. art of investigating the truth by discussion and logical argument. 2. process whereby contradictions merge to form a higher truth. 3. any situation or discussion involving the juxtaposition or conflict of opposites.

Discourse is commonly defined as speaking or writing at length about a subject and to discuss has the same Meaning as that word. People who wish to establish the Truth of the matter do compare different points of view about a subject using Reasoning and they do that alone or with other people. But according to the common definition of discourse the Qualities of the narrative do not impinge on the use of that word. The Qualities of Meaning are determined by the Point of Reference. And truth has been defined as whatever two or more people agree is true when they are or have consulted an Authority. Then the agreement becomes the Authority in their absence. But Truth is Not the Sayings of an Accepted Authority. The Construct of Peer Agreement is standing in the place where the Concept of Truth should be. Hearsay that has Not been Tested for Truth and Justification is a core Factor in peer agreement produced by Induction. Hearsay can be Tested for Truth and Justification. It is not Hearsay that is the problem. The problem is the lack of Testing for Truth.

Plato defined truth as the words of a god, spoken through an oracle. Plato created an Illusion of Truth and Aristotle paid Attention to the topic of Illusion. Some influential Ancient Greeks determined the course of Meaning and Action to the present day. That is an extraordinary feat. Perhaps the word god does apply to them! But Hades is defined as the underworld. And devil might be a suitable substitute for the word god in relation to the Fathers of the Formal philosophical position. Gods and devils have different Points of Reference. But let us regain our footing on firm ground for none of us actually believe in gods and devils. Or do we? If we do, we do not now Believe in multiples of them. Or do we? And does it matter what we Believe? Surely what matters is how we Function in the Real World in relation to other things. Well, all things are Related in Dynamic Systems. The Ancients had some understanding of the Structure of Meaning and that shows in their legacy. Beliefs have particular Meaning and Meaning is translated into Action. It does matter what we Believe. Belief in the Formal philosophical position has led to Systematic Malfunctioning in Social Dynamics. That is a Fact. But you can take that Fact as a proposition for further Testing. As a proposition that is Testable in part and as a Whole.

Sophistic Art gives to one's ignorance the outward Appearance of Truth where Truth is Not an Issue and where data processing is firmly under the Control of that which is Not Reasoning. You can clearly see where the Sophistic Art of the Ideology with the False Appearance of Logic has led Social Dynamics. If you cannot see and feel the effects of that Illusion in Qualitative Terms you can count the number of problems in Social Dynamics and the number of Verified Errors of Judgment, and look at the size of the number as it grows. You will never reach a point where counting can stop unless you create one Artificially.

Socrates did not leave a written legacy and I have dealt in part with Plato's Version. There is No Evidence for the Premise that Plato's Socrates favoured telling the Truth and there is Evidence that he did not. But somewhere in his narratives Plato may have recognised that Truth has the highest value and there are signs that Aristotle recognised that. The False Appearance or Semblance of Truth is the Core of Illusion. The fathers of Specialist philosophy also recognised that eliciting the Beliefs and Opinions of other people and searching for the Truth in them has value due to the Fact that Inductions do not have that Quality. The Socratic Method enables battles to be won. Surely telling the Truth is what actually matters, or at least finding the

Truth matters. A question is prompted by that: Must the Truth be hidden from predators?

Well that is necessary in Nature and that Necessity does give the Appearance of Justification for Deception. But it is Not Natural to be unable to distinguish between a friend and a predator; that is an infrequent happening across other species in the Natural World. In the Wild, predators hide but eventually they attract Attention and that happening is determined by cues to their presence. Escape from predators is not always possible but the necessity for doing so is recognised and there is an effort to do so. Hiding is also part of that effort – Action that has the same Appearance can serve different purposes and thus has different Meaning. Which purpose it serves gives particular Meaning to the Action. Meaning and Action are Intrinsically Related in type through the Point of Reference.

Now, as cues are reduced, thc difficulty of detecting the presence of a predator increases. Some human predators make an effort to look like something else and some do not need to bother doing that for their Actions are peer agreed and licenced in Human Social Systems. There is overwhelming Evidence that Discrimination is necessary. And I note that the word discrimination covers different types of Meaning and has been gifted a negative Emotional Association without showing the difference in type of Meaning. In the absence of seeing the difference, the Emotional Association Acts against both types of Meaning. There is a need to discriminate between things in terms of their Meaning and that requires Analysis of doings and happenings. Discrimination in terms of visible characteristics does not distinguish between things in terms of Meaning and leads to problems in Social Dynamics for groups and for Single Cases. And I note that the Method of Induction draws Attention to that which is visible and away from that which is invisible. That is a Fault.

Socrates was sentenced to death after his trial. Plato wrote at length about the last day of Socrates. I have read that narrative. But I will use a brief summary description of Meaning that Plato delivered in relation to the day from Nails (2010):

> "Socrates spent the day in philosophical conversation, defending the soul's immortality and warning his companions not to restrain

> themselves in argument, "If you take my advice, you will give but little thought to Socrates but much more to the truth. If you think that what I say is true, agree with me; if not, oppose it with every argument" (Phaedo 91b-c). On the other hand, he warned them sternly to restrain their emotions, "keep quiet and control yourselves". (Phaedo 117e)"

Sophists in the Image of Plato's Socrates control their display of Emotions, and in that sense they keep quiet. They are told to voice their Opinions and to argue against Beliefs and Opinions with which they do not agree; in that sense they do not keep quiet. Those who are Sophists in the other Ancient form readily display their Emotions – they do not keep quiet in either sense. Now, restraining the display of Emotions or giving a misleading display of Emotion, hides warning signals that there is opposition and hides the strength of the opposition. Thus, once Meaning is finally displayed through Action that Meaning can take the audience by surprise. Surprise is an effective strategy in warfare. Plato and Aristotle showed how to delay the Detection of a War in progress and how to fight a War in the Cognitive Domain. Rapp states:

> "It could .... be .... that rhetoric is only useful for those who want to outwit their audience and conceal their real aims, since someone who just wants to communicate the truth could be straightforward and would not need rhetorical tools. This, however, is not Aristotle's point of view: Even those who just try to establish what is just and true need the help of rhetoric when they are faced with a public audience. Aristotle tells us that it is impossible to teach such an audience, even if the speaker had the most exact knowledge of the subject."

Here we are told that Aristotle did not state it is impossible to teach a public audience the Truth, just that Rhetoric is needed as well. We are also told he stated that "people have a natural disposition for the true". Here it seems to be suggested that:

TRUTH IN CONJUNCTION WITH RHETORIC

IS MORE POWERFUL THAN EITHER FACTOR ALONE.

There is Evidence for that statement. Rhetoric is commonly defined as the art of effective or persuasive speaking or writing. It is difficult to teach an audience many Unrelated Facts when speaking to them. Memory is limited in its Capacity to receive Unrelated Data. But Meaning is not Unrelated Data. Rhetoric puts ideas into relations and that enables their reception and storage. Manipulating Emotions in conjunction with the ideas expressed is another Factor that helps the reception of Meaning by the audience and its embedding in Memory. Manipulating the type of Emotion can change the Meaning that is perceived for it affects the relations between parts. It is not difficult to teach an audience Meaning. Any type of Meaning can be taught. But True Meaning has Power that False Meaning does not have. Aristotle knew that Fact. So, Aristotle recognised that the Power of Rhetoric can be harnessed to deliver Meaning that is Just and True. But he Acted to remove that Source of Power from those who would do precisely that. Or at least, some agents saw that could be done and did that.

Aristotle did not advocate setting Truth as the Point of Reference for Rhetoric. Aristotle points to creating an Appearance of Truth. That is in line with Plato's Actions. Aristotle's Rhetoric is necessary to teach an audience False Meaning. That is misuse of Rhetoric, if that word is defined as public speaking. We need to understand the type of Action that Rhetoric is if we are to understand the Source of its Power. The Facts that enable us to do that have been available across the Centuries. Non-verbal tools have powerful effects on Attention and on Emotions and in public speaking, as in other public displays of Meaning, those tools are coupled with the tool of Natural Language. Tools can be used to serve different purposes. According to Rapp, (2010):

> Prior to Aristotle "…. theorists of rhetoric gave most of their attention to methods outside the subject; they taught how to slander, how to arouse emotions in the audience, or how to distract the attention of the hearers from the subject. This style of rhetoric promotes a situation in which juries and assemblies no

longer form rational judgments about the given issues, but surrender to the litigants."

Use of the word prior suggests that there was a change after Aristotle. The only Fault I can find with Rapp's description is the suggestion that the type of method actually changed. That description accurately represents the Meaning of Rhetoric to this day. Observe the Action in any assembly and you will see the fit. When Formal philosophical skill is exercised you will see attacks on the personal reputation of the opposition, the arousing of emotions in others and the channelling of their heat towards a chosen target. One effect of those Actions is to distract Attention away from the weak position that the attacker is defending. That is not exactly the Socratic Method, but does co-ordinate with that Method. The Socratic Method brings the Appearance of Calm and the Appearance of Reasoning into the equation. But the type of Meaning and Action does not change. The Appearance of Reasoning is False.

Now, recall Plato's words delivered through the mouthpiece of Socrates in his essay on the last day in the life of Socrates – "he warned them sternly to restrain their emotions; "keep quiet and control yourselves". The Appearance of Calm can also be False. Emotions can be suppressed. Once Emotions are suppressed the audience does not know they are there. But they are still in place and guide Responses. Thus, we are Deceived and taken by surprise when the Response finally shows. It is baffling when that happens, for when the Response is delayed the Active Relation between Stimulus and Response is not seen. But the Response is Dissociated in Appearance only.

The difference between Aristotle and the Sophists is Trivial. The Appearance changed and the type of Meaning did not. Dingle had been indoctrinated into a philosophical position that did not work and he had presumably glimpsed Pure Science as I had done. Given he had accepted and used the Method of Induction for argument he could not Verify the Meaning of what he had seen. His Emotions rose as he struggled to tell the difference between Science and pseudo-science as he Believed it to be. Having stated his own Belief, Milne kept quiet and controlled himself which was in line with the advice from Plato's Socrates. In doing so, Milne had the False Appearance of being Reasonable while his position was Not Reasonable. Dingle had not heeded the warning from Plato's Socrates and resorted to the old Sophist tradition which was also Not Reasonable. The Contrast is telling. Milne won the battle in terms of Appearances. But neither side won the War. Artificial philosophy does not provide Verification and predictions are rarely Falsified. But there is Pure Science hidden somewhere in Physics and it is Certainly visible in Chemistry, and the Certainty of Truth is possible. Milne's theory was Falsified. While Falsification is delayed, profit can be reaped from Dogma. Did Popper know that? We will never know if he did or if he did not. And we do not need to discover his Intention. The way his Actions and the Meaning he delivered Function can be Calculated while the presumption of No Malign Intent can remain standing.

The same tools of Persuasion are applied to large groups of people and to Single Cases. Rhetoric has an important place in the Art of Persuasion. It is a technique that is not confined to public speeches but is most often recognised in that Context. Rhetoric and Dialectic engage the Capacity of the audience for Emotion. The Action that defines those words disengages the Capacity for Reason. Emotion is a Complex Active Mechanism with many parts. In line with the Formal philosophical position, the Active Premise is that nobody knows what Emotion is. It is a Fact that much more is known than Appears to be known. Specialists in philosophy know what Emotion is. The Evidence is that they are Not in Denial of the Fact that they deny. That is Not a Contradiction and is Pure Deception. But we will accept that Specialists are Not Malicious and are Full of Good Intentions. The Premise may be that we are deceived for our own good. But does good result from Deception? That topic will have to wait for another occasion.

Emotions are manipulated with precision, not precisely in terms of Quantities but precisely in terms of type. The type of Emotion aroused makes a difference to the type of Meaning that is perceived and to the type of Action that follows. The fathers of Specialist philosophy did the groundwork for their followers. That work could be used well by those who aim to Stabilise Human Interactions and thus enable good Functioning of all parts and the Whole of Social Dynamics. That is the conclusion to which Reasoning takes me. But we need something different to Facts in order to use them well – we need Causal Knowledge. Once Facts are known in their Intrinsic Relations, we will have the ability to determine what happens. And of course, that conclusion must be taken as a proposition and Tested by Experiment. There will be Resistance to doing that. If testing is done in the Context of Inductive science then Uncertainty will prevail. Decisive Action is not taken when there is Uncertainty. Inductive science embodies the Resistance to Pure Science. But it is a Fact that Resistance can be overcome.

The need for alignment of Meaning and Action in the direction of Truth and Justification has always been there. The tools to enable that need to be met are those of Pure Science. Those tools were confined by the Ancients and have been closely guarded since then, preventing their escape into the World. That which was confined and kept hidden from view should now be released. The Evidence for the granting of Parole in this Case is overwhelming. In Fact, the Verdict of the Ancients should be overturned and thrown out. There was a Miscarriage of Justice. The Action of Pure Science in the future would benefit us all. There should not be a Pardon; there should be a Full Apology and Compensation.

I see no difference in type between Aristotelian Rhetoric and the Sophists. A difference has been introduced where there is no Real difference. Those who use Formal logics pick up and drop differences as a matter of convenience in service of particular purpose. Differences can be Tested for their Meaning. Meaning must be Disclosed if it is to become Testable. That is easily done. So how is the particular difference that Aristotle is supposed to have made Represented? Rapp continues:

> "Aristotelian rhetoric is different in this respect: it is centred on the rhetorical kind of proof, the enthymeme ……, which is called the most important means of persuasion. Since people are most strongly convinced when they suppose that something has been proven ……………, there is no need for the orator to confuse or distract the audience by the use of emotional appeals, etc. In Aristotle's view an orator will be even more successful when he just picks up the convincing aspects of a given issue, thereby using commonly-held opinions as premises. ……………".

The "convincing aspects of a given issue" are pre-existing Beliefs. If people can be Deceived into Believing that an Appearance of Truth is actually the Truth, there is no need to manipulate their Emotions. Premises work in that way. But in practice, Emotions inevitably change during the delivery of Meaning. Rapp describes the basic idea of the Enthymeme:

> "In order to make a target group believe that q, the orator must first select a sentence p or some sentences p1 … pn that are already accepted by the target group; secondly he has to show that q can be derived from p or p1… pn, using p or p1 … pn as premises. Given that the target persons form their beliefs in accordance with rational standards, they will accept q as soon as they understand that q can be demonstrated on the basis of their own opinions."

Beliefs are Not Rational and are rooted in Emotion. Here, the term "rational standard" is used for standards that do not involve the Capacity to Reason. And it is acknowledged that audiences are targets for Specialists in philosophy. They set their sights in line with the target. The audience already accepts their own Opinions and those form the foundations on which the orator builds different Meaning. The orator gives the Appearance that the ideas he is delivering are deductions from the existing Opinions of the audience. Bayes Theorem expresses that principle in Quantitative Terms. I will deal with that Theorem at length in a separate narrative. For now, you can see the Error in this form of Aristotelian logic. That approach is Not Logic at all. The word enthymeme is used for a Method that Constructs Meaning on the foundations of pre-existing Beliefs. The different emphasis did not replace the old ways and was additional to those. When the audience is not convinced that something has been proven then it is necessary to use the old ways, which involve the use of Visible Force. The new way which dates from Socrates and Plato and became classical philosophy uses Invisible Force. If the audience is convinced there is proof, then that is sufficient to complete the Trick of having False Meaning accepted by them. One way of doing that is to state a Belief and follow that up with the term "now why do I say that" and give a plausible reason to stand as proof where Proof should be. The order of the sequence of statements is Trivial. The proof can come early and the conclusion late or vice versa. Plausible reasons are one Factor in the creation of Illusions of Reasoning.

Aristotle enabled the Art of creating Illusions to be perfected. An Illusion of Truth and an Illusion of No Apparent Force are Powerful Factors that are used in conjunction. According to MOM Truth does speak for itself. I will repeat some Facts. Aristotle saw that the Appearance of Truth works just as well as the Real thing. The Appearance of Truth only works just as well under the condition that Truth can be prevented from making an Appearance. False Meaning departs when True Meaning Emerges into View. But True Meaning only Emerges when many different Points of Reference are applied to arrangements and the different possible Meanings are seen and Tested for Justification. Illusions are produced when angles of view are restricted. And angles of view are restricted by designers. While Truth guides the search for Meaning within the Process of Pure Science, particular Truths cannot be the Point of Reference until they emerge and they only emerge through Comparative Analysis. Dropping Points of Reference from view removes the Potential for Testing Meaning for Truth and Justification. Dropping the Point of Reference of Truth as a guide prevents its finding in relation to a particular happening. Truth is the star in the firmament on which we set our sights in order to reach particular Truths.

Rhetoric and Dialectic are part of the creation of Illusions. Rapp (2010) states that:

> "Aristotle does not hesitate to concede on the one hand that his art of rhetoric can be misused. But on the other hand, he tones down the risk of misuse by stressing several factors: Generally, it is true of all goods, except virtue, that they can be misused. Secondly, using rhetoric of the Aristotelian style, it is easier to convince of the just and good than of their opposites. Finally, the risk of misuse is compensated by the benefits that can be accomplished by rhetoric of the Aristotelian style."

The concession here is an acknowledgment of a Fact - Rhetoric can be misused. The statement quoted suggests there is only a possibility of misuse and that good use is possible. In Fact, here it is suggested that it is easier to put Rhetoric to good use. But Aristotle also took the view that, "...speeches...... are not able to encourage the many to nobility and goodness" and that would rule out that which he had ruled in. Aristotle reserves public speaking for those who have a different type of purpose. But he did not make the reservation clear. Now, Aristotle's Rhetoric cannot be well used in the service of good purpose if the purpose is not good. And, Aristotle's Rhetoric cannot deliver Truth when Truth is not found. The term "tones down" indicates the reduction of a Contrast between misuse and good use – Attention is taken away from the difference. I have not seen specified any Method of encouraging the many to nobility and goodness. Curiouser and Curiouser said I. The benefits of rhetoric that are said to be there need to be specified and their effects Tested.

I note that it is True of all goods except virtue that they can be misused. Virtue is a particular Quality that is defined at least in part by an absence of misusing things. I have seen the word virtue defined differently and misused. But I would pick up an absence of misusing things as a necessary part of a Concept of Virtue. And I would categorise the Process of Pure Science as a Virtue. Pure Science is a type of Action and is not goods. The products of Pure Science can be and are misused, but the Method is Pure and has Virtue.

Speeches use Natural Language to deliver Meaning to an audience, and an audience can be a large group of people or one person. If you talk to your reflection in a mirror then you are testing the effect of your Appearance on a person and its fit to the Meaning of the words you are speaking; you are pretending to be the audience. By Observing your own Actions and Reactions you can predict their effect on a different audience and adjust the delivery to fit your purpose. Any type of Meaning can be delivered using words in public and in private to an audience of any size and of any type. The size of an audience is irrelevant to the Quality of the narrative. And the type of audience is not relevant to the Quality of Meaning in the narrative and is relevant to aspects of its Representation, for that is the route for Communication. Meaning can be communicated accurately when the Representation is a good fit to the skills of the audience and the Representation can be separately adjusted to fit the level and type of skill of the audience.

But no audience is "intellectually deficient". Audiences are composed of different groups of people with different types of experience and training. The designer of the arrangement needs to adapt the form of the arrangement so that the perception of Meaning by the audience accurately fits the Meaning to be delivered. The Representation is a Mediating Variable that takes different forms. Written narratives have also been undermined as vehicles for the delivery of True Meaning to audiences. In Fact, the Concept of Truth and the Method of producing viable Constructs of Truth (which are related to particular Truths) have been undermined altogether. The Illusion that Natural Language cannot deliver Truth, or any particular Meaning at all, has been created and is part of the undermining of the Calculation and Communication of True Meaning.

More often than not, Rhetoric is used to produce Cognitive Illusions. You will not be surprised when I tell you that nobody knows what an Illusion is. The Appearance that Nobody Knows is False. It is a Fact that when you are subject to a Cognitive Illusion you have accepted Meaning that is False in relation to a doing and happening. Meaning that is False does not fit what actually occurs, is about to occur, or has occurred. However, it fits what could occur if False Meaning is accepted. The details of happenings depend on the conditions in which they occur and the conditions can change Naturally or can be changed by design. Conditions are the Structure that gives rise to happenings.

Public speaking is not Naturally Restricted to Aristotelian Rhetoric or to the Rhetoric of other Sophists, nor is it Naturally Restricted to the presentation of False Meaning. Aristotle imposed Artificial Restrictions on Actions and those Restrictions are Active against the Calculation and Communication of True Meaning, and against the development of Concepts. What chance of survival does goodness have against the combined onslaught of an arrangement of Factors Acting against it in Conjunction? There is little risk of goodness taking hold in many people across the World. I have given you a particular example that Verifies there has been Action to reduce the Risk of Goodness. That type of Risk Reduction is Extraordinary. That is Outrageous.

Yet designers of Structures that deliver Meaning show the Appearance of Surprise at the Disappearance of Goodness. Perhaps they are surprised by the Certainty of the Evidence that we have reached the stage where we are transitioning in very large numbers into being Not Decent at all. I cannot date the point at which we will all be in the State of Badness but it is approaching fast. And I do not know what surprises the Specialists in Formal logic or if they actually are surprised at the success of their efforts. But I do know they deny any responsibility for the effects and outcome of arrangements they design. The Ideology of Formal logic uses the Premise that nothing is Determined and everything is randomly distributed. The Appearance of No Causal Process is False. That is Certainly an Illusion. The Evidence in this narrative Verifies that and it is a Fact. But it is not peer agreed. And if there is any doubt preventing an agreement that doings and happenings are Determined, more Evidence can be produced. I have defined Truth in a different way to those who are Marked as Specialists in philosophy. And I note at this point that if people have been in the audience and a deficiency has been Installed, then when they are forced into Dialogue the deficiency becomes Active. They are no longer able to use Resources that they have, once they have been Stupefied. And when the "partner in dialogue" is a Socrates then they will not emerge unscathed. Nor will they emerge in a Harmonious Emotional State. There are many Sophists in this World and they have a Licence to Act on other people. That Licence should be withdrawn.

Beliefs are rooted in Emotions. Human beings have the Capacity for Reason. The two Capacities of Emotion and Reason are Intrinsically Related. They both deal with Meaning. Meaning from one part is brought to bear on Meaning from the other part and the Whole Meaning and Action that emerges from Intimately Conjoined Capacities is different to that of both parts when they are weakly related or disconnected. When the Capacity for Reason is reduced or disabled, Emotion rules the Actions of humans. Aristotle knew those Facts. Aristotle demonstrated how those Facts can be used to gain personal advantage. Rapp (2010) has read Aristotle's narratives and many interpretations of those. He informs us on that topic in this way:

> "How is it possible for the orator to bring the audience to a certain emotion? Aristotle's technique essentially rests on the knowledge of the definition of every significant emotion. Let, for example, anger be defined as "desire, accompanied with pain, for conspicuous revenge for a conspicuous slight that was directed against oneself or those near to one, when such a slight is undeserved." (Rhet. II.2 1378a31–33). According to such a definition, someone who believes that he has suffered a slight from a person who is not entitled to do so, etc., will become angry. If we take such a definition for granted, it is possible to deduce circumstances in which a person will most probably be angry; for example, we can deduce (i) in what state of mind people are angry and (ii) against whom they are angry and (iii) for what sorts of reason. Aristotle deduces these three factors for several emotions in the chapters II.2–11. With this equipment, the orator will be able, for example, to highlight such characteristics of a case as are likely to provoke anger in the audience. In comparison with the tricks of former rhetoricians, this method of arousing emotions has a striking advantage: The orator who wants to arouse emotions must not even speak outside the subject; it is sufficient to detect aspects of a given subject that are causally connected with the intended emotion."

So, it is acknowledged that Aristotle shows how to manipulate the Emotions of an audience. And it is acknowledged that he has that in common with those who are categorised as Sophists. The difference between them is not in the Point of Reference. The difference is not one of type. Some other Facts emerge from this description of Aristotle's work. It is acknowledged that Emotions can be Determined by Design. Chance is not involved in producing effects once Facts are known and Facts about Emotions are known. We can define Emotions. We do know what Emotions are. We know the details of their Appearance. We know the types of Action that accompany different types of Emotion. We can produce Emotions on demand to suit particular purposes in terms of type, but not precisely in terms of Quantities. Emotion is a Factor in the Perception of Meaning. Different Emotions have different effects on the type of

Meaning that is perceived. Emotion is an important Variable in terms of Functioning.

ARISTOTLE MUDDLED INDUCTION WITH DEDUCTION

I have disclosed some of the Action against Natural Deduction across the Ages. The muddling of deduction with induction is not new and has been maintained. The effects are systematic. Aristotle removed the Power of Natural Deduction to Act on the Perception by audiences of the Meaning of the Controllers in Social Dynamics. The Controllers in Social Dynamics are not actually living people and that is difficult Meaning to grasp. And it is important to understand that Fact. For that changes the Perception of how we should now Act and what we should Act upon. So, I will highlight that Judgment.

THE CONTROLLERS IN SOCIAL DYNAMICS ARE NOT LIVING PEOPLE.

AND

THEY ARE NOT GODS EITHER.

The type of Action that the fathers of Formal logic set in stone to determine the future following their departure from this Earth has already been described, but more than one description is useful to drive Meaning home. Rapp describes the Action in this way:

> "For Aristotle, there are two species of arguments: inductions and deductions (Posterior Analytics I.1, 71a5ff.). Induction (epagôgê) is defined as the proceeding from particulars up to a universal (Topics I.12, 105a13ff.). A deduction (sullogismos) is an argument in which, certain things having been supposed, something different from the suppositions results of necessity through them (Topics I.1, 100a25ff.) or because of their being true (Prior Analytics I.2, 24b18–20)."

There you see some Strings and Necessities in Formal logic. We recall that Aristotle advised building Meaning onto pre-existing Beliefs by creating the Appearance that different Meaning aligns with them. Induction proceeds from suppositions up to a universal which is also a supposition. The deductive part of Induction is Active on universal suppositions and ends in a conclusion that has the False Appearance of Necessity in terms of truth. Beliefs are suppositions. The universal supposition is merely a different Belief. The Quality of Truth is missing. Beliefs are changed to different Beliefs by Formal logic and acceptance of the different Beliefs is engineered through the Illusion of Necessity. Within Formal logic, inductions and deductions are one species of argument. Rapp continues thus:

> "The inductive argument in rhetoric is the example (paradeigma); unlike other inductive arguments, it does not proceed from many particular cases to one universal case, but from one particular to a similar particular if both particulars fall under the same genus (Rhet. I.2, 1357b25ff.). The deductive argument in rhetoric is the enthymeme ……: but when, certain things being the case, something different results beside them because of their being true, either universally or for the most part, it is called deduction here (in dialectic) and enthymeme there (in rhetoric)."

Recall the second part of Carnap's taxonomy of inductive inference: "the most important and fundamental kind of inductive inference…. includes the special case, known as singular predictive inference, in which the second sample consists of just one individual." In Rhetoric, the first sample, from which Meaning is taken, is just one individual, but though the speaker of Rhetoric is just one person the Meaning delivered is frequently from a group. Socrates was a composite. The second sample, to which Meaning is given, is a group of individuals, while in Dialectic both samples can be just one individual or a small group of individuals. Using the same type of Action, Meaning can be transferred from any thing, or any number of things, to any other thing, or any number of other things, under the condition that some resemblance between parts is visible during the transfer of Meaning between them.

Suppositions are Not True. Beliefs are Not True. We saw that the Beliefs of the audience can be taken as the starting suppositions. What is the "necessity" that leads from Beliefs to acceptance by an audience of different Beliefs that are Not Justifiable? That is called deduction and it is also called the enthymeme; they are one and the same type of thing. The enthymeme is deductive and has the False Appearance of Natural Deduction. Aristotle did not hide the tools by which Intellectual Resources are restricted. Intellectual Deficiencies are created in an audience and yet he attributes the Deficiencies to the audience. That is Clearly Unjustifiable but it has the Appearance of Justification for once the audience have the Deficiency, the Deficiency is clearly there in that place. You see the Deficiency that is there and you do not see how it got there. That is a Magic Trick.

Until and unless Causal Processes are understood, Deficiencies in Reasoning cannot be prevented or corrected. It is a Fact that the Action of Pure Science can lead us to Causal Knowledge. The Truth of that statement is Testable, has been Tested in this narrative and continues to be Testable. It is a Fact that once it is discovered Knowledge should be shared amongst us. And most important of all, the Method of discovering True Meaning should be revealed. Under the condition that those in the audience Calculate Meaning and Truth is their guide, then they can discover when Meaning is False and reject it, thus Deficiencies in Judgment are preventable. But that change is only part of the Action required. It is not as easy as that to change the Whole of Social Dynamics.

People do not React well to Rejection. For there to be Harmony there must be an Ecosystem and Social Dynamics is far away from that type of System. But the Solution is Simple as a Whole. The Point of Reference for the Action in Social Dynamics must be changed and applied to all parts. Those in Power do not recognise any Authority higher than them but it is a Fact that they are parts in the System too. There is a Higher Authority than them. Meaning governs the Actions of active agents. Concepts should govern us, not particular people. But particular people must develop the Concepts using the Process of Pure Science. Then a different type of Meaning can be passed on.

That which is Not Justified is Not Moral. But is that which is Justifiable necessarily Moral? Different Concepts have to be carefully Calculated and any differences in their Meaning drawn together. That will require yet another Concept and I am not looking that far ahead in this narrative. We know the name of that Concept already. The Concept of Proportion awaits development. A

Quantitative Construct of Weight stands firmly in the place where the Qualitative Concept of Weight should be. Aristophanes told us that. His Teachings were Ignored. That was a Fault and it is one we can remedy. Meaning can be Tested for Truth and Justification.

Let us Observe Aristotle's instructions for producing an audience that will not use Reasoning and the Appearance of Justification that is given to the instructions. Rapp tells us that:

> "Aristotle's rhetoric .... tells the orator how to stimulate emotions and how to make himself credible .... There are, he says (III.1, 1404a2f.), methods that are not right, but necessary because of certain deficiencies of the audience...."

It is not right to create Deficiencies in an audience and it is not right to use those Deficiencies to give the Appearance of Justification to the use of methods that are not right. Rapp describes the three technical means of persuasion specified by Aristotle:

> "A speech consists of three things: the speaker, the subject that is treated in the speech, and the listener to whom the speech is addressed (Rhet. I.3, 1358a37ff.). It seems that this is why only three technical means of persuasion are possible: Technical means of persuasion are either (a) in the character of the speaker or (b) in the emotional state of the hearer or (c) in the argument (logos) itself."

In relation to the character of the speaker, Rapp takes Meaning from Aristotle as follows:

> "..... how does the speaker manage to appear a credible person? He must display (i) practical intelligence (phronêsis), (ii) a virtuous character, and (iii) good will (Rhet. II.1, 1378a6ff.); for, if he displayed none of them, the audience would doubt that he is able to give good advice at all............... if he displays all of them, Aristotle concludes, it cannot rationally be doubted that his suggestions are credible. It must be stressed that the speaker must accomplish these effects by what he says; it is not necessary that

> he is actually virtuous: on the contrary, a pre-existing good character cannot be part of the technical means of persuasion."

That is astonishing. You might want to read that again for it is hard to grasp that this particular Meaning is clearly given. I will highlight part of that:

A PRE-EXISTING GOOD CHARACTER

CANNOT BE PART OF THE TECHNICAL MEANS OF PERSUASION.

Or at least that is so according to Aristotle and he is the father of the current philosophical position. And this Ancient Artificial philosophical position is an Ideology with many Believers. I am not one of those. Questions are the route through which Meaning can be found. So, we ask: Why cannot "a pre-existing good character" be "part of the technical means of persuasion"? Is that defeasible, or is that a Fact? We have seen the word credible before. You will have heard that word used by many Statesmen and by people in all Contexts where Formal logic has taken root. All Statesmen display the trappings that create the Appearance of their being a credible person. And they are Trappings. That is a Fact, and in describing the Fact in that way you might assume I have accepted the suggestion by Aristotle that the Appearance of Virtue is necessarily an Illusion. But it is not as Simple as that. And Aristotle suggested that Illusions should be created, which I do not accept as a Universal Rule. Suggestions must be Tested for Truth and they must be Tested in Single Cases. Differences in type of Illusion must be taken into account.

The word credible is defined to cover two types of Meaning. A credible person is someone who "is believable or worthy of belief". And now we see that, according to Aristotle, a credible person is believable and is not worthy of belief. We could change the definition and restrict the word credible to the Meaning that Aristotle gave. Or we could use the parts alone or in conjunction with the word credible to clarify the Meaning in use. There are many ways to clarify Meaning. Personally, I would relegate the word credible to history and stop using it due to the severity of the contamination. But

I do not have the Power to do that. A person is not of good character when the Appearance of Virtue is an Illusion. It is a Fact that an arrangement that is credible is not an arrangement with Meaning that is Tested and True. Statesmen display "practical intelligence, a virtuous character, and good will". That display makes them credible in the sense of being believable. That which is credible is Not Necessarily True and is good enough to produce an effect. But if a pre-existing good character cannot be part of doing that, the Appearance given by Statesmen of having those Qualities is Necessarily False. If that is True, then those who are defined informally or through a title in the political system as honourable are actually dishonourable. I am not saying they are all dishonourable and I am not saying that any particular Statesman is dishonourable. I am merely saying that is the conclusion when Natural Logic is applied to the Meaning Aristotle left behind.

And I am saying that differences in type of character have disappeared by design. Statesmen give speeches. We all speak to audiences. What applies to Statesmen applies to all those who use Artificial logics and only applies while they are Acting in an Artificially Induced way. There is an alternative to the Ideology of Formal logic. But Pure Science is not an Ideology. Pure Science is a type of Action and I have shown you that its Action has particular Meaning. Pure Science leads to ideas and to sets of them in particular configurations, and thus the products of Pure Science could be mistaken for Ideologies. So how is that word defined?

ideology: n. 1. ideas at the basis of an economic or political theory (Marxist ideology). 2. characteristic thinking of a class etc. (bourgeois ideology) [French: related to IDEA, -ology]

Negative Emotion has been attached to that word. That is part of the Action against Metaphysics, from which category peers have agreed to exclude Formal logics. The peers who have agreed that their Beliefs fall outside that definition are the ones who use Formal logics. Ideologies are defined as particular Formal types of Cognitive Structure. Pure Science is Not a Formal Structure and it leads to Meaning which also does not fit a Formal Structure. Thus, Pure Science and its products are not Ideologies. And Pure Science is a type of Metaphysics. Pure Science is a Natural Structure of Action and Cognition, which is harmonious and orderly. In line with Einstein we should forget about the –isms and –ology's. Pure Science leads to patterns of ideas that are different in type to those to which Beliefs lead.

The immediate problem when people are Selected is that we cannot tell the difference in character when the signs of good and bad character are the same. The Marks that should be distinctive, and which we presume and assume are distinctive of Meaning, are presented indiscriminately. Differences in type of Meaning are dropped. Remember that we in the audience are Intellectually Deficient. That which is credible is good enough to be accepted by those who have lost the Power of Reasoning through being Stupefied. But when we see the credible people in Action once they have gained positions of Power, we are roused to our Senses for the difference in Meaning is made clear when the Action is seen. It is too late then to mount a Defence. The existing Structure provides the time that it takes for those who have risen to positions of Power to produce the effects they desire. Those who use Formal logic have agreed the Structures that enable them to Act on the World as they choose to do and those Structures enable them to be Undisturbed while they do so. In the existing Construct of Democracy, when their time is up, they simply present for re-election and the choice is restricted by Selections to a few credible people in groups of the same type that are merely different in form. Round and Round they go, sharing Power between them and reaching agreements between them by the Quantitative Method. The Qualitative Concept of Democracy needs development and we need to do that at high speed.

It is our Misfortune to be guided by the philosophical legacy that Plato and Aristotle left behind. We cannot dismiss the Fact that we have been guided by Artificial logic for over Twenty-Five Centuries, or the Fact that Artificial logic is a Pretence of logic. We cannot dismiss the Fact that Meaning and Action are Intrinsically Related. The Artificial star launched by these particular Ancients into the firmament has led us to our current position on earth. We can all Sense that things are now changing in Social Dynamics. That star is now spent and its light is fading. Let us hasten its ending and move on by setting our sights in a different way on a different star.

ALL I HAVE DONE IS TO LOOK AT SPECIALIST LITERATURE

AND SHOW ALL OF YOU WHAT IS THERE.

That is what Specialists do in their published narratives; they describe what others have written or said.

BUT IS THAT ALL I HAVE DONE?

In selecting a particular Point of Reference, I have Selected the Meaning that Emerges from that literature and I also Selected the parts of the literature from which the Meaning Emerges. In doing so, I clarified particular Meaning and brought it to the foreground where it catches Attention. I drew conclusions using Natural Logic and I gave them to you, but you do not have to accept those conclusions and you should not do so without Testing their Truth. And if the Meaning that emerges is True then the only possible conclusion is that we need Statesmen who are guided by a different type of philosophy and who thus Act in a different way. Statesmen set the Rules in Social Systems. Formal logic as a general philosophy and a Method of Acting is agreed to be valid and is upheld by them. The Premise that there is no alternative to a type of logic that they admit is flawed has been set and taken hold. And it is said that Power Corrupts and that Absolute Power Corrupts Absolutely. That is plausible but that which is plausible could be an excuse for not changing the existing System. Reasoning leads to the conclusion that a pre-existing good character cannot be part of the technical means of Deception and that if we select those who practice Deception then we are selecting those who are not of good character. Deception is a particular type of Persuasion and there is a different type. Truth is Persuasive. We should try that.

Moral Concepts are Highly Significant to the development of Stability in Social Dynamics. Qualitative Testing is threaded throughout the production of Sound Judgments. When Attention is focussed on Functioning, we take into account both Meaning and its Representation, through Testing the fit between them during Action. If the Action does not fit the Meaning that is Represented then the Action should be stopped and should not be allowed to continue beyond the point at which the failure in fit was disclosed. That is not the current position.

It seems to follow from Aristotle's teachings that being a Statesman is a visible sign of bad character. But, do signs determine the Meaning they hold? If that is so then we do have an Insoluble problem for there must be a System of government and people to run that. But through Analysis of the Single Case of an argument on the topic of Abortion it became clear that signs do not determine the Meaning they represent. Meaning comes from a Source elsewhere. Additional Evidence was provided through Qualitative Analysis of more than one arrangement. I propose that problems that Appear to Defy Solution can be Solved when Induction is not the Master of Judgments.

I would not rule out the proposition that some existing Statesmen are of good character. That cannot be ruled out while we cannot distinguish those who are good from those who are not good. We need good Statesmen with experience and thus we need to be able to tell the difference between them in Qualitative Terms. Statesmen are people who are Selected as leaders, but people are driven by a philosophy so we should Select people on the basis of the type of philosophy they accept, for it is philosophy that leads us all. Meaning and Action are Intrinsically Related. Before we can tell the difference, the types of philosophy must be distinguished and Marked. We can distinguish one type of thing through its Reactions to other types of thing. And perhaps also some Statesmen are minimally decent and still in free fall to the bottom of Thomson's proposed Quantitative Standard. They will reach the bottom eventually unless their fall is stopped. Without a stopping point they will fall out of the weak dimension that lacks the strength to contain them.

Now, a good apple in a barrel of bad apples cannot spread its goodness to the others. It is the rot that spreads and it does so until the bad apples are removed. The problem with barrels is that they hold the good and the bad in close proximity. Thus, the spread of badness is accelerated and we soon find that there is a large quantity of bad apples. There must be room for air to circulate for that helps reduce the problem. But in addition, the bad apples must be identified and in the Natural World it is possible to do that. There are visible signs of rot in Natural things. And regular monitoring is necessary to maintain the Quality of the group within the barrel.

Politicians are Designers of Social Systems. And the Premise is that there are no bad people amongst specialists and designers, and that if there were the problem could be solved by identifying them and taking them out. That Solution is blocked. We cannot identify for Certain those whose Actions will result in harm and there is No Licence to take them out before harm results. As I have demonstrated by Analogy, that is Not Natural. But the Premise of No Malign Intent can stand as a General Presumption to guide our responses. I have identified a Method as the Fault. It is the Formal Inductive Method that is bad and must be removed from Power.

And I will make something else clear here. By engaging in the Process of Pure Science, the presence of Malign Intent can be Verified in some Single Cases. I agree with the removal of those who harm others from the place where they can do harm. We do not need to wipe them from the face of the earth to prevent further harm, though there are certain conditions under which that is necessary. And as Plato saw, the death of a Single Case does not prevent the same type of doing by other Single Cases or the happenings that follow from the doings. The doings of Plato and Aristotle were not stopped by their death. Meaning flows from one part of a Dynamic System to another part. In order to change doings and happenings it is necessary to change the type of Meaning that is flowing.

My guiding Principle is that it is Not Moral to do to them what they do to us. I will not take the time to Justify that position. Or perhaps that is simply done through the following statement. If we do to them what they do to us and to others, and if what they do is wrong, then it follows using Natural Logic that what we do is wrong. But it is a Fact that the same Action can have a different type of Meaning and Facts must be taken into account in Calculating Meaning. So, while that is an Aspiration on which we set our sights it is not always possible to meet that High Standard. Exceptions to a General Rule depend on the type of Meaning in the Single Case. And to prevent more doings of the same type takes more than Death in the Single Case. Or perhaps it takes something that is different to Death. After all, Meaning is Not Mortal and is passed on. Gosh, philosophy is far more fun than I ever thought it was! And puzzles are easy to create and difficult to solve. I presume that philosophers who have Specialist skills will easily Unsettle the general statements by focussing Attention on definitions, which as you will recall they have prepared in advance. I understand why Plato's Socrates was never going to stop.

But philosophy is easier and more fun to do when there is no particular person interfering with your Judgments. Aristophanes basket set aloft in the Clouds is more commonly visualised as armchairs where those who have wisdom sit and opine in splendid isolation. They are oracles who deliver truth from gods with whom they have no relation and from whose assumed existence they draw Power. But it is a Fact that clouds form from water that rises from the ground and that, when precipitated, water falls from them to the ground. Clouds do not travel with their contents up to the heavens. And it is a Fact that those who use Formal logic have a different definition of god to those who Believe in other types of Religion and to those who have accepted none of the Religions and do not use that particular word. When different people use the word god they are talking about different types of thing and some of them are talking about human beings. Human beings are Mortal and death does not end their Meaning. So, the Concept of Immortal Being requires further development. But No Human Being is Immortal. Humans are Mortal. That is a Fact we cannot change.

A Focus on Functioning is a Focus on Meaning for the Analysis is in terms of the relations between Actions and the changes that occur. Meaning is a Whole that emerges from parts. Active organisms cannot move safely in their environment without Sensing what is there. Active organisms Sense movement and Meaning is drawn from that in relation to other things. Testing involves Reasoning in relation to doings and happenings, and is a Constant Natural Activity. The Action to divert Attention and Block Perception of the type of thing that Natural Deduction actually is has been repeated across the Centuries. As I have said before, Natural Deduction was shackled and became a Slave of Induction. The wrong type of Processing became Master. Where Illusions are accepted, those who create them become Master of the Meaning Perceived by the audience and thus the Designer of the Illusion becomes Master of the Actions of the audience. That Fact is Denied. But the Fact is clearly stated in the literature together with the Instructions for Rising to Power. Aristotle tells of the necessity to control the State of the audience through setting the Emotional Conditions for them. That is a Fault and it is an Active Fault. The Power of Association is harnessed to determine what the audience feels about the Meaning it has been given through a description in some form. We saw what Thomson did with the Meaning she learned from her lessons. We should be careful what we teach. Ideas should be well related and soundly guided as they are Installed in minds. That is not what generally happens.

When the Art of Persuasion is in Operation, Emotional Conditions are carefully orchestrated to Align with Meaning in the arrangement. I quote again from Rapp:

> "The success of the persuasive efforts depends on the emotional dispositions of the audience; for we do not judge in the same way when we grieve and rejoice or when we are friendly and hostile. Thus, the orator has to arouse emotions exactly because emotions have the power to modify our judgments: to a judge who is in a friendly mood, the person about whom he is going to judge seems not to do wrong or only in a small way; but to the judge who is in an angry mood, the same person will seem to do the opposite (cp. Rhet. II.1, 1378a1ff.)."

Emotions can be manipulated for any particular purpose, including for the purpose of disclosing True Meaning. There are many ways of determining the Emotional Dispositions of an audience. The orator can easily arouse Emotions and can do so exactly. If that can be done then the Reaction of the audience is Certain and has been determined by the Composer of the arrangement. It is a Fact that that can be done. It is a Fact that that is done systematically. It is a Fact that Emotions have the Power to influence our Judgments. It is clear that Aristotle preached the use of psychological techniques. He did not encourage the use of Moral Reasoning. And I note that Reactions are Not Simple and are Complex. The Reactions of the audience to an arrangement are Certain and are determined by the Composer of the arrangement but one Reaction is Related to another and together they lead to a Whole that is different to the parts. The designer who does not Calculate Meaning can be taken by surprise at the outcome of arousing Emotions exactly. Look back at my Analysis of Thomson's Persuasion and you should understand that statement and its implications.

In the following quote from Aristotle there is a contradiction:

> "Now if speeches were in themselves enough to make men good, they would justly, as Theognis says, have won very great rewards, and such rewards should have been provided; but as things are … they are not able to encourage the many to nobility and goodness." (EN X.9. 1179b4–10)

Why not! I see no Justification for that Meaning. Aristotle picked Meaning up from Theognis and placed a sign that points those of good character firmly away from tools that enable the effective Communication of Meaning. That Action has been systematically repeated across the Centuries. Speeches arouse Emotions. If Emotions can be determined exactly by type and if the type of Emotion has the Power to modify our Judgments, then it follows that they can be aroused for different types of purpose and no purpose can be ruled out. Of course, speeches are not in themselves enough, but they relate to other things and are a Significant Factor in determining the Qualities of men. Most Orators deliver speeches that have previously been written and could instead have been read in private by the audience. Public speaking provides an opportunity for the use of non-verbal tools during the delivery of Meaning. But written narratives also have the Power to deliver and embed particular Meaning in minds. Written narratives arouse Emotions in a different way.

Different types of Meaning can be delivered using Natural Language. But we are told that is Not True. If any person aims to "encourage the many to nobility and goodness" then it looks as though they would be wasting their time. There is no point in trying to make men good if that cannot be done. The suggestion that it is impossible to encourage the many to nobility and goodness removes any obligation to try to do so. In equal measure, there is no point in trying to make men good if there will be a cost to you. Thomson clearly stated that there is no point in doing that. Even the Parable of the Good Samaritan delivers the Meaning that there is a cost attached to being good. There will be No Reward for the Effort. The Good Samaritan paid all the expenses himself. Curiouser and Curiouser said I. But, in the Context of the original parable that Makes Sense and that Does Not Make Sense in all Contexts. Generalisation is a Problem that leads to Error in Judgments.

The Ancients had some psychological Facts at their disposal but Knowledge is different to Facts and I presume they did not have that. Since it is a Fact that Emotions can be aroused accurately and a Fact that Meaning can be delivered accurately and a Fact that the two things can coincide precisely when designed to do so, then it is necessarily possible to make men good. So why is that not done? The pairing of goodness with the absence of wealth has been repeated across the Centuries. Moral Action is Not Rewarded. In Fact, there is Negative Reinforcement - there is a cost to the Action. That is a strange Association and Association is a powerful Mechanism. There is No Reward for Goodness and No Incentive to be good in the only terms of reward that we now understand - there is No Quantitative Reward and No Quantitative Incentive for Goodness. And Goodness has a low frequency of occurrence. That was predictable!

The idea that Goodness will be rewarded by a God in Heaven had supported the absence of reward in the here and now. But we no longer Believe in a God in Heaven and we do Believe in the Quantitative Method and Representation by Numbers. And the Right to immediate gratification of our personal desires has been given the Appearance of Justification. There was a necessity to Calculate the Meaning of all these things. That was never done as far as I can see. The failure to Calculate True Meaning is a Fault. If the Meaning of all these things was Calculated and foreseen then we would have to abandon the presumption of No Malign Intent. And if that is actually so, then nothing will change and the human species will continue along the path to Extinction by removing that which is needed to sustain all life to satisfy the desires of a few people. If there is Malign Intent then the Action that is necessary to produce change is different in type to the Action that is effective when They Know Not What They Do. That Fact Emerged from extensive experience in the Analysis of Single Cases in a range of different Contexts. I presume that Fact was known to the Ancients but that is Not a Testable Proposition. They are The Departed. But it is a Fact that the difference in type was removed from view and the Premise of No Malign Intent was fixed in position. That Premise supports the Premise that only one type of Action is necessary to produce change. Those Premises are False.

I could hang on to the possibility that the individual man who was Socrates was among a number of human beings who existed and attempted to do good, and to the possibility that his efforts were not wasted but were overwhelmingly subject to misuse and could be used well. The tale of Bridgman and the monster that Frankenstein created is Evidence for that as a possibility. But we will never know the Truth of that, for the proposition is Not Testable. When people across the Centuries have tried to encourage the many to nobility and goodness there has been concerted Counter-Action. Many of those who aimed to develop the Forces for Good in Social Systems have been shackled, imprisoned and executed. But they have not been forgotten and some of their work survives. Metaphysics has been Selectively Disabled. The Power of Metaphysics to do Good can be enabled. The Power of Metaphysics to do Harm is still enabled and could be disabled. Points of Reference can be changed. The need to do that is now plain to see.

We can tell the difference between Malfunctioning and Good Functioning when we have a Method to Calculate Meaning, for that type of Action leads to Knowledge. And at this late stage in my narrative I will propose an amendment to the Ancient Mantra that I picked up from the founding fathers of the Christian religion and used as a Chorus. Human beings are conscious of much of what they do for they have the Capacity for Awareness, but:

THEY KNOW NOT WHY THEY DO WHAT THEY DO.

And thus, they cannot give us the Causal Knowledge we seek in relation to their Actions, nor can they or others gain Control over their doings and happenings without that. Only the Action of Pure Science can access Causal Knowledge.

PURE SCIENCE IS MORAL REASONING and MORAL REASONING IS PURE SCIENCE

You will not have been Stupefied by reading my narrative and should be in a condition that enables you to Test the Meaning I have Represented for its Truth. It is impossible to draw conclusions about the Relations between things and Test them for Truth while the Meaning of the things that are to be Related is Not Specified. The Quality of Testability is removed when a definition is not supplied. In conclusion I need to clarify the Meaning of the word moral. We have seen the word moral being used in different ways. I will show more of the common Meaning from the Pocket Oxford Dictionary. I will first note the definition of that word as a noun:

moral: n. 1. moral lesson of a fable, story, event, etc. 2. (in pl) moral behaviour.

That does not tell us the Meaning of the word moral and can cover any type of story, event, or behaviour. I have already picked up the differences between what is moral and what is not moral through telling stories. It is clear that only some stories, events, etc. deliver a moral lesson so this definition does not clarify Meaning and we need to Calculate that. I will pick up the word fable in order to explore the Meaning of moral and the Action on its Meaning by the designers of definitions. The word fable is related through definitions to the word moral and a fable is a particular type of story categorised as Fiction. However, part of the definition of the word fable is puzzling and I will highlight that part:

fable: n. 1 a. fictional, esp. supernatural, story. b. moral tale, esp. with animals as characters. 2. legendary tales collectively **3 a. a lie. b. thing only supposed to exist.**

Here you see that Attention is Focussed on the Appearance and not on the Meaning of the story. The Premise that Meaning Represented in Fiction is Not True is Active in this definition. In part 3 we have the term moral tale associated with the word lie and with a thing only supposed to exist. There is Confusion here. Or is there a clear direction? The direction is not one I would follow. The characters in legends are imaginary but we know that, so there is no lie there. The characters are merely Representatives delivering Meaning in the form of a Moral Lesson. So, what type of Meaning do such lessons deliver? I will call on the definition of the word moral as an adjective in the dictionary.

moral: adj. 1 a. concerned with goodness or badness of human character or behaviour, and with the distinction between right and wrong. b. concerned with accepted rules and standards of human behaviour. 2 a. virtuous in general conduct. b. capable of moral action ....

This definition shows the dropping of the difference between a moral lesson and a lesson of a different type, together with the Mechanism that has that effect. The familiar Quality of the Meaning of moral is included as part 1 a. and part 2 a. That Meaning can be automatically transferred through the Mechanism of Association, unless the transfer is Blocked, to the term "accepted rules and standards of human behaviour" at part 1 b for it is in close proximity and being given the same part Point of Reference, which is the word moral. The transfer of Meaning between parts is Not Blocked in this definition and is Enabled. Points of Reference are designer products. Here, that which is good and bad, right and wrong, and virtuous conduct becomes subject to peer agreement through the Association of Meaning in different types of part within the Structure of the definition. Specialists in philosophy agreed that Thomson's story is a moral story. But the Evidence that I provided through Comparative Analysis leads to a different Judgment. I tell a different story. Is mine a Moral Story?

Through movement in the definitions, a lie becomes moral. Moving back to the definition of a fable, the word lie does not fit the Meaning in a Moral Lesson but there it is associated with that. A Moral Lesson distinguishes between goodness and badness in Qualitative terms. The definition of fable is clearly in Error. But if you accept the definitions that we have been gifted then to lie is moral Action. This is part of the Action against Moral Actions which are properly defined as virtuous and which rely on distinguishing between goodness and badness of human character or behaviour, and between right and wrong. I have shown you sufficient data to Verify that the Attack on Moral Action has been systematic over a long period of time. Yet we would not suspect that such Action is occurring for it makes No Sense at all. We have seen that the word lie does fit parts of the Meaning in Thomson's story and that the word moral was applied to that. The definition backs up the misuse of the word moral. Her peers have reached agreement that Thomson's Actions in her argument are legitimate and moral, and they cannot be faulted for their description is warranted by previous agreements. But none of the agreements are Justifiable.

Now, there is one thing I do know for Certain after a lifetime of being baffled by that which has the Appearance of Non-Sense. The Appearance is Deceptive.

THAT WHICH APPEARS TO MAKE NO SENSE AT ALL

WILL MAKE SENSE EVENTUALLY

Moral Reasoning was nowhere in my mind when I decided to write a book and began to focus my research on doing that. My Attention finally narrowed to a search for the Mark of Science. I had no reason to think the two things were related because I had perceived the Premises that science is a different type of Meaning and Action to Metaphysics, that science is objective, and that being objective means there is No Bias of any type. I had also perceived the Premise that science is one particular type of thing without fully understanding the type of thing that it is said to be or seeing that there is a difference between Inductive science and Pure Science that is Highly Significant. I had accepted Premises after trying and failing in my attempts to Test them for Truth and Justification. But Doubt remained in my mind. And I read and heard many narratives over the years in relation to many different happenings of various types. In my work I focussed on Single Cases and attempted to Calculate Meaning in relation to those. I began to discover that Premises give rise to Illusions. That is so obvious that I cannot Believe it took me so long to work out what is actually happening in Social Dynamics. I take comfort from some Facts. Illusions have Power over the

Perception of Meaning. No human being is immune from the spell of an Illusion until and unless the Appearance of Meaning is Tested for Truth. Testing the Appearance of Meaning for Truth has been made very difficult to do. Having True Meaning accepted as being True is even more difficult to do. But once found, Truth is difficult to dismiss and for me it is impossible to dismiss True Meaning once it is found through Experiment.

I have defined Truth as what actually happens in a World. And I have shown that Meaning is Relative. What actually happens in one World relates to what actually happens in a different World. I have defined the Natural World as the Real World. Within the Boundaries of Pure Science, Functioning in relation to the Natural World is the Standard for Judging the Qualities of the Artificial World of human design.

I have highlighted the Significance of Differences in Type of Meaning and that those differences are disclosed by Reactions when things come into Conjunctions. I have noted that all Wholes are composed of parts in Relations and that Wholes are Qualitatively Different in Meaning to the parts of which they are composed. And I have shown that different Whole things Relate to other Whole things and are parts in larger arrangements.

I have defined Pure Science in part as the Operation of Natural Deduction engaged in the Discovery of True Meaning in the Natural World, and included the Action of Experiment as part of Natural Deduction. I have moved the starting point for Natural Deduction back to Sense Data where it fits the purpose of Making Sense of the Real World. And I have noted that the Calculation of Meaning is necessarily flexible and can begin anywhere. Meaning can be Tracked backwards and forwards from any part within a Dynamic System.

I have specified that Natural Deduction is Reasoning and vice versa. Experiments are threaded through the Calculation of Meaning. Meaning is Abstracted from Sensations, Justified through Testing by Experiment, and stored ready for use as Attention turns again to Focus on the Meaning of different, particular Single Cases in the environment. When the Process of Pure Science is engaged, Meaning is Tested from every point of view until Verification or Falsification is reached, then Testing is redirected. Particular Deductions begin and end with particular Facts or propositions.

There are No Premises in Pure Science. I have identified a Moral Dimension to the Action in Pure Science. That is only a beginning in understanding the Process of Pure Science and its particular Bias. But that is the ending of this particular narrative.

AN ENDING IS MERELY THE BEGINNING OF SOMETHING DIFFERENT

REFERENCES

Aristophanes (405 BC) “The Frogs” Source Penguin Classics “The Wasps, The Poet and the Women, The Frogs” translation by David Barrett 1964.

Aristophanes (423 BC) “The Clouds” Source Penguin Classics “Lysistrata and Other Plays” translation by Sommerstein A H 1973, 2002

Bartha, P. (2013) "Analogy and Analogical Reasoning", The Stanford Encyclopedia of Philosophy

Brown, J R and Fehige Y (2011) "Thought Experiments", The Stanford Encyclopedia of Philosophy

Chang (2009) “Operationalism” The Stanford Encyclopedia of Philosophy

Crupi, Vincenzo (2014) "Confirmation", The Stanford Encyclopedia of Philosophy

Douven, I. (2011) "Abduction", The Stanford Encyclopaedia of Philosophy

Gale (2011) “Cosmology: Methodological Debates in the 1930s and 1940s”
The Stanford Encyclopedia of Philosophy

Garrett B L 2012 “Convicting the Innocent: Where Criminal Prosecutions Go Wrong”
Harvard University Press

Green (1992) “Of Immortal Mythological Beasts: Operationism in Psychology” Theory & Psychology, 2, 291-320.

Hardcastle, G. L. (1995) "S.S. Stevens and the Origins of Operationism" Philosophy of Science, 62, 404-424.
Hempel (1966) "Philosophy of Natural Science"

Hasok Chang (2009) "Operationalism" The Stanford Encyclopaedia of Philosophy

Howard D A December 2005 "Albert Einstein as a Philospher of Science" Physics today American Institute of Physics S-0031-9228-0512-010-9

Howard, D. A. (2017) "Einstein's Philosophy of Science" The Stanford Encyclopaedia of Philosophy

Kelly T 2008 "Evidence" The Stanford Encyclopedia of Philosophy

Klima G 2008 "The Medieval Problem of Universals" The Stanford Encyclopedia of Philosophy

Lakatos, I., 1976, "Proofs and Refutations" Cambridge University Press.

Lakatos 1973 Lecture Science and Pseudoscience (transcript) Copyright © LSE 2013

Lamond, G (2006) "Precedent and Analogy in Legal Reasoning" first published on Tue Jun 20, 2006. Last modified on Feb 11, 2014.

Margolis E and Laurence S (2012) "Concepts", The Stanford Encyclopedia of Philosophy

Nails Debra (2010) "Socrates", The Stanford Encyclopedia of Philosophy

Plato Five Dialogues Euthyphro, Apology. Crito, Meno, Phaedo Translated by Grube G M A Revised by J.M. Cooper Second Edition 2002 Hackett Publishing Company, Inc.

Pojman P 2011 "Ernst Mach" The Stanford Encyclopedia of Philosophy

Popper K R (1963) Conjectures and Refutations

Popper, K R. 1935. The Logic of Scientific Discovery, 3rd ed. 1963 London: Hutchinson.

Portoraro, F. (2014) "Automated Reasoning", The Stanford Encyclopedia of Philosophy

Rapp, C. (2010) "Aristotle's Rhetoric", The Stanford Encyclopedia of Philosophy

Ribes-Iñesta (2003) What is Defined in Operational Definitions? The case of Operant Psychology Behavior and Philosophy, 31, 111-126 (2003). © 2003 Cambridge Center for Behavioral Studies

Richardson, H. S. (2013) "Moral Reasoning", The Stanford Encyclopedia of Philosophy

Russell 1923 "Vagueness" Read before the Jowett Society, Oxford, 25 November 1922. First published in The Australasian Journal of Psychology and Philosophy, 1 (June 1923)

Russell (1946) "History of Western Philosophy" Routledge (First published 1946 George Allen & Unwin Ltd.)

Ryan S. (2013) "Wisdom", The Stanford Encyclopedia of Philosophy

Smith R. (2019) "Aristotle's Logic", *The Stanford Encyclopedia of Philosophy*

Thomson J J 1971 "A Defense of Abortion" Philosophy and Public Affairs Vol 1. No 1 (Autumn 1971) pp 47-66

Thornton S 2011 "Karl Popper", The Stanford Encyclopedia of Philosophy (Winter 2011 Edition), Edward N. Zalta (ed.)

Uebel T (2012) "Vienna Circle", The Stanford Encyclopaedia of Philosophy

Vickers, J. (2012) "The Problem of Induction" The Stanford Encyclopaedia of Philosophy

REPORTS AND COURT JUDGMENTS IN DATE ORDER

923: - Appeal on FRYE v.UNITED STATES. 293 F. 1013 (D.C.. Cir 1923)

1975 : - R. v. TURNER (1975) Law Reports, Queen's Bench (Court of Appeal), 834-843

1984:- Bonython (1984) 38 SASR 45, 46 to 47

1993 and 1994 :- Daubert v Merrell Dow Pharmaceuticals Inc., 509 U.S.579, 113 S. Ct 2786 (1993) and Daubert v Merrell Dow Pharmaceuticals Inc United States Court of Appeals, ninth Circuit. No 90-55397 Argued and submitted March 22 1994.—January 04 1995 Before Kozinski and O'Scannlain, Circuit Judges and McNamee, District Judge

1993 : - Ikarian Reefer Lloyds Law Reports 1993 Vol 2

1999 : - Kumho Tire Co. v. Carmichael, 526 U.S. 137 (1999)

2004:- appeal : REGINA - v - ANGELA CANNINGS Neutral Citation Number: [2004] EWCA Crim 1 Case No: 200201711 D3

2005: - Law Reform Commission Report 109 (2005) New South Wales - Expert Witnesses A report relating to expert witness testimony in civil law proceedings

2005 :- Parliamentary Office of Science and Technology Postnote Science in Court October 2005 Number 248

Biscoe Justice Peter (2006) EXPERT WITNESSES: RECENT DEVELOPMENTS IN NSW A paper delivered to the Australasian Conference of Planning and Environment Courts and Tribunals by a Judge of the Land & Environment Court of NSW on 16 September 2006.

2006: - Risk v Northern Territory of Australia [2006] FCA 404 per Mansfield J at [469]

2007: - Supreme Court of Florida No. SC06-118 [November 21, 2007] Appeal on Marsh v. Valyou, 917 So. 2d 313 (Fla. 5th DCA 2005)

2010 : - The UK Guidance manual for Expert Witnesses (2010)

2012 :- Dina Yehia SC EXPERT EVIDENCE Dina Yehia SC NSW Public Defender
August 2012

Made in the USA
Middletown, DE
05 November 2026

36611312R00275